Medical Data Sharing, Harmonization and Analytics

Medical Data Sharing, Harmonization and Analytics

Vasileios C. Pezoulas

Themis P. Exarchos

Dimitrios I. Fotiadis

ELSEVIER

ACADEMIC PRESS

An imprint of Elsevier

Academic Press is an imprint of Elsevier
125 London Wall, London EC2Y 5AS, United Kingdom
525 B Street, Suite 1650, San Diego, CA 92101, United States
50 Hampshire Street, 5th Floor, Cambridge, MA 02139, United States
The Boulevard, Langford Lane, Kidlington, Oxford OX5 1GB, United Kingdom

Library of Congress Cataloging-in-Publication Data
A catalog record for this book is available from the Library of Congress

British Library Cataloguing-in-Publication Data
A catalogue record for this book is available from the British Library

ISBN: 978-0-12-816507-2

For information on all Academic Press publications visit our website at https://www.elsevier.com/books-and-journals

Publisher: Mara Conner
Acquisition Editor: Chris Katsaropoulos
Editorial Project Manager: Ali Afzal-Khan
Production Project Manager: Punithavathy Govindaradjane
Cover Designer: Miles Hitchen

Typeset by TNQ Technologies

Contents

Preface

This book provides a framework for comprehending the fundamental basis of data sharing, data harmonization, cloud computing, machine learning, and data analytics in the clinical domain. The rationale behind medical data sharing combined with the unmet needs in chronic diseases are described first, along with popular frameworks and global initiatives in the field. Data protection legislations are discussed to state the problem of the ethical, legal, and privacy issues, which are involved in medical data sharing, and to discuss potential solutions provided by global initiatives in the domain. The concept of data harmonization is then described. Cloud infrastructure and security protocols are discussed to enable data sharing and at the same time ensure the privacy of the data. Emphasis is given in data mining, deep learning, machine learning, and visual analytics tools and frameworks to analyze the harmonized data. After presenting several case studies which combine the concepts of medical data sharing, harmonization, and analytics, the book concludes with future trends in medical data sharing. More specifically, per chapter:

- Chapter 1 is an introductory chapter which aims to familiarize the reader with the fundamental principles and concepts behind medical data sharing, data protection, data harmonization, cloud infrastructure, and data analytics toward the establishment of a federated cloud platform to deal with the clinical unmet needs in various diseases.
- Chapter 2 presents the most common types and sources of medical data along with data collection standards for each type. Various types of medical data, such as biosignals, medical images, omics, and laboratory tests, are extensively described along with the sources of such types of medical data including patient registries, health sensors, electronic health records, genome registries, cohorts, clinical trials, and claims, among others. The evolution of big data in medicine is also discussed along with different cohort study designs.
- Chapter 3 presents the fundamental concept of data sharing as the backbone of any healthcare cloud platform toward the interlinking of medical data from different sources. Methods for enhancing the quality of medical data in terms of accuracy, relevance, and completeness, such as outlier detection and deduplication, are also presented along with methods for data standardization as a preliminary step for data harmonization. Emphasis is given on the description of existing data sharing frameworks and ongoing global initiatives along with barriers that hamper the vision of data sharing, such as the misuse of the shared data, among others.
- Chapter 4 offers the basis for understanding the fundamental aspects of popular data protection regulations, such as the General Data Protection Regulation (GDPR) in Europe and the Health Insurance Portability and Accountability Act (HIPAA) in the United States. The legal and ethical issues, which are posed during the sharing of medical data, are identified along with the common characteristics of international data protection regulations, global initiatives,

principles, and guidelines toward the establishment of the legal and ethical compliance of the cloud computing platforms in health care.

- Chapter 5 aims to present the latest technological advances and methods for medical data harmonization including lexical and semantic matching approaches toward the identification of lexically similar terms, as well as terms that share a common concept between two heterogeneous datasets. The concept of ontologies is introduced as a reference model to describe the domain knowledge of a disease of interest. The importance of the ontologies is further highlighted for semantic matching during the data harmonization process. Emphasis is given on global data harmonization initiatives and frameworks.
- Chapter 6 presents the current advances in the overwhelming field of cloud computing technology in health care along with the related challenges. Popular cloud computing vendors in health care are described along with cloud computing architectures including the Infrastructure as a Service (IaaS), the Provider as a Service (PaaS), the Software as a Service (SaaS), and the Data as a Service (DaaS). Emphasis is given on international security protocols to ensure the legal and ethical compliance of a cloud computing platform, as well as on different types of data storage topologies in the cloud, such as the centralized, the distributed, and the decentralized (blockchain).
- Chapter 7 presents methods to effectively preprocess and analyze medical data to deal with the unmet needs in various diseases including the development of robust patient stratification models, the identification of biomarkers and/or the validation of existing ones, and the selection of therapeutic treatments for effective disease monitoring. Methods for data preprocessing, including data discretization and feature selection, are presented along with supervised and unsupervised algorithms for classification and clustering. Emphasis is given on the distributed learning strategy for the application of machine learning models across data in distributed databases. Popular machine learning frameworks are presented along with applications in the medical domain.
- Chapter 8 summarizes the fundamental cohort studies on the promising field of medical data harmonization across various medical domains including cohort studies on aging, autoimmune diseases, cancer, phenotypes and epidemics, personality scores, and obesity.
- Chapter 9 summarizes the key points of the previous chapters and presents the latest trends in medical data sharing, harmonization, and analytics.

This book is intended for undergraduate and graduate students in the field of medicine, computer science, computer engineering, data science, and biomedical engineering. The book may also be beneficial for professionals in those fields. This work was carried out at the Unit of Medical Technology and Intelligent Information Systems (MEDLAB) at the University of Ioannina, whose research excellence in the field of biomedical engineering is internationally acknowledged.

We would like to thank our team in the Unit of Medical Technology and Intelligent Information Systems for their scientific and emotional support during the time of the writing of this book and the team of the HarmonicSS (harmonization and integrative analysis of regional, national and international cohorts on primary Sjögren's syndrome toward improved stratification, treatment, and health policy-making) project funded by the European Commission (grant agreement No. 731944 and from the Swiss State Secretariat for Education, Research and Innovation SERI under grant agreement 16.0210). We are also grateful to the editorial team for their valuable guidance throughout the publishing process. We also express our sincere gratitude to our families who contributed to the final realization of this work through the continuous motivation and inspiration they provided to us.

Vasileios C. Pezoulas
Themis P. Exarchos
Dimitrios I. Fotiadis
University of Ioannina, Ioannina, Greece

Terminology list

Data sharing: The process of interlinking sensitive medical data from different medical databases, worldwide, fulfilling all the necessary ethical and legal requirements for data protection.

Data curation: The computational process of enhancing the quality of the clinical data through the identification of outliers, incompatible and inconsistent fields, missing values, etc. The data curation workflow also includes functionalities for data standardization and thus serves as a preharmonization step.

Data harmonization: The computational process of homogenizing medical databases with heterogeneous structure and value ranges under a common medical domain usually through a reference schema (e.g., an ontology). Data harmonization can be accomplished using lexical and/or semantic matching in a semi-automated manner through the detection of lexically identical or similar terms, as well as terms that describe a common concept.

Ontology: A high-level data presentation model where the data are described in the form of entities and object properties, where the entities are defined as classes and subclasses and the object properties are defined as the relationships between them.

Semantic matching: The process of identifying terminologies that share a common conceptual basis (e.g., belonging to the same class or subclass) between two heterogeneous ontologies.

Lexical matching: The process of identifying lexically identical terminologies between two heterogeneous ontologies, i.e., terminologies with common lexical blocks or synonyms, using string similarity measures.

Stringent harmonization: A simple case of data harmonization which involves the harmonization of heterogeneous medical data that have been collected under a specific data collection protocol.

Flexible harmonization: A challenging case of data harmonization which involves the harmonization of heterogeneous medical data that have been collected in the absence of a specific data collection protocol.

Reference model: A set of parameters that efficiently describe the domain knowledge of a medical condition or disease including classes (e.g., laboratory tests, medical conditions, demographics, lifestyle, interventions) and additional subclasses and variables as well. This set of parameters is usually determined by a team of clinical experts in the related medical field.

Patient stratification: A clinical unmet need in across several medical domains which involves the application of machine learning models for the identification of a subgroup of patients who are highly likely to develop a specific medical condition or disease.

Biomarker(s): A set of one or more prominent features according to a given target feature. This set of features is usually identified through a feature selection or a feature ranking method.

Precision medicine: "An emerging approach for disease treatment and prevention that takes into account individual variability in genes, environment, and lifestyle for each person" (the definition has been provided according to the Precision Medicine Initiative).

Health impact assessment: A multidisciplinary process which involves the assessment of health policies in terms of evidence, where a health policy is defined as "the decisions, plans, and actions that are undertaken to achieve specific healthcare goals within a society."

Cloud computing: "A model for enabling convenient, on-demand network access to a shared pool of configurable computing resources (e.g., networks, servers, storage, applications, and services) that can be rapidly provisioned and released with minimal management effort or service provider interaction" (the definition has been provided according to the National Institute of Standards and Technology).

Federated healthcare platform: A federated cloud computing environment, where multiple cloud computing systems/models interact under a common purpose in healthcare. It can support the provisioning and management of the cloud infrastructure in multiple cloud computing systems/models by standardizing the interactions between the individual cloud environments (federated cloud management).

Supervised learning: A machine learning approach which involves the application of a machine learning algorithm on a set of annotated clinical data for classification purposes (the target feature is predefined), e.g., for the development of prediction models.

Unsupervised learning: A machine learning approach which involves the application of a machine learning algorithm on a set of clinical data without annotation (the target feature is absent) for clustering purposes, e.g., for the detection of features with similar patterns within a clinical dataset.

Feature selection: The extraction of a specific subset of features which are highly correlated (highly dependent) with a target feature and less correlated (highly independent) with the rest of the features.

Feature ranking: The ranking of features within a set of clinical data according to their association with a given target feature.

Big data: Massively accumulated sets of daily generated medical data which are characterized by four dimensions, namely the volume, the velocity, the veracity, and the variety.

Batch processing: The strategy of processing big data as smaller partitions/subsets by fetching them into the memory in a sequential manner.

Online learning: The process of additively adjusting a continuous data model (e.g., a machine learning model) on upcoming data streams (or batches).

Deep learning: The process of identifying hidden patterns across large subsets of clinical data (big data) for classification purposes. This process is usually conducted by deep learning artificial neural networks, such as the convolutional neural networks and the recurrent neural networks.

Incremental learning: The process of incrementally adjusting a continuous data model (e.g., a machine learning model) on subsets of a large dataset.

List of abbreviations

APEC	Asia-Pacific Economic Cooperation
API	Application programming interface
AUC	Area under the curve
BOLD	Blood oxygen level dependent
CA4GH	Global Alliance for Genomics and Health
CAGE	Cap analysis gene expression
CAMP	Cloud Application Management Protocol
CASB	Cloud Access and Security Broker
CBC	Complete blood count
CBPR	Cross-Border Privacy Rules
CC	Creative common
CCD	Charge-coupled device
CCM	Cloud Controls Matrix
CCSK	Certificate of Cloud Security Knowledge
CCSP	Certificate of Cloud Security Professional
C-CDA	Consolidated Clinical Document Architecture
CDA	Clinical Document Architecture
CDEs	Common data elements
CDMI	Cloud Data Management Interface
cDNA	Complementary DNA
ChIP	Chromatin immunoprecipitation
CJEU	Court of Justice of the European Union
CNNs	Convolutional neural networks
CoE	Council of Europe
CPIP	Cloud Portability and Interoperability Profile
CPU	Central processing unit
CRF	Case report form
CRP	C-reactive protein
CSA	Cloud Security Alliance
CSF	Cerebrospinal fluid
CT	Computerized tomography
DaaS	Data as a Service
DAST	Dynamic Application Security Testing
DataSHaPER	Data Schema and Harmonization Platform for Epidemiological Research
DataSHIELD	Data Aggregation Through Anonymous Summary-statistics from Harmonized Individual Level Databases
DBMS	Database management system
DCC	Data controller committee
DFA	Direct fluorescent antibody
DICOM	Digital imaging and communications in medicine
DIF	Differential item functioning
DLP	Data loss prevention
DLT	Distributed ledger technology
DNA	Deoxyribonucleic acid

DOC	US Department of Commerce
DoS	Denial of Service
DPAs	Data Protection Authorities
DPIA	Data protection impact assessment
DPO	Data protection officer
DPPA	Data Privacy and Protection Agreement
DSA	Digital Signature Algorithm
DTI	Diffusion Tensor Imaging
DWI	Diffusion weighted imaging
eCRF	Electronic case report form
ECC	Elliptic curve cryptography
ECDSA	Elliptic Curve Digital Signature Algorithm
ECG	Electrocardiography
EMG	Electromyography
ENISA	European Network and Information and Security Agency
EOG	Electrooculography
ECoG	Electrocorticography
ECS	Electrical cortical stimulation
ECHR	European Convention on Human Rights
EEG	Electroencephalography
EHRs	Electronic Health Records
ELISA	Enzyme-linked immunosorbent assay
ENG	Electronystagmography
EPI	Echo-planar imaging
FBP	Filtered backprojection
FDA	Food and Drug Administration
FDG	Fluorodeoxyglucose
FFT	Fast Fourier transform
FBP	Filtered backprojection
FCBF	Fast correlation-based filter
FDAAA	Food and Drug Administration Amendments Act
FDAP	Federal Act on Data Protection
FISMA	Federal Information Security Management Act
fMRI	Functional magnetic resonance imaging
fNIRS	Functional near-infrared spectroscopy
FOAM	Framework for Ontology Alignment and Matching
FOV	Field of View
FTC	Federal Trade Commission
GDPR	General Data Protection Regulation
GI	Gini Impurity
GGI	Gain in Gini Index
GLFA	Generalized linear factor analysis
GLM	Generalized linear model
GLS	Generalized least squares
GWAS	Genome-wide association studies
GUI	Graphical user interface
HCP	Human Connectome Project
HCT	Hematocrit

HGB	Hemoglobin
HHS	US Department of Health and Human Services
HIPAA	Health Insurance Portability and Accountability Act
HITECH	Health Information Technology for Economic and Clinical Health
HMMs	Hidden Markov Models
HQMF	Health Quality Measures Format
HR	Hazard ratio
HRIC	Health Research and Innovation Clouds
HTC	High-throughput technology
HTTP	Hypertext transfer protocol
IaaS	Infrastructure as a Service
ICH	Immunocytochemistry
IDA	Integrative data analysis
IDPC	International Data Protection Commissioner
IDSS	MIT Institute for Data, Systems, and Society
IEEE	Institute of Electrical and Electronic Engineers
IG	Information gain
IHC	Immunohistochemistry
IoT	Internet of Things
IP	Internet Protocol
IPC	Inter-process communication
IQR	Interquartile range
IRT	Item response theory
JSON	Javascript Object Notation
KOGG	Kyoto Encyclopedia of Genes and Genomes
KVM	Kernel-based virtual machine
LASSO	Least absolute selection and shrinkage operator
LCS	Longest common subsequence
LFA	Linear factor analysis
LHS	Learning health system
LOF	Local outlier factor
LR	Likelihood ratio
LRR	Local research repository
LSC	Life Sciences Consortium
mRNA	messenger RNA
MAD	Median absolute deviation
MCD	Minimum covariance determinant
MDL	Minimum description length
MEG	Magnetoencephalography
MFA	Multifactor authentication
MI	Mutual information
ML	Maximum likelihood
MNFA	Moderated nonlinear factor analysis
MRI	Magnetic resonance imaging
MS	Mass spectrometry
NAS	Network-attached storage
NCBI	National Center for Biotechnology Information
NCBO	National Center for Biomedical Ontology

NCI	National Cancer Institute
NEWMEDS	Novel Methods leading to New Medications in Depression and Schizophrenia
NIH	National Institutes of Health
Nifti	Neuroimaging Informatics Technology Initiative
NIST	National Institute of Standards and Technology
NGS	Next-generation sequencing
NLM	National Library of Medicine
NMR	Nuclear magnetic resonance
NSA	National Security Agency
OASIS	Organization for the Advancement of Structured Information Standards
OCR	HHS Office for Civil Rights
OCT	Optical coherence tomography
OECD	Organisation for Economic Co-operation and Development
OLS	Ordinary least squares
ONC	Office of the National Coordinator for Health Information Technology
OR	Odds ratio
OVF	Open Virtualization Format
OWL	Web Ontology Language
P2P	Peer to peer
PaaS	Platform as a Service
PCG	Phonocardiography
PCR	Polymerase chain reaction
PET	Positron emission tomography
PHI	Protected health information
PLT	Platelets
PPI	Protein–protein interactions
PPRs	Patient-powered registries
PRs	Patient registries
RBC	Red blood cell
RCTs	Randomized clinical trials
RD	Risk difference
RDBM	Relational database management
RDF	Resource Description Framework
RDFS	Resource Description Framework Schema
REST	Representational state transfer
RFE	Recursive feature elimination
RIM	Reference Information Model
RF	Radiofrequency
RLS	Recursive least squares or regularized least squares
RNA	Ribonucleic acid
RNA-Seq	RNA sequencing
ROC	Receiver operating characteristic
ROIs	Regions of interest
RPC	Remote procedure call
RR	Risk ratio
RSA	Rivest–Shamir–Adleman
SAML	Security Assertion Markup Language

SAN	Storage Area Network
SNPs	Single-nucleotide polymorphisms
SaaS	Software as a Service
SAGE	Serial analysis gene expression
SAML	Security Assertion Markup Language
SAMV	Sparse asymptotic minimum variance
SAST	Static application security testing
SDV	Synthetic Data Vault
SIIF	Standard for Intercloud Interoperability and Federation
SMUAP	Single-motor-unit action potential
SNOMED-CT	Systematized Nomenclature of Medicine - Clinical Terms
SOAP	Simple Object Access Protocol
SORTA	System for ontology-based re-coding and technical annotation of biomedical phenotype data
SPECT	Single-photon emission computed tomography
SQL	Structured Query Language
SFF	Standard Flowgram Format
SQUIDs	Superconducting Quantum Interference Devices
SSL	Secure Sockets Layer
SVMs	Support Vector Machines
TEU	Treaty on European Union
TFEU	Treaty on the Functioning of the European Union
TLC	Thin Layer Chromatography
TLS	Transport Layer Security
TOSCA	Topology and Orchestration Specification for Cloud Applications
VLANs	Virtual Local Area Networks
VM	Virtual Machine
VMM	Virtual Machine Monitor
VPC	Virtual Private Cloud
VPN	Virtual Private Networks
W3C	World Wide Web Consortium
WBC	White Blood Cell
WLS	Weighted Least Squares
WSNs	Wireless Sensor Networks
WGS	Whole Genome Sequencing
XML	eXtensible Markup Language

CHAPTER

Introduction

1

Chapter outline

1.1 Origin of medical data

In medical research, cohort, case–control, and cross-sectional studies are three special types of observational studies [1]. A clinical cohort study comprises data from a group of people who share common disease occurrences and medical conditions (e.g., experience a common type of a chronic disease) and are useful for measuring the disease occurrence and progress [1]. A cohort study design can be either prospective or retrospective. In a prospective study, the cohort data are expected to be updated within the duration of the study, whereas in a retrospective study, the patient data are predefined. In prospective studies, the existence of individual follow-up time points is necessary to keep track of the upcoming data. On the other hand, cross-sectional studies measure the disease occurrence at one particular time point and thus are not able to capture the relationship between the occurrence and the progress of a disease. To understand the meaning of a cohort, it is necessary to understand the fundamental types and sources of medical data.

Laboratory results comprise a widely known source of medical data. Laboratory tests include a large number of biochemical tests [2], such as (i) hematological tests that measure the oxygen levels in the blood flow, urine tests that are usually used to detect kidney, liver disease, and diabetes, (ii) serological tests that are blood tests that seek for antibodies (e.g., to detect rubella, fungal infections), (iii) coagulation tests that are used to detect thrombophilia and hemophilia, (iv) histological tests that are employed to examine different types of tissues (e.g., muscle, nervous, epithelial), etc. Laboratory results combined with valuable information from medical conditions and medications can offer a powerful basis for (i) understanding the progress of a disease, (ii) dividing sensitive populations into subgroups (i.e., patient

Medical Data Sharing, Harmonization and Analytics. https://doi.org/10.1016/B978-0-12-816507-2.00001-3

stratification), and (iii) evaluating existing and/or proposing new treatments, in large-scale population studies. Other common parameters that can often be found in clinical datasets include demographic information (e.g., age, gender, socioeconomic factors), vital parameters (e.g., heart rate, blood pressure), medications (e.g., antibiotics, antiseptics) and medical conditions (e.g., Alzheimer's, Parkinson's), physical and mental conditions, nutrition habits, and environmental and lifestyle factors [3], among others.

Other sources of medical data include medical images that are obtained by a variety of diagnostic imaging modalities or systems, such as, computed tomography, magnetic resonance, optical topography, ultrasound, positron emission tomography, single-photon emission computed tomography, etc. Advances in surface-rendering and volume-rendering methods have led to three-dimensional medical image visualization that has significantly improved the quality of image interpretation. Moreover, the rapidly increasing spatial resolution of such systems combined with the technical advances in medical image processing (e.g., reconstruction, fusion) can significantly enhance the diagnostic accuracy and the consistency of the image interpretation by doctors in a variety of diseases ranging from heart failure, osteoporosis, and diabetes to Alzheimer's disease and cancer [4]. Undoubtedly, computer-aided diagnosis comprises one of the major computer-assisted technologies for medical diagnostics.

Biosignals comprise another domain of medical data including a variety of biomedical signals, such as (i) electroencephalography (EEG) and (ii) electrocorticography, which capture the electrical fields that are produced by the activity of the brain cells, (iii) magnetoencephalography that captures the magnetic fields that are produced by the electrical activity of the brain cells, (iv) electrocardiography that records the electrical activity that arises from the depolarization and repolarization activity of the heart, (v) electromyography that records the electric potential that is generated by the muscle cells, (vi) electrooculography that records the electric potential generated by the cornea and the retinal activity, etc. Biosignals provide high temporal information about a disease's onset and progress and have been employed in a variety of diseases ranging from epilepsy and schizophrenia to heart failure and muscle atrophy [5]. Biomedical signals are usually combined with medical imaging systems (e.g., EEG and MRI) to provide both high spatial and temporal information for more effective diagnosis and treatment. The advances in biomedical signal processing have made signal manipulation much easier.

The field of genetics constitutes a vast domain of medical data. Genetic data can be generated from high-throughput (next-generation) DNA and RNA sequences. The outrageous number of these sequences has created the well-known field of genomics. Genetic data are generally of more complex form than the aforementioned types of medical data because they require the use of multiple processing pipelines with unique input. This complexity arises from the different formats of the genetic data, such as the FASTQ files used for RNA sequence analysis, the haplotypes for haplotype analysis, etc. In the last decade, genetic data generated from genome-wide association studies have led to thousands of robust associations between common single-nucleotide polymorphisms (SNPs) and common diseases ranging from autoimmune diseases to psychiatric disorders, quantitative traits, and genomic traits [6,7].

The recent advances in omics technologies [8,9], such as genomics (the study of genomic information), transcriptomics (the study of all the RNA transcripts of an organism), proteomics (the study of proteins and their interactions), lipidomics (the study of lipids, i.e., biomolecules with structural diversity and complexity), and metabolomics (the study of the multitude of metabolites) have increased the demand for properly annotated and well-preserved biospecimens, which has led to the development of the biobanks [10]. Biobanking involves the (i) collection, (ii) processing, (iii) storage, and (iv) quality control of the biological samples along with their associated clinical information. Biobanks have been widely used for meeting scientific goals in genetic and molecular biology due to their long-term sustainability [10].

1.2 Toward medical data sharing and harmonization

Cohort, case–control, and cross-sectional studies are capable of resolving crucial scientific questions related to predictive modeling of a disease's onset and progress, the clinical significance of genetic variants, the adequate identification of high-risk individuals, and the effective patient selection for clinical trials, among others [1]. However, the fact that these cohorts are dispersed withholds important clinical information and leads to small-scale studies with reduced statistical significance and thus poor clinical importance. In addition, the rapidly increasing gap between healthcare costs and outcomes obscures the evolvement of a sustainable healthcare system that is able to adapt on the technological advances of our era [1]. Traditional clinical epidemiology poses several obstacles on the development of clinical decision support systems, public health policies, as well as medical research in general, as it is conducted by individual researchers who do not share any common research interests, a fact that often leads to indigent manipulation of the available clinical data and hampers health research and innovation.

On the other hand, the current technological advances in biomedical and health research significantly increase the generation of digital data, leading to vast amounts of data in a variety of disciplines, varying from finance to medicine [11]. This kind of data is widely known as big data. In the healthcare sector, big data have many sources varying from large-scale clinical trials, clinical registries, and electronic health records to medical imaging and genetic data. Medical big data comprise a powerful tool toward the establishment of an expandable and "smart" healthcare system that is able to improve the existing healthcare quality using machine learning to extract actionable knowledge. However, several technical challenges lie behind the concept of big data in healthcare—the heterogeneity of the protocols among clinical centers, the lack of tools to interpret and visualize such a large amount of data, and the dissimilarity and incompleteness of the dataset structures, to name but a few, are issues that need to be addressed by a modern healthcare system.

All these clinical needs promote the existence of a medical data sharing and data governance framework that is capable to effectively address the needs for (i) sharing medical data across international heterogeneous cohorts, (ii) assessing the quality of the data, and (iii) overcoming the heterogeneity among the cohorts, toward the establishment of a secure federated cloud system [12,13]. Such a framework will not only be able to interlink heterogeneous medical cohorts but also to lead to more accurate studies of rare diseases, i.e., studies with high statistical power. Indubitably, federated analysis must comply with all the legal, ethical, and patient privacy issues under the technical requirements and challenges of our era. These challenges involve the imposed regulations of the General Data Protection Regulation (GDPR) in Europe [14–16] and its effect on the existing medical infrastructures, its relationship with data protection regulations in other continents, such as the Health Insurance Portability and Accountability Act (HIPAA) in the United States [17], and finally novel methods for medical big data manipulation, visualization, and analytics for general purposes.

Data sharing is a complex procedure that faces several obstacles related to the heterogeneity of ethical and legal issues across different countries all over the world. Data sharing reduces the duplication of studies and provides cheaper and more transparent infrastructures for data curation and storage and thus more efficient infrastructures for conducting clinical research. However, the underlying fear for data abuse and the loss of data control comprise the main aspects that obscure the sharing of medical data. Moreover, data sharing is effective only when the scope of sharing is well defined, a fact that is not always taken into consideration in research studies. According to the former EU's Data Protection Supervisor, Mr. Peter J. Hustinx, privacy and protection is defined as "the right to respect for private life and the right to protection of one's personal data—are both fairly expressions of a universal idea with quite strong ethical dimensions: the dignity, autonomy and unique value of every human being" [14]. Toward this direction, strict data protection regulations provide legal barriers to privacy issues for avoiding any data breach, embezzlement, and misuse.

In the United States, the HIPAA of 1996, Public Law 104–191, which is part of the Social Security Act, aims to protect the healthcare coverage of individuals who lose or change their jobs, as well as embraces the sharing of certain patient administrative data for promoting the healthcare industry [17]. Toward this direction, federal data protection and legal obligation rules have been developed concerning the security and privacy of electronic health transactions to ensure the confidentiality, integrity, and security of health information. More specifically, the US Department of Health and Human Services (HHS), considering the fact that the technological advances could potentially erode the privacy of health information, developed two fundamental rules under the HIPAA, namely the HIPAA Privacy Rule and the HIPAA Security Rule. The former establishes national standards for the protection of healthcare electronic transactions including medical records and other personal health information that are conducted by healthcare providers. Moreover, the HIPAA Privacy Rule applies also during the development of any kind of small-scale and large-scale health plans. On the other hand, the HIPAA Security Rule sets national standards for the protection of personal health information that is created, received, used, or maintained by a covered entity [17]. To evaluate whether the security rules of a covered entity are fulfilled or not, the HHS Office for Civil Rights and the Office of the National Coordinator for Health Information Technology offer a security risk assessment tool.

In Europe, GDPR [14–16] imposes rules on the use of big data, the nature of biobanks, and research infrastructures in general, as well as the purposes of data processing, legitimate and liability exemptions, and data manipulation. In this book, the actors of the new data protection regulation are described from a medical point of view due to the nature of the book's context. The data controller is authorized to manage the patients' data, whereas the data processor can apply any kind of (authorized) preprocessing on the data. Great emphasis is given on the data owner (i.e., patient) who is the core of the data protection regulation and lies at the intersection of these two roles with respect to a final role, the one of the Data Protection Officer (DPO) who is responsible for supervising the compliance of the processes that involve the sharing and processing of private data. The data controllers and data processors shall prepare codes of conduct with respect to (i) the collection of private data, (ii) the pseudonymization of data, (iii) the legitimate interests pursued by the data controllers, (iv) the transparency and fairness in data processing, and (v) data minimization, among many others. More specifically, to conduct medical research, the data controllers must collect a variety of documents, varying from signed consent forms, the purpose of processing, and the contact details of the corresponding DPOs to data protection impact assessment (DPIA) reports and data protection guarantees. The patients are the real owners of their medical data and have the right to request any kind of processing on their data and even request these data, i.e., the right for data portability and the right to be forgotten.

The data protection compliance, as well as the quality of the data (e.g., data completeness), must be carefully assessed by the clinical experts before data sharing. The data governance framework is responsible for the evaluation of data quality and data completeness by taking into consideration existing public health policies. The ensuing framework is responsible for the evaluation of performance metrics (e.g., DPIA), the organizational structure of the clinical data, and information management, as well. As part of the data quality assessment process, the data curation involves methods for fixing potential missing and duplicate values, incomplete terminologies, and any related misinterpretations within the data, therefore enhancing the overall quality assessment process [18]. After assessing the quality of the involved data, harmonization takes place to deal with the dissimilarity of the data sources.

The structural heterogeneity of the clinical data across different countries is another crucial factor that introduces biases during the analysis of medical data. Data harmonization is a data-driven approach that aims to overcome the heterogeneity of medical cohorts worldwide by converting the heterogeneous datasets into homogeneous ones (e.g., with similar structure and terminologies), with minimum loss, based on the contents of each dataset [19]. Harmonization involves several mechanisms including dataset transformation to a common format, dataset description, similarity detection, terminology detection, and alignment. According to the literature, there are two types of methods for matching heterogeneous datasets, namely the semantic matching and the lexical matching [20–22]. The latter method simply seeks for string matches among the fields of the datasets using various distance measures, such as the Euclidean distance, by mapping each term to a point in the two-dimensional space. On the other hand, semantic matching does not only involve simple terminology matching but also considers for further associative relationships between the entities and the classes of the heterogeneous datasets, i.e., their ontological representations.

An ontology summarizes the entities, classes, terminologies, and vocabularies (knowledge) of a dataset's domain and can be usually expressed using a Resource Description Framework Schema format [23]. Except from the data themselves, the knowledge representation of the data can be also available in the form of ontology languages, i.e., OWL (Web Ontology Language) [24], so that the technical experts can further examine the elements used within each specific dataset. A semantic matching algorithm uses a source ontology and a target ontology as input, and the output is an aligned version of the source ontology based on the target ontology. In fact, the output is a set of proposed associations between the terms of the heterogeneous datasets where a domain expert is necessary for validation. Although the semantic matching procedure is more effective than lexical matching, it is semiautomatic, i.e., involves the manual mapping of each dataset to an ontology. A knowledge base is a promising solution to this limitation. The knowledge base is a collection of existing ontologies and vocabularies related to the domain knowledge of interest. Thus, the knowledge base can be combined with new ontologies to train a "smart" system that will be able to execute automatic semantic matching. To assure the privacy of the patient data in large-scale studies, data harmonization should be performed locally on each clinical center's private repository.

The fact that the data should be processed in a distributed manner introduces the concept of decentralized analysis according to which the data can never move out from the hospital. Data aggregation is mandatory before data sharing. Patient data must be already pseudonymized at the clinical databases, and unique identifiers must be constructed per patient. However, for example, in case that the patient data are anonymized, GDPR (in Europe) does not stand and no prospective analysis is supported. Cloud architectures should take into consideration the cost scalability over security, and online data collection has been characterized as a "killer application." In addition, the definition of a common structural data model (including structure terminology, terminology description, and terminology mapping) is a technical limitation as it requires time and large effort by the clinicians. The complexity of the hospital IT infrastructures by means of the clinical data formats is also a technical limitation before data harmonization. Finally, the nature of the primary data collectors and the secondary analysts must be clearly defined before the implementation of

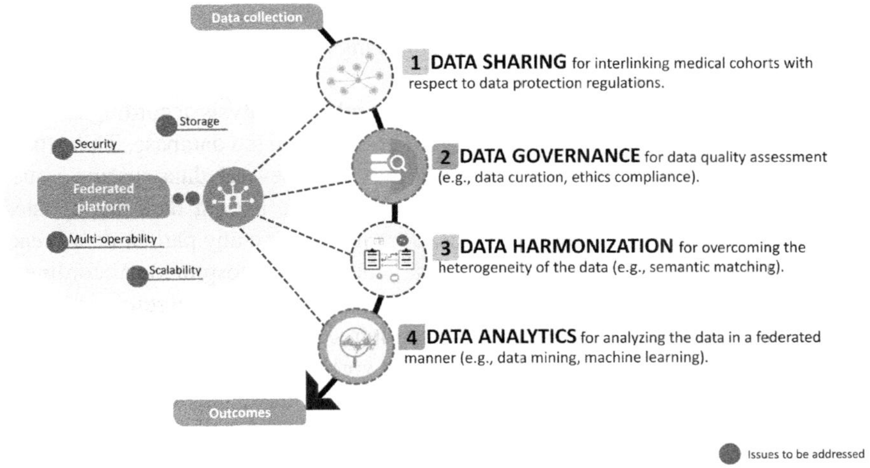

FIGURE 1.1

Steps toward federated analysis of clinical data.

a distributed data analytics architecture for federated analysis. The major steps toward the establishment of a federated cloud platform's architecture are presented in Fig. 1.1.

1.3 Distributed data processing architectures

Data analytics services consist of novel supervised and unsupervised learning algorithms for constructing data models, as well as testing these models on new data using various performance evaluation measures. These data models can be used, for example, to propose a drug to be tested on appropriately selected patients, construct accurate predictors for a disease's progress (e.g., risk stratification), reduce the risk of producing unsatisfactory results in clinical trials by employing expensive drugs, and therefore improve the patient diagnosis and screening procedure. A complete architecture for big data analytics should provide services for data preprocessing, knowledge extraction, and performance evaluation. Data preprocessing includes functionalities for (i) feature discretization, (ii) correlation analysis and statistical analysis to select clinical features that appear to be significant (feature selection), (iii) dimensionality reduction methods for extracting significant features from genetic data (e.g., transcriptomic), (iv) image preprocessing methods (e.g., reconstruction, segmentation, sharpening), (v) data modeling that involves the application of data mining algorithms on preprocessed medical data (e.g., laboratory measures) for pattern recognition, etc. Data mining consists of supervised (e.g., artificial neural networks) and unsupervised (e.g., k-means) methods for developing data models that are able to predict medical outcomes. The performance of these models, however, needs to be validated, and thus, cross-validation methods are necessary to evaluate the prediction accuracy of the data mining algorithm on random subsets of the input data.

The most common type of analysis is the centralized analysis according to which the data are gathered in a common database, i.e., a centralized database. This type of analysis, however, is prone to data breach and poses several data sharing issues, especially in the case of prospective studies where the patient data are updated. The distributed analysis concept (Fig. 1.2) does not involve any patient data breach as the data never move out from the clinical centers (e.g., hospitals). According to this concept, the initial data model is distributed from an authorized reference center (i.e., an external database) to each clinical center's local database for training and testing purposes (e.g., assume a Bayesian network model that is distributed and executed separately on each clinical center). In fact, the data model is executed on each clinical center (i.e., in a parallel way), and the individual results are returned to the reference center where they are finally combined and distributed to all involved clinical centers.

The distributed analysis scenario has many similarities with the Hadoop architecture [25]. The private data repositories in the clinical centers can be visualized as

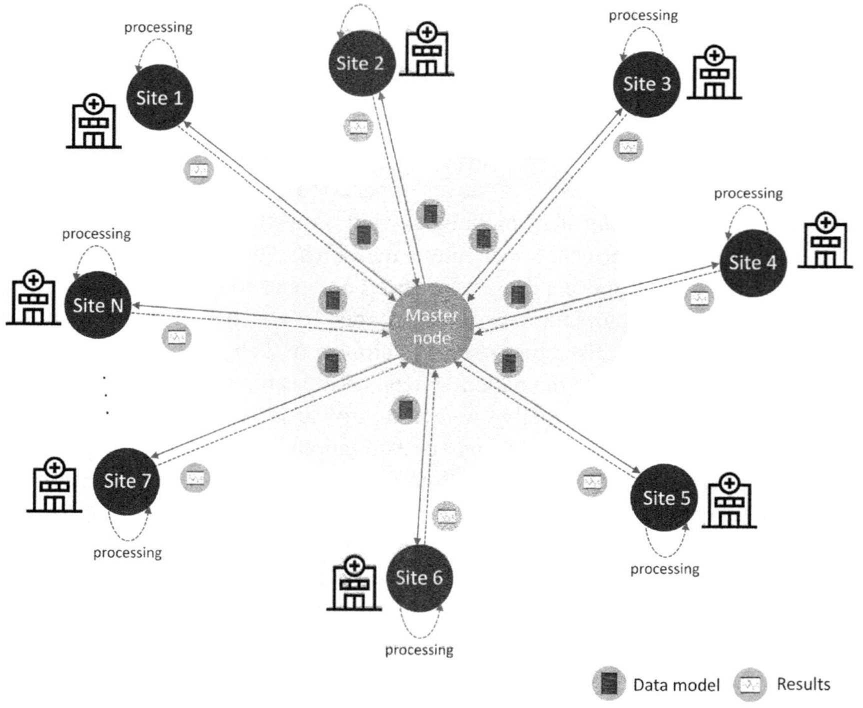

FIGURE 1.2

The concept of distributed analysis.

computational nodes that lie within a virtual environment. A master node distributes the model to each node, and each node applies the model to the corresponding data in a parallel execution manner. The results from the computational nodes are finally returned and combined to the master node. Through this way, the analysis is secure as the patients' privacy is ensured. However, biases are often introduced into the results due to (i) the heterogeneous structure of the data among the cohorts and (ii) the "nature" of the data mining algorithm (e.g., linear or nonlinear). The first factor can be effectively resolved by harmonizing the involved data. As far as the second factor is concerned, emphasis should be given to the development of appropriately defined distributed machine learning algorithms (e.g., sparse support vector machines) and distributed cross-validation methods, which are currently an ongoing research field of interest.

The nature of the data mining algorithm determines its ability to be adapted to a distributed environment. Emphasis must be given on the way that the results from the individual executions will be combined to produce the final result. So far, the largest portion of distributed systems for decentralized analysis is focused on extracting descriptive statistics (e.g., mean, variance, and histogram) from data in different sites. A distributed learning system requires the application of machine learning algorithms on decentralized data. The scenario of simply averaging the results across the computational nodes applies only when linearity stands (e.g., descriptive statistics, linear regression), which is a technical limitation as the majority of the data mining algorithms are nonlinear. Another challenge could be to seek for a way to parse the algorithmic parameters from one cohort to another, for example, when a clinician wishes to train a lymphomagenesis prediction model (i.e., a machine learning algorithm that is trained to predict lymphoma outcomes, using laboratory measures, age, gender, medical conditions, etc., as input) in two cohorts and evaluate its performance on a third cohort. Such technical challenges have been already addressed by the scientific community, and some of them are currently under investigation [26–28] due to the fact that they pose significant barriers toward the realization of a federated platform.

Another limitation toward the realization of distributed learning in decentralized facilities lies on the complexity of the hospital infrastructures. Installing software in medical facilities demands too much time for receiving the green light from the hospital's board members. In fact, a hospital can be seen as a local database that receives commands from an external database. A web portal is thus necessary to send the commands that need to be executed (e.g., a data mining algorithm to predict lymphoma outcomes like before). This web portal must guarantee that no malicious software will enter the facility of each hospital and must fulfill all the necessary requirements (e.g., appropriate licenses). A more conventional and rather innovative approach is the cloud. Each hospital can maintain a highly remote private cloud space that lies within the cloud where the data can be securely uploaded and stored. Then, the processing can be performed on each hospital's private cloud space, and the results can be combined as described in the first approach. A cloud, however, is still more sensitive to data breach as it has to deal with technical challenges related to the security protocols for data access and management. Moreover, the cloud administrator must provide appropriate contracts to the hospital's board that guarantee the privacy of the hospital's data, an agreement that is legally binding.

It is also suggested to make use of synthetic data for assessing the performance of the algorithms before the final application on the clinical data. The cloud administrators can also use synthetic data to assess the impact of the employed data obfuscation/aggregation methods for anonymizing patients' data, as well as evaluate various multilevel hacking attempts against the system to enhance the sustainability and overall safety of the cloud. These synthetic data (also referred to as artificial data) are the result of machine learning algorithms operating within global statistical models, such as the Synthetic Data Vault (SDV) system that has been launched by the MIT Institute for Data, Systems, and Society [29]. The SDV system automates the modeling and synthetic data generation for relational databases and is able to (i) operate on different relational databases (generalizability), (ii) allow users to define input parameters and synthesize their data (usability), and (iii) generate synthesized data that can realistically replace the original data (accuracy) [29]. As a result, researchers can make use of synthetic data to test disease models, without compromising the patients' privacy, and in fact "use machine learning to enable machine learning."

1.4 Scope and contribution to the state of the art

Medical data sharing, harmonization, and analytics aim to extensively describe the advances in medical data harmonization and analytics, as well as crucial aspects and challenges of the GDPR in Europe and its effect on the existing medical data sharing protocols, as well as its association with the HIPAA and other related data protection regulations in the United States. Medical data sharing is undoubtedly the most fundamental part of a federated database to promote data analytics services. As a matter of fact, it is important to investigate the data protection regulations across different countries to reveal the mechanisms that enable data sharing. Of course, such an attempt is difficult as each country has its own data protection regulations, especially the member states in the United States. One of the goals of this book is to shed light into these mechanisms by giving the reader the ability to comprehend the concept of cross-border data sharing along with the related technical challenges that lie behind data sharing. Data protection issues will be part of this attempt, highlighting the necessity to preserve the patients' privacy and rights before any data manipulation. The reader will be able to understand the scope of data protection regulations and be familiar with widely known terms, including data minimization, data pseudonymization, and its difference from anonymization, data standardization, data curation, and data with expiration date, among a variety of other ethical and legal obligations. The clinical needs of a federated platform will be stated along with all the related functionalities and requirements to accomplish them. Moreover, technical challenges posed by the scientific community will be stated including the necessity to establish standard cloud protocols that fulfill the data protection regulation, education on the power of big data in medicine, definition of common data models (e.g., structure terminologies), formation of official ethics group, legal advisory boards, etc.

In addition, this book presents clinical data harmonization methods for overcoming the heterogeneity among different clinical cohorts. All the recent advances in data harmonization techniques, varying from semantic matching and lexical matching to multivariate regression modeling and statistical matching, will be extensively presented. The reader will also become familiar with technical challenges and limitations in data harmonization. Realistic examples of lexical and semantic matching methods will be presented toward the homogenization of heterogeneous biomedical ontologies. Emphasis will be given in existing data sharing frameworks that include data harmonization as part of their overall strategy. Data sharing combined with harmonization is a compelling consolidation that offers the ability to collect and analyze heterogeneous data from different clinical cohorts to extract homogeneous data structures that aim to increase the statistical power of the clinical studies. The reader will understand the importance of a semiautomatic framework for prospective and retrospective data harmonization, as well as how it can be extended to (i) be able to coexist with data analytics in distributed environments and (ii) be as much automated as possible using knowledge bases that are enriched with existing ontologies. The importance of the knowledge base will be further highlighted to promote automatic semantic interlinking approaches so as to effectively reduce the clinician's involvement during harmonization.

This book goes one step further and extends this federated strategy by providing the latest advances in data analytics, including machine learning and artificial intelligence, which can be employed to analyze the harmonized data. Data mining approaches can be applied on the latter with the purpose of (i) extracting useful information concerning a disease's progress, (ii) evaluating the performance of a patient stratification model for effective patient treatment, and (iii) seeking for significant clinical parameters that will serve as biomarkers or validating any existing ones. Toward this concept, emphasis will be given in the description of technical issues related to novel big data mining and machine learning tools and most importantly how these tools can be combined to realize the concept of distributed analysis. Distributed frameworks comprise the second main area that this book intends to cover to fulfill the novel idea of a federated platform that is capable of analyzing homogeneous data in a decentralized manner by taking into account the data sharing regulations (i.e., ethical and legal), as well as the patients' data protection protocols employed by different clinical cohorts worldwide. The technical challenges and the clinical importance of the decentralized analysis will be further stated along with recent methodological guidelines and practical examples from existing as well as ongoing research studies.

Recent technological advances in distributed learning algorithms combined with existing data mining algorithms will offer the reader the opportunity to (i) better understand the fundamental basis and functionalities of a distributed framework,

(ii) meet the technical challenges and requirements toward the establishment of a distributed environment, and (iii) assess the concept of distributed frameworks in medical research followed by federated studies that make use of such methods. In addition, the reader will be able to understand the clinical need of data analytics in the medical field. These needs are mainly related to patient stratification for the early identification of high-risk individuals according to several clinical laboratory predictors, as well as to the identification of significant clinical features that can serve as potential predictors for the progression of a specific disease. The reader will understand how it is possible to address these needs using a virtual environment where all the tasks are performed in a parallel way and the individual results are combined and converged under certain criteria. In addition, the reader will become familiar with conventional and novel computational methods for analyzing sequential and transcriptomic genetic data, including SNPs identification and validation, correlations of existing SNPs with certain disease subphenotypes, haplotype analysis, differential expression analysis, case−control associations, and case−case associations, among others. All these methods will be followed by appropriate descriptions of state-of-the-art tools for depicting the analysis results so as to provide a complete overview of data analytics and big data visualization methods.

Indubitably, a federated plan needs to be somehow realized, and to accomplish that, a physical space is needed. Health research innovation clouds comprise a novel solution to this by providing all the necessary infrastructures for data collection and data analytics services toward the establishment of a federated database. A federated database is a high-level mechanism according to which the individual medical data from different medical cohorts are collected into private cloud spaces and thus ensures the patients' data protection. As might be expected, several technical challenges and limitations tend to obscure the attainment of such a platform. This book accounts for all these issues to offer the reader the ability to (i) comprehend the basis behind a cloud system, (ii) understand the importance of health clouds in medical research, (iii) meet the technical challenges related to the expandability and scalability of the cloud platform, and (iv) understand the way that data analytics services are executed in a cloud system.

Security issues concerning cryptographic primitives (e.g., symmetric and asymmetric key) for ensuring the secure transfer of the sensitive data will be presented along with the blockchain technology, an ambiguous field that is recently emerged and has been increasingly receiving attention nowadays with recent applications in biomedical engineering, mainly for medical data sharing. At this point, it is important to note that the blockchain-based federation of health innovation clouds can significantly increase (i) the findability, (ii) the accessibility, (iii) the schematic interoperability, and (iv) the reusability of data. The fact that the impact of all these mechanisms will be finally evaluated in clinical case studies covering a vast domain of pathological medical conditions, ranging from diabetes to cancer, enhances the benefaction of this book toward the enhancement of the reader's scientific domain knowledge. In a genuine manner, the eminence of this book lies on the fact that its contribution is concentrated on the intersection of five major and rapidly evolving

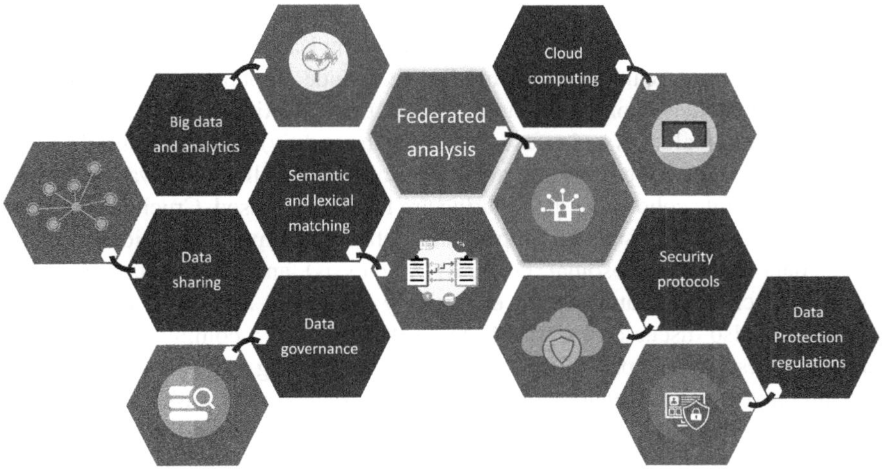

FIGURE 1.3

Contribution to the state of the art.

scientific fields, namely data sharing, data governance, data harmonization, (health) cloud infrastructures, and data analytics, as it is depicted in Fig. 1.3.

1.5 Organizational structure

Chapter 2, "Types and sources of medical and other related data," discusses the different types of medical data (e.g., mobile health, virtual populations, cohorts), the data acquisition protocols, as well as the sources of medical data. Emphasis is given on cohort studies and how they can be used in large-scale studies. The types and sources of medical and other related data (e.g., prospective, retrospective) are stated as crucial factors before any analysis of the medical data. The fact that the clinical cohorts do not follow a common type of data acquisition protocol hampers their analysis, as biases are introduced to the results. For example, hematological examinations do not always follow the standard clinical requirements among different clinical centers. Therefore, the importance of a clinical reference model that is able to address these minimum requirements is highlighted. The same stands for clinical images as different scanners produce images with different scale, intensity values, etc. In addition, guidelines are provided to deal with the complexity of the hospital IT infrastructures to (i) agree to a common format and (ii) convert the clinical datasets into that common format. A standard format should be well-defined by each research team before data sharing as a first step toward data harmonization is to deal exactly with the heterogeneity that lies among different clinical cohorts. Finally, the opportunities and challenges that yield from the power of big data in medical research are stated.

Chapter 3, "Medical data sharing," comprises the core of federated analysis. The rationale behind data sharing is discussed, which envisages to deal with the unmet needs in the current healthcare systems. These include (i) patient stratification, (ii) identification of new biomarkers and/or validation of existing ones, (iii) patient selection for clinical trials and effective treatments, and (iv) health policies development, among others. Opportunities and methodological challenges toward the interlinking of multiple cohorts for data sharing are further discussed. Crucial ethical and legal topics related to sharing of personal data (e.g., physical, socioeconomic), genetic data (e.g., chromosomal, DNA, RNA analysis), and health data (e.g., medical history, clinical treatment) are extensively discussed. Several standardization methods are also discussed for transforming the data into a common format. Data curation is described as part of the generalized data sharing and governance framework that involves functionalities for (i) automatic detection of missing values (bad features), (ii) extraction of new features, (iii) outlier detection based on classification algorithms, (iv) outlier removal, (v) detection of duplicate fields and highly correlated features, (vi) attribute identification and grouping, and (vii) automatic fill of missing values based on various methods (i.e., data imputation). Existing and ongoing data sharing frameworks are also described, such as the responsible sharing of genomic and health-related data [30] and the DataSHIELD framework for enabling statistical analysis of individual-level data without pooling them [31]. Several data sharing initiatives are finally discussed, such as the BioSHaRE platform [32], along with valuable guidelines and practical solutions against the misuse of the shared medical data.

Chapter 4, "Data protection," comprises the fundamental basis of data governance. The significant challenges posed by data protection are discussed in detail including the legal and ethical barriers, patient privacy issues, and technical limitations, among others. Emphasis is given on the GDPR (Regulation [EU] 2016/679) that aims to replace the existing Directive 95/46/EC in Europe and introduces new advancements in personal data protection within the European Union, affecting different areas of life, especially in healthcare systems. The two main roles of the GDPR are presented, namely the data controller and the data processor with respect to the data owner (i.e., patient). The role of DPO is also described for evaluating the compliance of all the processes that are executed within a healthcare platform. The impact of the European data protection directives on the existing ones in the United States is also discussed and further evaluated, including the Federal Trade Commission Act, HIPAA and legislation, and HIPAA security rules. The data governance framework that provides a complete mechanism for data quality assessment and distribution of the data analytics services is also described. Before data harmonization, it is necessary to assess the quality of the clinical data (e.g., the level of detailed description of the attributes).

Chapter 5, "Medical data harmonization," describes novel approaches that are able to deal with the structural and domain heterogeneity of disease-related datasets that originate from different clinical cohorts. The main idea behind this concept lies on the fact that data harmonization can increase the statistical power of the clinical studies as it is capable of combining heterogeneous data from different cohorts, in a homogeneous way, and thus yields a more effective analysis with high statistical power. Harmonization techniques, such as lexical matching methods for mapping lexical terms (strings) into points in a space using various distance measures, such as n-gram, Euclidean, and Levenshtein distance, are presented along with methods for semantic matching (which is also known as ontology mapping or ontology alignment). The lexical and semantic matching methods are followed by performance comparisons and appropriate pseudocodes. As far as semantic matching is concerned, guidelines to construct an ontology for describing the domain knowledge of a specific type of disease are also presented, including tools toward this direction, such as Protégé [33]. The need of a disease-specific reference model is also stated by means of an ontology that meets all the minimum requirements of the disease under examination. At this point, it is important to note that semantic matching is usually a semiautomatic procedure as the domain expert's assistance is necessary to define the domain type of each attribute. Toward this direction, emphasis is given on the existence of a knowledge base to train machine learning algorithms that will be able to automatically match ontologies using knowledge from existing disease-related ontologies, such as structures and vocabularies. Such a system will be further discussed and evaluated based on the current advances in harmonization. Several examples of software tools that are often employed for data harmonization are finally presented. Examples of these tools include the Opal software [22], the system for ontology-based recoding and technical annotation of biomedical phenotype data (SORTA) [20], the S-Match tool [34] for semantic matching to overcome the semantic interoperability problem, and the BiobankConnect software [21] for semiautomatic ontological and lexical indexing, among others.

Chapter 6, "Cloud infrastructures for data sharing," aims at describing popular cloud architectures and provides guidelines for secure cloud protocols for medical data sharing. The concept of cloud computing is presented along with its contribution in sharing medical data. Popular cloud architectures, such as the Infrastructure as a Service, Platform as a Service, Software as a Service, and Data as a Service, are further discussed. Security protocols including the Cloud Security Alliance, Institute of Electrical and Electronic Engineers standards, European Network and guidelines such as the Information Security Agency guidelines, and National Institute of Standards and Technology are also described followed by the related technical challenges. Encryption protocols for secure information transfer within private networks, such as the Secure Sockets Layer/Transport Layer Security, are also discussed. In addition, methods for implementing Representational State Transfer services under Virtual Private Networks are also presented along with popular security frameworks for sharing data, such as the rapidly evolving field of the blockchain technology. The latter can facilitate the consolidation of multiple data modalities for attaining a reliable standardization of data availability as well as of transactions that are dealing with heterogeneous big data. Crucial technical challenges are finally addressed including a cloud's (i) multidimensional interoperability, (ii) expandability,

(iii) scalability, (iv) storage, (v) user and data monitoring, (vi) automated error recovery, etc.

Chapter 7, “Machine learning and data analytics,” provides a complete overview of the state-of-the-art topics in medical data analytics. It enables the reader to better comprehend the importance of big data analytics in medicine along with the latest machine learning and data mining approaches employed by a variety of clinical studies. Modern machine learning and big data mining strategies are discussed and compared against their effectiveness and computational costs, as well as their ability to be adapted in distributed analysis environments. Methods to extend the functionalities of existing widely known data mining algorithms, such as the Support Vector Machines and the Naïve Bayesian networks, will be described and further evaluated to be able to work in a distributed way. The latter is important to create a framework for analyzing data from several sources without actually pooling them together at all. Of course, before recruiting any data mining algorithm for clinical data analytics, it is of great importance to preprocess the data. A simple data preprocessing mechanism allows for (i) automatic feature selection for biomarker identification (e.g., wrappers, filters, and embedded methods), (ii) the computation of new features, (iii) feature discretization, and (iv) sample selection under a predefined condition (e.g., an appropriate query that helps clinicians to select patients according to predefined criteria) and thus prepares the clinical data for effective analysis. The concept of distributed learning is then introduced to enable decentralized analysis. The latter is based on the fact that the clinical data never leave from the clinical centers, thus ensuring patients’ privacy. Popular data mining frameworks are further described including Apache Spark (e.g., MLlib) for the recruitment of data mining algorithms, Apache Hadoop for the implementation of virtual environments to enable distributed analysis, and other Python packages that are widely used for data mining (e.g., scikit) and deep learning (e.g., Keras, Tensorflow, Theano), as well. Algorithmic tools and software packages for visual analytics (e.g., Data-Driven Documents) and genetic data analytics (e.g., Haploview, Biopython) are finally presented. Emphasis is given on the opportunities and challenges that are posed by the implementation of distributed learning frameworks for ensuring patients’ privacy according to the rapidly involving concept based on the “bring the analysis to the data” design.

Chapter 8, “Case studies”, provides the reader with the ability to discover the latest clinical studies that combine medical data sharing, harmonization, and data analytics. These studies may vary from obesity and aging to cancer, studies on phenotypes, and studies on diabetes. Through this chapter, the reader can gain an overview of the clinical studies that combine all the aforementioned items into a variety of clinical fields and thus consummate this scientific book in a constructive way.

Chapter 9, “Conclusions and future trends,” summarizes the major outcomes of the literature on medical data sharing, harmonization, and analytics, along with the overall guidelines toward the design of a federated platform that takes advantage of distributed data analytics with respect to data protection regulations. The future trends toward the establishment of a modern healthcare system are finally discussed.

References

[1] Song JW, Chung KC. Observational studies: cohort and case-control studies. Plast Reconstr Surg 2010;126(6):2234–42.

[2] Nesar A, editor. Clinical biochemistry. 2nd ed. Oxford University Press; 2016.

[3] Egger G, Binns A, Rössner S, Sagner M. Introduction to the role of lifestyle factors in medicine. Lifestyle medicine. 3rd ed. Academic Press; 2017 [Chapter 1].

[4] Farncombe T, Iniewski K, editors. Medical imaging: technology and applications. CRC Press; 2017.

[5] Semmlow JL, Griffel B, editors. Biosignal and medical image processing. 3rd ed. CRC Press; 2014.

[6] Visscher PM, Brown MA, McCarthy MI, Yang J. Five years of GWAS discovery. Am J Hum Genet 2012;90:7–24.

[7] Visscher PM, Wray NR, Zhang Q, Sklar P, McCarthy MI, Brown MA, Yang J. 10 years of GWAS discovery: biology, function, and translation. Am J Hum Genet 2017;101(1): 5–22.

[8] García-Cañas V, Cifuentes A, Simó C. Applications of advanced omics technologies: from genes to metabolites, vol. 64. Elsevier; 2014.

[9] Karczewski KJ, Snyder MP. Integrative omics for health and disease. Nat Rev Genet 2018;19(5):229–310.

[10] Vaught J, Kelly A, Hewitt R. A review of international biobanks and networks: success factors and key benchmarks. Biopreserv Biobank 2009;7(3):143–50.

[11] Lee CH, Yoon HJ. Medical big data: promise and challenges. Kidney Res Clin Pract 2017;36(1):3–11.

[12] Liaqat M, Chang V, Gani A, Ab Hamid SH, Toseef M, Shoaib U, et al. Federated cloud resource management: review and discussion. JNCA 2017;77:87–105.

[13] Esposito C, Castiglione A, Choo KKR. Encryption-based solution for data sovereignty in federated clouds. IEEE Cloud Comput 2016;3(1):12–7.

[14] Hustinx P. EU data protection law: the review of directive 95/46/EC and the proposed general data protection regulation. Collected courses of the European University Institute's Academy of European Law, 24th Session on European Union Law; 2013. p. 1–12.

[15] Regulation (EU) 2016/679 of the European Parliament and of the Council of 27 April 2016 on the protection of natural persons with regard to the processing of personal data and on the free movement of such data, and repealing Directive 95/46/EC (General Data Protection Regulation). Off J Eur Union 2016;L119:1–88.

[16] De Hert P, Papakonstantinou V. The proposed data protection regulation replacing directive 95/46/EC: a sound system for the protection of individuals. Comput Law Secur Rep 2012;28(2):130–42.

[17] Atchinson BK, Fox DM. From the field: the politics of the health insurance portability and Accountability Act. Health Aff 1997;16(3):146–50.

[18] Stonebraker M, Beskales G, Pagan A, Bruckner D, Cherniack M, Xu S, et al. Data curation at scale: the data tamer system. In: CIDR; 2013.

[19] Fortier I, Raina P, Van den Heuvel ER, Griffith LE, Craig C, Saliba M, et al. Maelstrom Research guidelines for rigorous retrospective data harmonization. Int J Epidemiol 2017;46(1):103–5.

[20] Pang C, Sollie A, Sijtsma A, Hendriksen D, Charbon B, de Haan M, et al. SORTA: a system for ontology-based re-coding and technical annotation of biomedical phenotype data. Database (Oxford) 2015;18:2015.

[21] Pang C, Hendriksen D, Dijkstra M, van der Velde KJ, Kuiper J, Hilege HL, et al. BiobankConnect: software to rapidly connect data elements for pooled analysis across biobanks using ontological and lexical indexing. J Am Med Inform Assoc 2014; 22(1):65–75.

[22] Fortier I, Burton PR, Robson PJ, Ferretti V, Little J, L'Heureux F, et al. Quality, quantity and harmony: the DataSHaPER approach to integrating data across bioclinical studies. Int J Epidemiol 2010;39:1383–93.

[23] Manola F, Miller E, McBride B. RDF primer. W3C Recomm. 2004;10:6.

[24] Bechhofer S. OWL: web ontology language. In: Encyclopedia of database systems. Springer; 2009. p. 2008–9.

[25] White T. Hadoop: the definitive guide. O'Reilly Media, Inc.; 2012.

[26] Jochems A, Deist TM, van Soest J, Eble M, Bulens P, Coucke P, et al. Distributed learning: developing a predictive model based on data from multiple hospitals without data leaving the hospital – a real life proof of concept. Radiother Oncol 2016;121(3): 459–67.

[27] Deist TM, Jochems A, van Soest J, Nalbantov G, Oberije C, Walsh S, et al. Infrastructure and distributed learning methodology for privacy-preserving multi-centric rapid learning health care: euroCAT. Clin Transl Radiat Oncol 2017;19(4):24–31.

[28] Brisimi TS, Chen R, Mela T, Olshevsky A, Paschalidis IC, Shi W. Federated learning of predictive models from federated electronic health records. Int J Med Inform 2018;112: 59–67.

[29] Patki N. The synthetic data vault: generative modeling for relational databases. Massachusetts Institute of Technology; 2016. Doctoral dissertation.

[30] Knoppers BM. Framework for responsible sharing of genomic and health-related data. Hugo J 2014;8(1):3.

[31] Gaye A, Marcon Y, Isaeva J, LaFlamme P, Turner A, Jones EM, et al. DataSHIELD: taking the analysis to the data, not the data to the analysis. Int J Epidemiol 2014; 43(6):1929–44.

[32] Doiron D, Burton P, Marcon Y, Gaye A, Wolffenbuttel BHR, Perola M, et al. Data harmonization and federated analysis of population-based studies: the BioSHaRE project. Emerg Themes Epidemiol 2013;10(1):12.

[33] Protégé: a free, open-source ontology editor and framework for building intelligent systems, http://protege.stanford.edu/.

[34] Giunchiglia F, Autayeu A, Pane J. S-Match: an open source framework for matching lightweight ontologies. Semantic Web 2012;3:307–17.

CHAPTER

Types and sources of medical and other related data

2

Chapter outline

Medical Data Sharing, Harmonization and Analytics. https://doi.org/10.1016/B978-0-12-816507-2.00002-5

2.1 Overview

In our rapidly advancing technological area, the large volumes of accumulated data, on a daily basis, yield many benefits in different areas of our everyday lives including finance, medicine, and industry, among others [1–3]. These large-scale datasets are referred to as big data. The big data are characterized by four dimensions, namely the volume, the velocity, the veracity, and the variety [1–6]. The speed of the daily generated data, the amounts of collected data, the different types of collected data, and the biases that are introduced during the data collection process are the fundamental characteristics of the big data against the traditional datasets, which are only characterized by one dimension, i.e., their volume. The big data in medicine can improve the patient care through the enhancement of the clinical decision-making process, as well as enhance the statistical power of the clinical research studies yielding more accurate outcomes and powerful prediction models [1–6]. Furthermore, the big data can further enhance the development of effective patient stratification methods toward the identification of sensitive population subgroups, as well as provide better insights on large population groups toward the development of new public health policies and targeted therapeutic treatments.

There are many types of big data in medicine. These types of data vary from biosignals and medical images to laboratory tests and omics data. The biosignals are produced by the electrical activity that arises from the biological function of the organs in the human body. Examples of the most common types of biosignals include the electrocardiogram (ECG) [7], which records the electrical activity as a result of the heart's depolarization and repolarization function, the electroencephalogram (EEG) [8], which records the changes in the electrical activity as a result of the neural activation (i.e., the electrical field from the extracellular currents), along with the magnetoencephalogram (MEG) [9], which measures the changes in the ensuing magnetic field (from the intracellular currents), the electromyogram (EMG) [10], which records the changes in the electrical activity as a result of the muscles contraction, the electrooculogram (EOG) [11], which records the corneoretinal potential as a result of the eye movement, etc. The biosignals yield high temporal information regarding a disease's onset and progress, with numerous applications in medical conditions and diseases that vary from amnesia and schizophrenia to heart failure, myopathy, and Parkinson's disease [5].

The medical images comprise another type of medical data with significant importance in clinical diagnosis and screening procedures. Computerized tomography (CT) [12] scans, and magnetic resonance imaging (MRI) [13] scans, can provide detailed insight on the anatomic and tissue characteristics of different body parts, yielding high spatial information, and are useful in the detection of malignancies and other disorders. Furthermore, the positron emission tomography (PET) [14] scans, the single-photon emission tomography (SPECT) [15] scans, and the functional magnetic resonance imaging (fMRI) [16] scans provide additional information regarding the biological and physiological operations, i.e., the metabolic processes, at a molecular level, as well as the brain activations under specific physical and mental tasks. Furthermore, ultrasound [17] and photoacoustic [18] images are fast, nonionizing, real-time methods that are based on acoustic properties, having numerous applications in echocardiography, obstetric ultrasonography (US), intravascular US, and duplex US, among others. Spectroscopy-based methods, such as the functional near-infrared spectroscopy (fNIRS) [19], can also shed light into the measurement of the metabolic rate of oxygen consumption, which indicates a neural activation, similar to the fMRI.

The field of omics constitutes another vast domain of medical data with numerous subfields, such as the fields of genomics [20], lipidomics [21], proteomics [22], metabolomics [23], microbiomics [24], epigenomics [25], and transcriptomics [26], among others. The omics data can be generated from high-throughput next-generation sequencing (NGS) technologies [27], such as ribonucleic acid (RNA)-sequence analysis [28], mass spectrometry (MS) [29], and thin-layer chromatography (TLC) [30], which are able to analyze the proteins, lipids, transcriptomes, metabolic profiles of the biological cells, microorganisms in the tissues, pathological factors, and even whole human genome. The RNA-sequence analyzers are able to capture all the single cell—based (or even group-based) RNA molecules (i.e., the whole transcriptome). In addition, MS technology is able to reveal the structural and functional characteristics of proteins, as well as identify the lipids and their involvements in cell functionality. Omics can be used to study a variety of molecular-level functions, including the examination of bacteria and fungi on the tissues and organs, the interactions between the proteins, the detection of pathological factors and metabolic effects in degenerative and chronic diseases, and gene expression analysis, among others.

The laboratory tests along with the medical claims and the subscribed medications can offer a powerful basis for understanding the underlying mechanisms of a virus and detecting various pathological conditions in tissue specimens. The most common laboratory tests include the hematological tests, the serological tests, the skin tests, the histopathological tests, the immunological tests, the endocrine function tests, and the coagulation tests, among others. Straightforward methods, such as microscopic analysis [31], fluoroscopy [32], immunocytochemistry (ICH) [33], and immunohistochemistry (IHC) [34], are used to analyze the tissue and blood samples. Each test offers a unique insight on a medical condition or a disease toward the detection of blood clotting disorders, tumors, anemia, diabetes, fungal infections, autoimmune disorders, skin cancer, allergies, inflammatory disorders, and endocrine dysfunctions, among many others.

The sources of medical data are many. With the growing number of large volumes of daily generated data from health sensors, medical images, laboratory tests, electronic patient records, patient registries (PRs), clinical and pharmaceutical claims, and genome registries, the estimated amount of data is expected to overcome the zettabyte (10^{21} gigabytes) and even the yottabyte (10^{24} gigabytes) [1,35]. The medical data acquisition process is often conducted according to international standards and protocols for each type of medical data. For example, in signal acquisition, well-known international standards are used for the placement of surface electrodes, such as the

12-lead placement [7,36] for ECG signal acquisition and the International "10–20" system (and "10–5" system) [37] for EEG signal acquisition. In laboratory tests, hemodynamic, coagulation, serological, and immunoassay analyzers are most commonly used for measuring biochemical (e.g., blood pressure, blood clotting time) and pathological factors (e.g., the presence of antigens in the antibodies), as well as analyzing tissue specimens (e.g., for skin cancer, endocrine disorders), under different measurement units. Medical image acquisition protocols are also used for the reconstruction of MRI, CT, fMRI, PET, and SPECT images, such as the filtered back-projection (FBP) algorithm [38], the family of the iterative reconstruction algorithms, such as the algebraic reconstruction algorithm [39], and the iterative sparse asymptotic minimum variance (SAMV) algorithm [40], as well as the universal backprojection algorithm [41] for ultrasound imaging reconstruction, toward the examination of tissues and organs for tumors and other disorders. In the field of omics, standard methods, such as the microarray analysis [42], the RNA-sequencing analysis [28], MS [29], TLC [30], along with the high-throughput NGS technology [27], are widely used in omics to study the proteins interactions, the genetic profiles, and metabolic effects of different viruses, lipids, whole transcriptome, and genetic profiles of the human microbiome, among many others.

A research-oriented source of medical data with high clinical significance is the cohorts. Cohort studies are special types of observational studies [43] that are used to examine a disease's origins and the effects of the population characteristics [44]. The longitudinal cohort studies are observational studies that involve the repetitive collection of patient data over long (or short) periods of time and are able to provide deeper insight on the disease progress over time with increased accuracy, overcoming recall biases [43,44]. In general, a cohort study can use either retrospective or prospective data. The retrospective cohort studies make use of data that have been already collected with the purpose of identifying the association between the causes (symptoms) and the disease's outcomes [45]. On the other hand, the temporal dimension that is introduced by the prospective cohort studies (i.e., the follow-up data) can reveal significant associations between the disease's outcomes and the causes of the disease, as well as the effects of various prognostic factors on the outcomes over time. The risk ratio (RR) and the hazard ratio (HR) are mainly used to quantify the associations between the drug exposure and the outcomes, as well as the frequency of death, as a ratio between the exposed group and reference (or control) group [46,47]. The former includes the subjects that are exposed on a specific drug, whereas the latter consists of healthy individuals. Cohort studies are able to overcome several limitations that are present in traditional clinical trial studies by (i) measuring patient-specific outcomes from large population groups, (ii) keeping track of follow-up patient data, and (iii) being less expensive than large-scale clinical trials [48]. An example of the clinical importance of a cohort study lies on the fact that it can address the unmet needs in the special case where the exposure is a rare condition, such as an autoimmune disease [43,44]. In practice, a well-designed cohort study can provide deep insight into the underlying mechanisms of a disease's onset and progress.

2.2 Types of medical data

Below we provide some details on the sources of data mentioned above.

2.2.1 Biosignals

The biosignals are produced by the electrical activity that arises from the biological activity that takes place within different tissues and organs of the human body. The most common types of methods that are currently used to record biosignals in clinical research are presented below along with a brief description of their functionality and related clinical applications.

- EEG signals: these types of signals are produced by the electrical activity of the brain cells. When a neuron fires, an action potential is generated as a result of the exchange of ions that occurs inside and outside the neuron's cell [8]. This causes an alteration in the electrical charge from negative to positive and thus generates an ionic current (extracellular current) that is then propagated through the neuronal axons to other neurons, and as a result an electrical field is generated. This field is propagated throughout the brain and can be recorded by electrodes that are placed around the scalp. EEG signals consist of various brain rhythms (brainwaves) including delta (0.3–4 Hz), theta (4–8 Hz), alpha (8–14 Hz), beta (14–30 Hz), and gamma (>30 Hz), with each one having a clinical importance in disease and pathological diagnosis [49]. EEG signals have been extensively used in clinical research to study potential fluctuations under specific events (i.e., event-related potentials [ERPs]) [50], as well as in various pathologies including epilepsy [51], schizophrenia [52], dyslexia [53], and dementia [52].
- MEG signals: these types of signals are produced by the magnetic fields that are generated by the electrical activity of the brain cells. The electrical activity that is generated by the neuronal triggering produces an extremely weak magnetic field (as a result of intracellular current flow) that can be only recorded by powerful magnetometers, known as superconducting quantum interference devices (SQUIDs) [9]. SQUIDs are usually placed in liquid helium and are able to capture the extremely small alterations in the brain's magnetic field ($\sim 10^{-15}$ T), when the Earth's magnetic field varies between 10^{-4} and 10^{-5} T [54]. For this reason, the MEG examination is performed inside magnetically shielded rooms to shield out the inference of outside magnetic fields [9]. The main advantage of MEG against EEG is that the former is not affected by the electrical field's distortion during its propagation through the skull, scalp, and cerebrospinal fluid. Thus, the MEG yields both higher spatial and temporal resolution [9]. However, the MEG equipment is very expensive due to its superconducting technology and is often subject to high noise levels. MEG has been used for the examination of neocortical epilepsy regions due to its high spatial resolution [55], amnesia [56], etc.

- EMG signals: these types of signals are produced by the electric currents that are generated by the muscle contraction. The depolarization and repolarization of the skeletal muscle produces a difference in the electrical potential within the muscle cells (i.e., an electrical field), which propagates throughout the muscle fiber [10]. The electrical activity of the selected muscle is detected by surface electrodes. A needle is often used to stimulate the skeletal muscles, yielding the single motor unit action potential (SMUAP) with an amplitude of 300–400 μV. EMG signals are used to detect anomalies in the activity of the muscles, including myopathy and neuropathy [57], as well as in biomechanics for the development of body prosthetics [58].
- EOG signals: these types of signals are produced by the electric potential that is generated by the cornea and the retinal activity during eye movement. A typical EOG records the electrical field that is produced by the difference between the cornea's positive potential and the retina's negative potential, i.e., the corneoretinal potential, with an amplitude from 0.4 to 1 mV [11]. EOG has been used as a method for removing ocular artifacts in other biosignals, such as EEG [59], as well as for studying the eye movement in human–computer interaction systems [60]. Other relevant procedures include the electronystagmography [][61] that records the eye movement during nystagmus.
- ECG or EKG signals: these types of signals record the electrical activity that arises from the depolarization and repolarization activity of the heart [7]. A typical ECG records the P wave as a result of the right atrium's activation (<80 ms), the QRS complex (<120 ms) as a result of the depolarization between the left and the right ventricles, the T wave (<160 ms) as a result of the repolarization of the right and the left ventricles, and finally the U wave as a result of the repolarization of the interventricular septum [7]. The intervals between the waves can be used as indications of abnormal heart activity, e.g., a prolonged PR interval from the atrial activation to the beginning of the ventricular activation might indicate heart failure [62] and a wide QRS complex might denote both left and right bundle branch block [62]. Furthermore, ECGs have been widely used for studying arrhythmias [63], coronary artery disease [64], and other heart failure conditions.
- Phonocardiography (PCG) signals: these types of acoustic signals record the sounds that are produced by the heart's beat and the blood flow (murmurs) between the heart valves [65]. PCGs have the same origins as ECGs, and they have been used to study abnormalities on heart sound for the detection of heart defects (e.g., cardiomyopathy) [66], as well as for biometric identification [67].
- Electrocorticography (ECoG) signals: these types of signals can directly capture the extracellular currents that are produced by the electrical activity of the brain cells within the cerebral cortex [68]. ECoG signals have been widely used to localize epileptic zones before epileptic surgery with very high precision [69] and for the localization of activated brain regions using motor- or somatosensory-evoked potentials through a procedure that is known as electrical cortical stimulation [][70]. ECoG signals yield high temporal resolutions, and their invasive

nature, however, involves surgical operation procedures, a fact that makes it difficult to obtain with the exception of heavy medical conditions.

In all types of biosignals, the sampling frequency (or sampling rate) and the recording duration are directly proportional to the size of the acquired data and the speed of data acquisition process. For the majority of the biosignals, the modern recording systems use sampling frequencies that may vary from 50 to 500 Hz to 1 kHz and even 10 kHz (i.e., 10,000 samples per second). According to the signal recording time (e.g., on an hourly basis or on a daily basis) and the number of bits used for encoding the samples (e.g., 8 bits), the data size may vary from MB (megabyte) to even GB (gigabyte) of digitized signal data. For example, the ECG biosensors for patient monitoring can record the activity of the heart for hours or weeks with sampling rates that vary in 50–250 Hz, yielding huge amounts of accumulated data (e.g., a system with a sampling rate 250Hz can record more than 21 million samples per day).

2.2.2 Medical images

The medical images can be obtained by a variety of diagnostic imaging modalities or systems. Widely used sources of medical images in clinical research include the computerized axial tomography (CAT), MRI, fMRI, PET, SPECT, the optical coherence tomography (OCT), fNRIS, the US imaging, and the photoacoustic imaging, where each type of medical image has a unique clinical meaning. The origins and applications of these types of medical images in clinical research are summarized below:

- X-ray CT or CAT: a medical tomographic imaging technique that uses X-rays (ionizing radiation) to construct highly detailed cross-sectional images (slices) of different body parts [12]. An X-ray tube produces X-rays that are passed through the patient's body parts (in parallel, fan, or cone shape), yielding radiation intensities (projections) that carry information regarding the structure of the examined body part [12]. The whole system is placed into a gantry that is able to generate projections around the body, which are captured by the X-ray detectors as electrical signals. The structural image is finally reconstructed using a special computer software. Multiple images can be obtained according to the scanner's ability to generate multiple cross-sectional images (i.e., the slice thickness). CT scanners have high spatial resolution and are able to detect fluctuations in small area units yielding clinical images with high anatomic detail [12]. Contrast agents can be also injected into the body to observe changes between normal and abnormal tissues (lesions) [71], as well as track the arterial blood flow (angiography) to examine for blood vessel diseases, such as aneurysms [72]. However, the radiation level that is followed by a CT scan is large enough and shall not exceed certain levels. CT scans have been widely used in clinical research for the three-dimensional (3D) reconstruction of the arteries, heart, and other organs [73] and cancer treatment [74], among others.

- MRI: a nuclear medical imaging technique that uses strong electromagnetic fields to excite the hydrogen atoms in the body and produce cross-sectional images highlighting the structure of the tissues in different planes [13]. The hydrogen atoms are excited using radiofrequency (RF) pulses that are emitted in a specific frequency range (i.e., the Larmor frequency), yielding an energy release from the nuclei of the hydrogen atoms in the form of RF signals [13]. The latter are recorded by RF detectors. As the hydrogen atoms have different magnetic characteristics in different tissues, the MRI scanner is able to capture various pathological conditions inside the organs of the body in the complete absence of ionized radiation. Echo-planar imaging is usually used to obtain multiple slices with a single RF excitation pulses, in almost milliseconds, overcoming motion artifacts [75]. Popular types of contrast MR images include the T1-weighted images that are produced using a short repetition time between the RF pulses, yielding images with good anatomic information, and the T2-weighted images that are produced using a long repetition time between the RF pulses, yielding images with high tissue detail, which are useful for the detection of tumors [76]. A practitioner can configure the contrast of the MR scanner (i.e., the T1/T2 ratio) to construct images that can be used to detect benign or malignant tumors [77], neoplasms [78], inflammatory myopathies [79], etc. An emerging MRI-based technique is diffusion weighted imaging, which is based on the motion of the water molecules in the human body [80]. A similar approach is the diffusion tensor imaging, which enables the examination of the brain's white matter tracts and fiber tracking by computing the diffusion tensors that describe the diffusion anisotropy per pixel [81].
- fMRI: a straightforward medical imaging technique that is based on the principles of MRI but offers an additional new temporal dimension that accounts for the changes in the blood's oxygen levels, i.e., the blood oxygen level–dependent contrast mechanism [16]. When a neuron fires, the blood flow on the neuron's region is increased, a fact that alters the concentration of oxygen in the hemoglobin (HBC). The change in the oxygenation state of HBC alters the strong surrounding magnetic field that is applied. The fMRI BOLD contrast mechanism is able to capture these fluctuations in the magnetic field and thus provides functional information regarding the activation of different neuronal regions according to different tasks. Although the fMRI is both spatially and temporarily limited, it can enable the mapping of different activities of the human body, such as sensory, motor, processing, memory, cognition, emotion, speech, and language tasks with brain regions [82], as well as for the detection of brain regions that are activated during epilepsy [83].
- PET: a molecular-level medical imaging technique that makes use of positron-emitting radioactive isotopes to detect the photons that are produced as a result of the radioactive decay of the positrons during their interaction with an important biological molecule, such as amino acids [14]. The radioactive decay is recorded by the PET detectors. Thus, the PET scan can track the position of the isotopes and provide significant clinical information regarding the biological and

physiological operations, i.e., the metabolic processes, at a molecular level, from different angles. Thus, the PET scan can record the temporal evolution of undergoing molecular-level interactions, e.g., the degree of the nucleic acid uptake and absorption as well as from other tracers too, a process known as compound labeling. PET scans are extensively used in clinical oncology to reveal the biochemical characteristics of tumors using the flurodeoxyglucose as a radioactive isotope, based on the fact that the glucose molecules are highly susceptible by tumors [84]. PET scans can be also combined with CT scans yielding the hybrid PET-CT scan that can provide images that combine both anatomic and functional information regarding biochemical and physiological conditions [85].

- SPECT: a medical tomographic imaging technique that is based on the principles of PET to detect gamma ray photons that are emitted by radionuclides using a scintillation camera to produce a 3D radioactivity distribution [15]. The SPECT scan records the distribution of the radionuclide uptake within the body, i.e., the radioactivity distribution, from different angles and thus provides functional information from multiple projections regarding the uptake of the radionuclides from the organs. SPECT uses typical radionuclides that do not emit positrons in the case of PET where highly expensive, short half-life radioisotopes are used, which are generated inside radionuclide generators. SPECT can be combined with CT to reduce the noise levels that are often introduced by the scattering of the gamma ray photons [86]. SPECT clinical applications include myocardial perfusion [87], tumor imaging [88], and cerebral blood flow imaging [89].
- OCT: an optical medical imaging technique that produces cross-sectional images of the internal structure of biological tissues and organs by capturing the back-scattered light from the tissue [90]. The produced images are of high quality (100 times larger than the ultrasound), on the micron scale (1–15 μm), and can be obtained in real time, making OCT an emerging diagnostic imaging modality. OCT has many applications in ophthalmology for detecting glaucoma and macular edema [91], in arterial pathologies for studying plaque development [92], in gastric and esophageal cancer for identifying lesions using catheters/endoscopes [93], etc.
- fNIRS: a medical imaging technique that measures the oxygen consumption in the blood flow within different types of tissues based on spectroscopy [19]. The NIRS measures the absorption of oxygenated and deoxygenated HBC by the living tissue in the near-infrared electromagnetic spectrum, as well as the oxygen delivery in the living tissue, and thus it is able to record the metabolic rate of oxygen consumption in different tissues. fNIRS is based on the principles of the BOLD contrast mechanism (the basis of fMRI), by measuring the hemodynamic response in different brain regions, where an increased hemodynamic response denotes an activation in the region of interest (ROI). fNIRS is a relatively fast technology that uses fiber optics to transfer the changes in absorption of the light that is emitted by sources onto the head's surface. A similar technique is the optical topography that measures the changes in the blood concentration using

near-infrared light [94]. fNIRS is easily portable and can be used to assess the functional recovery after stroke and other traumatic brain injuries [95], as well as to measure the cortical blood flow and activation during physical activities, e.g., sensorimotor tasks [96].

- US: a real-time, medical imaging technique that is based on the principles of ultrasound to produce ultrasonic images that capture the echoes that are produced by soft tissues when the probes (transducers) generate sound waves with high frequency (>20 kHz) [17]. In an ultrasound imaging system, a probe is usually used to generate sound waves that travel into the tissue. The sound waves are reflected or backscattered from the internal tissue structures having different acoustic properties. The frequency of the sound waves can be reconfigured to yield better resolution of the tissues or more tissue depth. Although ultrasound imaging does not make use of ionizing radiation and is easily portable with real-time evidence, it does not provide detailed anatomic information regarding the structures behind the bones and sometimes it is difficult to interpret the outcomes [17]. US has many clinical applications including the obstetric US that is used for examining the fetus during pregnancy [97], the echocardiography that is used to examine the functionality of the heart valves [98], the intravascular US [99] that is used to visualize the blood flow within the blood vessels using catheters with attached probes to detect coronary artery lesions, and the duplex US [100] that makes use of the Doppler effect to display the tissue movement and blood flow to detect stenosis in carotid arteries and intracerebral arteries, among many others. The B-mode US has been also proposed, enabling the visualization of anatomic details in the ultrasound images for atherosclerotic progression [101].
- Photoacoustic imaging: a real-time, medical imaging technique that is based on the principles of the photoacoustic effect [18]. A photoacoustic imaging system consists of a nanosecond pulsed laser that generates pressure waves that are emitted into biological tissues that produce photoacoustic pressure waves that can be recorded by ultrasonic transducers yielding an image of the tissue. During the pulse emission process, the biological tissues absorb the light energy and convert it into heat. This causes the biological tissue to expand its size due to the changes in its thermoelastic properties (the thermoacoustic effect) and to produce sound waves (in the frequency range of MHz) that propagate and are recorded by the ultrasound transducers. Photoacoustic imaging is a hybrid method that combines the spectral characteristics of optical imaging with the spatial information of ultrasound. Photoacoustic imaging has numerous applications in the field of medical diagnostics including cardiovascular diseases [102], cancer detection, diagnosis and treatment [18], and brain function [103], among others.

The number of acquired images may vary from several hundred to even thousands of images per scan depending on the slice thickness technology that is used. The majority of the medical imaging methods generate a series of two-dimensional (2D) image/slices that are organized into a 3D structure. In the majority of the imaging acquisition protocols, the size of a 2D slice/image may vary from 256 × 256 pixels and 512 × 512 pixels to even 1024 × 1024 pixels (in the modern scanners). The size of a voxel depends on the spatial resolution (slice thickness) and the field of view of the scanner, where a voxel is a pixel in the 3D space (i.e., a volume element). As the modern scanners are characterized by high spatial resolution, they are able to capture very small fluctuations in pixels with extremely small dimension, such as 0.5×0.5 mm^2, and in voxels with size (width, height, depth) $= 0.5 \times 0.5 \times 0.5$ mm^3 (depending on the slice thickness), yielding images that consist of millions of volume elements. Therefore, the smaller the pixel size, the higher the resolution of the images. In the majority of the cases, a voxel is usually stored as 1 byte (i.e., 8 bits) according to the Digital Imaging and Communications in Medicine (DICOM) standard (see Section 2.3.3), where in each imaging type, the voxel represents a unique information (see Section 2.3.3 for the medical imaging standards). For example, the voxel in a CT scan represents the absorption rate of the X-rays from a specific part of the body. In 3D ultrasound imaging, the voxels represent the volumetric flow rates and densities of the organs. In MRI scans, the voxels represent the tissue characteristics as a result of the hydrogen excitation in the water molecules. A scanner that generates 1000 images/slices with 512 × 512 voxels per slice, in the 3D space, can produce images with a total size 256 MB (512 × 512 voxels/slice x 1000 slices × 1 byte/voxel). In all cases, the image size is equal to the header size (which includes patient metadata and imaging acquisition specifications) plus the pixel data size, which is calculated by multiplying the number of columns with the number of rows, the pixel depth (i.e., the number of bits that are used to encode the pixel's information), and the number of slices (or frames).

2.2.3 Omics

The field of omics consists of a large number of areas/types of omics, with each one having a specific unique clinical importance toward understanding the underlying mechanisms behind the cellular interactions. Examples of omics data types include the following:

- Genomics: the study of structural and functional genomic information. Structural genomics involves the coding of whole genomes, whereas functional genomics involve the investigation of the genome's behavior during the gene's development process, as well as under environmental circumstances [20]. A genome consists of genes, i.e., RNA or deoxyribonucleic acid (DNA) sequences, which carry valuable genetic information regarding the formation, development, and functionality of the biological cells. The DNA and RNA are nucleic acids that comprise the fundamental genomic materials of any living organism along with the lipids, proteins, and polysaccharides [104]. The clinical applications in the field of genomics include chromosome mapping (the assignment of genes to chromosomes, i.e., DNA molecules that carry portions of genetic material), coding unannotated genes, and mapping genetic variants to phenotypes (i.e., traits of an organism) and haplotypes (i.e., genetic variants that are grouped together) using single-nucleotide polymorphism (SNP) genotyping arrays to study the implication of genetic variants in different diseases [105]. In contrast to genomics, the genetics is a smaller field that studies the behavior of specific genes instead of the whole genome.

- Transcriptomics: the study of all the RNA transcripts of an organism. The RNA is a nucleic acid that is involved in different biological processes including gene expression and gene coding and decoding [26]. The transcriptome refers to all the RNA molecules that are present in a single cell or in a group of cells, where the messenger RNA (mRNA) serves as a transient intermediary molecule that transfers the genetic information (i.e., the nitrogen-containing nucleobases: guanine, uracil, adenine, and cytosine) that is used to synthesize the proteins. High-throughput RNA-sequencing and microarray technologies can capture the whole transcriptome, yielding important details regarding the gene regulation and the gene expression in different tissues and organs of the human body and thus being able to capture the genetic profile of an organism. This is the reason why transcriptomic technologies, such as RNA-sequencing [28], are often employed to annotate previously unannotated genes and understand the underlying cellular mechanisms of diseases, such as cancer [106].
- Proteomics: the study of the proteins, their topological characteristics, and further interactions between them [22]. High-throughput technologies, such as MS [29], are used to detect proteins from the peptides and thus shed light into the structure and functionality of the proteins in specific types of cells. Proteins are involved in a variety of biological processes including the cell growth and the intracellular communication, among others, and thus it is important to understand their expression profile in terms of how they function and interact with other proteins. Toward this direction, the analysis of protein−protein interactions (PPIs) using concepts from graph theory is able to simulate the complex interactions between the proteins and thus reveal functional associations between them under a given biological process [107]. Furthermore, the topological characteristics of the PPI networks can reveal hubs, i.e., proteins with increased participation in different biological processes, as well as identify specific groups of proteins (clusters) that exhibit common functional characteristics [107].
- Lipidomics: the study of lipids, i.e., biomolecules with structural diversity and complexity. Lipids are used to address the cellular needs for energy storing, signaling, and regulation and are highly engaged in preserving the balance within a biological cell [21]. Any attempt to disturb this balance has a direct effect on the chemical properties of the lipids. Therefore, the lipids can reveal important information regarding pathophysiological conditions, such as inflammatory and neurodegenerative diseases [108]. TLC [30] and MS [29] are the two most

commonly used methods for identifying the lipids and understanding their functions in any biological cell.

- Metabolomics: the study of the multitude of metabolites (e.g., amino acids). Metabolomics involves all those processes that involve the characterization of the metabolic profiles of cells according to a genetic variation [23]. Technologies such as nuclear magnetic resonance and MS are widely used to detect the metabolic changes in cells under different metabolomics experiments (e.g., under a drug administration). Metabolomics can provide valuable information related to the metabolic effects of cancer cells in the presence of drugs, as well as assess the chemical toxicities of a proposed drug in different organs and tissues of the body during preclinical drug development [109].
- Microbiomics: the study of the effect of microorganisms in the human physiology [24]. These effects include the negative implications of bacteria, fungi, and other related microorganisms in the tissues and organs. These microorganisms are usually referred to as microbiota. The genes of the microbiota form the microbiome. The high-throughput NGS technology [27] can be used to generate genetic profiles of the human microbiome and thus identify microbiota variations [109,110] for various human diseases, including heart disease, diabetes, and cancer, among many others.
- Epigenomics: a field that focuses on studying the effects of pathological factors, such as degenerative diseases, cancer, and metabolic disorders to the whole-gene expression [25]. These effects usually yield alterations in DNA methylation, proteins, enzymes, and other epigenetic marks. In fact, epigenomics aims at exclusively identifying gene expression implications "on" the genome. The most commonly used method for epigenomics analysis is the chromatin immunoprecipitation (ChIP), which is combined with high-throughput sequencing (HTS) [111]. This enables the genome-wide epigenetic analysis so as to study for the effect of the isolated DNA from chromatin-associated proteins that have been shown to play an important role during cancer development [112].

The majority of the omics data, including genomic, proteomic, and transcriptomic, among many others, are generated by the NGS analysis, a straightforward technology that uses high-throughput methods for molecular profiling, such as transcriptome sequencing (RNA-Seq) [28]. The raw data that are generated by the NGS technology, for an individual sample, may vary from 10 gigabytes (GB), e.g., in the case of microRNA sequencing (miRNA-Seq), and 20 GB, e.g., in the case of RNA sequencing (RNA-Seq), to even more than 200 GB, e.g., in the case of whole-genome sequencing (WGS) [27]. The size of the generated sequencing data (raw data) is proportional to the number of samples under examination. The raw data are usually stored in a FASTQ format (see Section 2.3.4) that is the fundamental input of an NGS analysis workflow that is executed by parallelized, high-performance computing pipelines. The output of the pipeline is stored in the SAM format (see Section 2.3.4).

2.2.4 Laboratory tests

The laboratory tests comprise a widely known type of medical data. Some examples include the following:

- Hematological (or haematological or blood) tests: perhaps the most common types of laboratory tests that are used for the analysis and examination of the hemic system [113]. These types of tests make use of microscopes and hematologic analyzers to examine the concentration of HBC in the blood flow (i.e., the oxygen levels in the blood flow), white blood cell (WBC) count, and the red blood cell (RBC) counts, as well as the number of platelets (PLTs), the iron concentration, and the number of erythrocytes and leukocytes. Furthermore, the hematological tests are also used to evaluate certain types of diseases and monitor different kinds of disorders through the measurement of the prothrombin time and the thrombin time, the hematocrit (HCT), the blood sedimentation, the blood coagulation time, the fibrin clot lysis time, and the bone marrow, among others [113]. In general, the hematological tests are widely used for the examination of inflammatory and blood clotting disorders and genetic defects, as well as for monitoring an administered drug's progress (e.g., the plasma levels under an anticlotting medication), the diagnosis of anemia, the detection of blood cancer, the measurement of the cholesterol levels and the blood pressure, and the detection of blood loss and blood clotting, among many others [114].
- Urine tests: tests that examine the urine through a procedure that is known as urinalysis [115]. Urinalysis uses chemical screening tests and microscopes to study and analyze the urine flow, the urine gravity, the urine color levels, and the occurrence of any bacteria and cellular fragments. The urine tests, along with the hematological tests, comprise the two most common types of laboratory tests that are conducted during a typical health examination process (check-up) and during a pregnancy test. Urine tests are widely used for the detection of kidney and liver disorders, as well as for the diagnosis of diabetes and prostate cancer and the detection of urinary tract infections and prostatic hypertrophy [116].
- Serological tests: special types of blood tests that seek for the infection of antibodies under a specific virus [117]. More specifically, microscopic analysis is used to examine the blood samples toward the detection of fungi, bacteria, and other parasites that are present in the blood cells, as a result of a virus. In fact, people who are infected by a virus exhibit different symptoms that weaken the immune system. For this reason, the serological tests are widely employed to understand the underlying mechanisms behind the virus infection and the origins of the symptoms and are used for the detection of rubella, fungal infections, and autoimmune disorders and for finding appropriate therapeutic treatments [118].
- Coagulation tests: special types of blood tests that examine the blood clotting process. Coagulation tests are performed through fast coagulation analyzers and

microcoagulation systems that measure the blood's coagulation speed, prothrombin time, complete blood count (CBC), and the overall blood PLT levels [119]. Coagulation tests are used for the detection of blood clotting disorders, such as thrombophilia and hemophilia, as well as for liver disorders that include excessive bleeding and the evaluation of any drug-related medication that involves blood clotting as a negative implication [119].

- Histopathological (histological) tests: these tests are used to examine the different types of tissues (e.g., muscle, nervous, epithelial) that reflect the disease's type [120]. Histopathological tests are primarily used in cancer diagnosis. A sample tissue is initially obtained, on the least invasive way, for the purposes of a biopsy test, where the size of the collected tissue sample varies according to the tissue area that is under examination. Microscopic analysis combined with cytological and histopathological principles, such as ICH [33] and IHC [34], as well as gene profiling techniques, is then used to analyze the tissue specimens yielding a histopathological report. The latter indicates whether the tissue sample was found malignant (or metastatic malignant) or benign according to the "behavior" of a specific marker that is used for tumor detection (e.g., like the p53 tumor protein). In fact, the marker binds into the sample causing the antigens or proteins of the sample to be marked or not (i.e., invokes a chemical release). Moreover, the histopathological report provides information regarding the tumor type, the hormone susceptibility, and similar tumor indicators to define a more targeted drug treatment in the case of a malignancy (or metastases) [121].
- Immunological tests: these tests examine the functionality of the immune system. Immunological tests share a similar basis with the histopathological tests, where markers are used to identify pathogenetic characteristics on tissue specimens [122]. These markers can take the form of either antibodies or antigens according to the type of immunological analysis. Immunological tests are used to diagnose chronic diseases, such as autoimmune diseases, where the immune system attacks itself (e.g., like in systemic lupus erythematosus, multiple sclerosis, rheumatoid arthritis), as well as to detect primary and secondary immunodeficiency disorders, where the immune system is not able to function properly and thus it is more susceptible to viruses and parasites (e.g., like in HIV/AIDS, neutropenia, etc.) [123].
- Skin tests: tests that examine the changes in the skin. These changes might be a result of an allergy, a skin disorder, or might even denote skin cancer. In the latter case, a skin biopsy is conducted to examine the suspected skin tissue sample by removing the skin tissue sample and applying microscopic analysis (similar to the histopathological tests) [124]. Skin tests are widely used to detect skin redness as a result of widened blood vessels, nonblanching hemorrhages (like purpura and palpable purpura), skin carcinoma, and skin lesions that might lead to skin cancer, as well as to detect allergies through a skin prick test [125]. The latter involves the insertion of allergens that trigger allergic reactions (through the release of chemical compounds), which might cause symptoms,

such as fever and dermatitis. Allergies can be originated by different sources including food (food allergies), drug substances (drug reaction allergies), etc.

- Endocrine function tests: tests that examine the functionality of the endocrine system [126]. The latter is vital for the maternal health and consists of endocrine glands that can be found in different parts of the human body including the hypothalamus, thyroid, pancreas, liver, and adrenal, among others. These glands release chemical compounds, which are known as hormones, which are transferred to the hormone receptors of the human body through the bloodstream. Hormones are involved in a variety of biological functions including growth, fertility, metabolism, and energy consumption, etc. Endocrine function tests are used to detect gland dysfunctions [127], where a hormone imbalance is observed. The hormone imbalance might be related to a hormone release by an endocrine tumor, a virus infection, a hyperplasia (e.g., enlarged salivary gland), a sexual dysfunction, etc. The early detection of an endocrine gland dysfunction is crucial as a dysfunction in one gland can cause a direct or indirect reaction to another [126].

2.3 Medical data acquisition

2.3.1 Biosignal acquisition standards

The ECG signal acquisition devices use surface electrodes that are placed to specific points in the chest and the limbs based on the standard 12-lead ECG placement [36]. According to the 12-lead ECG placement, four electrodes are placed symmetrically on the limbs, namely on the right-hand wrist, the left-hand wrist, the right ankle, and the left ankle, where one of this lead usually serves as the neutral lead. Alternatively, these leads can be placed close to the hips and shoulders. Then, six electrodes are placed on the chest, i.e., the precordial electrodes, starting from the intercostal space at the right of the sternum to the midaxillary line, according to the angle of Louis method. Additional placements involve the 3-electrode system and the 5-electrode system where the electrodes are limited to the chest on equal distances from the heart. In any case, the electrodes record the electrical activity that occurs due to the depolarization and repolarization activity of the heart's valves, providing a lateral and inferior view of the ventricular and articular activity. The contraction activity is recorded as a potential difference that is transformed into electrical signals with peak amplitudes that vary from 3 to 7 mV [7].

For EMG signals, there are both invasive and noninvasive methods for the data acquisition process. In the latter case, surface electrodes are placed on the skin around the muscle with a conductive gel to enhance the signal's quality. In the invasive case, a needle is used to detect the electrical signals directly from the muscle. In both cases, electrical potentials are recorded as part of the muscle's contraction activity with peak amplitudes varying from 0 to 10 mV [10]. Depending on the muscle under observation, the EMG potentials might be higher than 10 mV [10].

The EOG signal acquisition process involves the placement of six electrodes to record the activity of the muscles that are responsible for the human eye movement (i.e., the medial, lateral, and superior rectus and the superior and inferior oblique), where one electrode is neutral (on the forehead). These muscles are usually referred to as extraocular muscles. The EOG signals are acquired through horizontal or vertical eye movement. The EOG potentials occur as a result of the cornea and retinal electrical activity (corneoretinal potential) during eye movement with peak amplitudes that vary on the interval [0.05,3.5] mV [11].

The EEG signal acquisition process involves the placement of 16–32, 128, and even up to 256 surface electrodes on the human brain (i.e., scalp electrodes), which are usually incorporated into appropriate head caps. The scalp electrodes are most commonly placed on the head cap according to the International 10–20 system [37], where the distances between the electrodes are either 10% or 20% the distance from the nasion to inion or from the left side of the head to the right side. Conductive gels are also placed for improving the quality of the recorded signals. For obtaining a high-resolution analysis, more sensors need to be placed. These additional sensors can be placed according to the International 10–5 system [37], where the distances between the adjacent electrodes are smaller, i.e., the area between the adjacent electrodes in the 10–20 system is filled. The EEG electrodes capture the changes in the electrical field with peak amplitudes that vary on the interval [10,20] μV and convert them to electrical signals that are digitized and stored in a computer device for further analysis [8]. In the case of the ECoG signal acquisition, the amplitude of the recorded transcranial signals is 1000 times higher, as the signal recording approach is invasive [68].

In MEG signal acquisition, the sensors are coils (and more specifically superconducting magnets) that are already placed inside a dewar. The powerful magnetometers are usually placed inside liquid helium to reduce the heat levels that are produced by the superconducting technology that captures the extremely weak changes in the magnetic field that occur during the neural activation. The whole system is placed into an electromagnetically shielded room for reducing the noise levels from the outside magnetic fields. These changes in the magnetic field are recorded from the sensors and finally digitized for further manipulation. The extremely small alterations in the brain's magnetic field are approximately equal to 10^{-15} T [9].

Biosignals can be also acquired using real-time health monitoring devices that record the physiological activity, such as the heart rate, the respiratory rate, and the blood pressure. These measures are transferred through wireless sensor networks (WSNs) to computer devices and can be monitored on a daily basis.

2.3.2 Laboratory tests standards

Hematological analyzers are used to conduct hematological tests. The blood sample is first collected either through venipuncture sampling or through arterial sampling and sometimes fingerstick sampling. The hematological analyzers are able to differentiate and measure the different types of blood cells using the traditional electrical impedance method [128], according to which a change in the electrical impedance between two electrodes (among which the blood flows) under an applied electrical field is proportional to the cells volume. Thus, the volume of the WBCs, RBCs, and PLTs can be differentiated and measured. Another widely used qualitative and quantitative method is the flow cytometric technology [129], where the blood sample is first treated with fluorescent antibodies that bind in the sample and then a laser beam is applied on the blood cells and appropriate photodetectors are able to capture the amount of light that is absorbed by the blood cells. Together with the light that is scattered by the blood cells from different angles, they can provide useful information regarding the chemical compounds of the blood and the cells morphology. A standard biochemical analysis involves the measurement of WBCs, RBCs, PLTs, HBC, HCT, C-reactive protein, creatinine, glucose, cholesterol, potassium, sodium levels, and the CBC as an overall index, among others. The measurement units vary according to the laboratory where the analysis takes place, with the typical measurement units usually being in mg/dL or mmol/L.

Coagulation analyzers are used to analyze the blood coagulation. The coagulation analyzers are based on an optical detection method according to which a laser beam hits the blood sample [119]. The scattered light is recorded by photodetectors that covert the optical signals into electrical signals. The intensity of the scattered light is proportional to the coagulation time and the blood clotting time. When thrombolytic agents are applied on the blood sample, the analyzer can measure the thrombin and prothrombin time.

In histological analysis, microscopes are used to analyze the tissue specimens for detecting pathological conditions. A tissue sample is usually obtained from the area under examination (e.g., from epithelium, endothelium, mesothelium), at the least invasive way. The sample is then placed on a glass microscope slide for analysis. The microscope that is used to analyze the sample can either be optical or electronic. The optical microscope uses light to magnify the sample using photodetectors, whereas the electron microscope uses a beam of electrons that is generated using high voltage, instead of the light beam that is used in the optical microscope, to reveal the morphological characteristics of smaller objects, with higher magnification ratio, through electrostatic lens [130]. The IHC technique [34] is usually used to detect specific proteins or antigens in the tissue sample using specific primary antibodies that bind in the tissue sample where the sample is obtained through a section, against the ICH technique where the cells have not been isolated by the tissue sample (remain intact). In endocrine analysis, the functionality of the endocrine glands is examined using gene expression analysis [131] to detect the activity of proteins that are affected by the activity of the hormones. Endocrine tumors usually cause a hormone imbalance that affects the activity of the proteins in the cells.

Standard urine test strips are used to detect pathological conditions in the patient's urine [115]. The urine test strip is a plastic strip with a colored scale that indicates the presence of leukocytes, pH, protein, glucose, specific gravity, and HBC levels, among others, on the urine sample. Each pad on the plastic strip consists of a specific chemical compound that reacts with the urine compounds yielding a color that indicates the presence of the pad's chemical compound on the urine sample. A urine test strip is able to detect metabolism disorders, urinary deficiencies, liver and kidney disorders, dehydration, and drug abuse, among others. Another widely used method is the optical microscopy, according to which the sample is magnified and examined. Laboratory measurements can also be obtained, e.g., for sodium, potassium, and calcium levels, with the typical measurement units usually being in mmol/24 h. Blood test–related measures, such as the number of erythrocytes and leukocytes, can also be obtained.

Immunoassay analyzers are used to examine a patient's collected samples for bacteria, parasites, and other disease-related substances. First, reagents are bound into the collected sample, and then a light beam (of specific frequency) is applied on the molecules of the sample. An emitted light from the molecules indicates that the collected sample is positive to the reagent. The light signals that are captured by the photodetectors can be converted to electrical signals. The most commonly used reagents in immunoassay analyzers are fluorescent substances. A prominent technology that is currently used in immunoassay analyzers is based on chemiluminescence [132], according to which light is emitted as a result of a chemical reaction.

Serological analyzers are used to analyze the blood samples for infected antibodies. A popular test in this area is the direct fluorescent antibody test [133], also referred to as direct immunofluorescence, where fluorescent antibodies are used to detect antigens that exist in specific bacteria and fungi. The emitted fluorescence (of specific wavelength) confirms the presence of the antigen under examination. Other widely used tests include the complement fixation test that tests the presence of antigens in the serum and the enzyme-linked immunosorbent assay (ELISA) test [134], where a microtiter plate is used to test the presence of antigens using antibodies with tethered enzymes that bind with the antigens and cause a chemical reaction that appears in the form of a color change (chemiluminescence).

2.3.3 Medical imaging acquisition standards

The traditional image reconstruction process involves the application of the inverse fast Fourier transform (FFT) to transform the data from the *k*-space (the projection space) where the spatial information lies in Cartesian coordinates to the original space (the image space). This is known as the inverse problem in tomographic imaging. A widely used, computationally efficient algorithm for tomographic image reconstruction is FBP algorithm [38], which is based on the inverse of the Radon transform. The FBP algorithm applies a one-dimensional filter on the raw data from different projections before the backprojection process to the image space to recover the image [38]. These methods, however, are often inadequate, especially in the case where non-Cartesian coordinates are available or when the FFT can no longer be applied due to the nonlinearity introduced [135]. To face this problem, iterative reconstruction algorithms have been proposed using statistical methods (e.g., by minimizing a regularized least square cost function) to provide object estimations from projections, in multiple iterations [135]. Examples of these algorithms include the algebraic reconstruction algorithm [39] and the iterative SAMV algorithm [40]. These algorithms provide better reconstruction but are more complex than the previous approaches.

In MRI, the RF detectors capture the RF pulses that are generated by the hydrogen atoms when the Larmor frequency is applied. The RF signals are converted to electrical signals that are digitized. The data are recorded in the *k*-space where the spatial information exists. Then, tomographic reconstruction algorithms are applied to convert the raw data from the projections space into the image space. The pixel's brightness in an MRI slice denotes the mean attenuation of the hydrogens' emitted RF signals by the tissue, bones, organs, and fat of the human body. In fMRI, the MRI slices are obtained along with additional temporal information regarding the BOLD responses that denote increased or decreased hemodynamic response (activation) in the ROIs. fNIRS shares a common basis with fMRI, where the hemodynamic responses are recorded by measuring the near-infrared light attenuation based on the neurovascular coupling. For fNIRS 3D image reconstruction, methods similar to the diffusion optical tomography can be used, such as boundary element methods for near-infrared absorption and scatter estimation [136]. In PET scans, the metabolic processes at a molecular level are recorded in the form of coincidence events [14]. PET image reconstruction is analogous to CT image reconstruction methods, yielding images with poorer quality, however, due to the photon scattering and coincidence overlaps. The pixels in a PET image quantify the absorption of nucleic acid as a result of the molecular-level interactions with the administered radioisotope (similar to the SPECT). In CT, the X-ray detectors capture the ionizing particles that pass through the human body and finally convert them to electrical signals. A pixel's brightness in a CT image depends on the mean attenuation of the tissue in terms of the radiodensity [12]. In general, the size of the pixel is disproportional to the spatial resolution, with higher spatial resolution yielding smaller pixels, with high detail. The 2D dimensions of imaging data may vary from 128×128 pixels to 256×256 pixels or 512×512 pixels and the number of slices varies based on the slicing technology.

Examples of standard image formats include DICOM [137] and the Neuroimaging Informatics Technology Initiative (Nifti) [137,138]. The former constitutes a standard medical imaging format that is adopted by the majority of the manufacturers and clinical centers worldwide. According to the DICOM standard, a DICOM file format (the format is recognized by the ".dcm" extension) consists of (i) a header that includes the patient metadata (e.g., patient identification number, age, gender) and the specifications of the image acquisition protocol (e.g., the scanner parameters) and (ii) the pixel data, where a pixel value is represented as an 8-bit, 16-bit, or 32-bit integer value through a linear transformation with a certain slope and intercept [137]. The DICOM standard supports image compression that is useful in the case where a series of DICOM images is produced by the scanner. On the other hand, the Nifti file format constitutes the backbone format for storing neuroimaging data (e.g., in brain studies), where the header and the pixel data are stored together under a ".nii" file format [137,138]. The most common method that is used by the Nifti format for storing neuroimaging data is through a rotation and translation matrix, which is used to transform the voxel coordinates to the desirable volume space. According to the Nifti format, the images are encoded as 16-bit integer values although the most recent Nifti-2 file format supports 64-bit integer values, thus significantly increasing the size of the image space.

In the ultrasound imaging systems, the sound waves produced by the piezoelectric transducers are scattered from the tissues and organs and finally returned to the transducer in the form of echoes. The echo time and the intensity of the echo are used to form the ultrasound image, where a pixel's brightness is proportional to echoes' intensity. In photoacoustic imaging systems, the universal backprojection algorithm [41] is used to reconstruct the image from the photoacoustic waves that are reflected by the tissues or organs denoting the optical absorption in these regions. In OCT, the optical scattering is measured. The images can be formatted either using single-point scanning based on depth information from two lateral dimensions or using charge-coupled device cameras for parallel scanning [139].

2.3.4 Omics acquisition standards

DNA sequencing is an approach that is used to detect the canonical structure of the DNA in terms of the four nucleotides, i.e., the adenine (A), cytosine (C), guanine (G), and thymine (T). According to the helix model, the "A" is always paired with "T" forming one strand, and the "C" is always paired with a "G" forming another strand. The strands are linked using hydrogen atoms. Popular DNA sequencing methods include the Sanger sequencing [140], where the four deoxynucleotides are used along with the DNA polymerase for four reactions where a DNA primer binds with the fragments, as well as the Maxam—Gilbert sequencing [141], which is more complex and uses radioactive labeling for sequencing DNA fragments, and thus it is less used. HTS technology is the key technology toward the generation of millions of sequences simultaneously [27]. The high-throughput NGS (also referred to as second-generation) method aims at enhancing the traditional DNA sequencing methods using parallel sequencing to achieve the WGS through genome fragmentation [27]. Examples of straightforward HTS methods include the pyrosequencing [142] and the Illumina sequencing [143], which are based on advanced polymerases and reversible dye-terminators for the detection of nucleotides (using fluorescence and luciferase) yielding millions of detected nucleotides, among others. In addition, ChIP-NGS [111] is used to examine the interactions between DNA and proteins. Common standard data formats include the Sanger FASTQ file format [144] that is used to store sequence information and the standard flowgram format (SFF) that is used to encode sequence reads [145].

RNA sequencing is used to examine the transcriptome, i.e., the set of transcriptomes in a biological cell [28]. RNA sequencing can provide great insight on the RNA molecules that are responsible for the gene regulation, function, and coding, as well as understanding the underlying mechanisms of a disease. Emphasis is given on the mRNA molecules that transfer information regarding the encoding of proteins, i.e., they act as an intermediate between the genes and the proteins [28]. Additional groups of RNA molecules have been recently emerged including the miRNA and the piwi-interacting RNA that is involved in the regulation of the gene expression [146]. A standard RNA-sequencing analysis procedure involves the execution of four steps [28]: (i) RNA isolation from the biological sample (e.g., from a biological cell), (ii) reverse transcription of RNA to complementary DNA (cDNA), (iii) amplification through polymerase chain reaction, and (iv) cDNA sequencing. The conversion to cDNA is performed due to the fact that the majority of the next-generation sequencers make use of DNA libraries for sequencing. The serial analysis gene expression [][147] and the cap analysis gene expression [][148] methods are often used to examine the mRNA molecules for transcriptome fragments, where the number of RNA molecules, for a mammalian cell, is more than 500 million [149].

The tissue microarray technology is used to detect tumors and other pathological characteristics in many tissue specimens that are placed on a tissue microarray block [42]. The tissue specimens are obtained using sectional methods that manage to isolate specific parts of the tissue for conducting a biopsy test. The block is then divided into hundreds of smaller sections that can be placed either on an optical or an electron microscope for analysis, similar to the IHC technology in histological analysis, where specific proteins or antigens are detected in the tissue samples using fluorescent antibodies that bind in the tissue sample. These proteins can lead to the identification of new biomarkers or the validation of existing ones, for different types of cancer, such as breast cancer and colorectal cancer, among others. Furthermore, MS are used to quantify the mass of a molecule using an ionizer that ionizes the molecule, then sorts the ions, and finally separates them based on their mass-to-charge ratio to quantify the mass of the molecule using electron multipliers or ion-to-photon detectors [29]. The output is a distribution of the detected ions versus the mass-to-charge ratio. MS is widely used for the characterization of the proteins' morphological characteristics and to study the function of the lipids in lipidomics [21].

TLC technology is widely used to monitor chemical reactions and discriminate mixtures [30]. The standard procedure involves (i) the solid phase, where a thin glass plate is covered with aluminum oxide or silica gel, and (ii) the mobile phase, where a solvent is applied on the mixture. The flow of the solvent can reveal the molecular structure of the mixture compounds through spots that remain on the plate and have different colors depending on the compound characteristics. Finally, the retardation factor is used for quantitative analysis that is defined as the distance of the spot from the starting point [30] and depends on the solvent, the layer's thickness, etc. These spots can also be visualized using fluorescent compounds that make them visible under a blacklight, as well as iodine vapors among others [30]. TLC is also widely used in lipidomics for identifying the lipids concerning different disorders, such as inflammatory and neurodegenerative diseases [108].

2.4 Sources of medical data

2.4.1 Patient registries

The PRs are research-oriented collections (records) of health information from patients for clinical or scientific purposes in the form of an observational study [150]. The term "health information" refers to standardized types and formats of medical data, including patient history, laboratory tests, medical images, and additional sources, such as patient and clinician/patient questionnaires, disease-related information, and reported patient outcomes (or clinician claims) under health-related events [150]. A PR is usually used by clinicians for the long-term (or short-term) collection, storage, and evaluation of health information for a specific group of individuals who exhibit a particular type of disease or medical condition. Therefore, the PRs can be used in clinical research to (i) manage patient data, (ii) recruit patients for clinical trials, (iii) develop drugs for chronic or rare diseases, including cardiovascular diseases, diabetes, cancer, and arthritis, among others, and (iv) evaluate the subsequent patient outcomes. The PRs, however, do not focus on patients and are purpose-specific as the majority of the PRs are mainly constructed for clinical research purposes (research-oriented registries) although efforts have been made toward the establishment of the patient-powered registries[][151] (patient-oriented registries) to promote the collaboration between the clinicians and the patients.

2.4.2 Health sensors

The health sensors are intelligent wearable or implanted devices that are used for health monitoring and disease prognosis. The current advances in wireless communications combined with the rapid development of "smart" biosensors have enabled the low-cost, continuous monitoring of a patient's health condition for home rehabilitation purposes, as well as the secure transfer of individual-related health information through wireless networks, directly to a central data repository for clinical evaluation and further analysis. The emerging biological microelectromechanical systems (Bio-MEMS) with organic substrates make use of mechanical and electrical components to record antigen interactions, nucleic acid interactions, and other biological fluctuations through analytes that interact with the biosensors to invoke fluctuations that are recorded by the transducers in the form of electrical (e.g., through electrochemical reactions), mechanical (e.g., through stress sensing), or optical signals (e.g., through fluorescence) [152]. Furthermore, the rapid advances in nanotechnology have enabled the construction of nanostructured thin films and nanomaterials [153] (e.g., polymer nanoparticles, gold nanoparticles) for constructing highly sensitive biosensors, in the nanoscale, having a larger number of bioreceptor units than the typical Bio-MEMS (in the micron scale).

In the majority of the electronic health (e-Health) systems, the patient data are transferred from the biosensors and through WSNs to a central processing node (system) for remote monitoring [154]. Furthermore, the health information can be also displayed to the user's mobile through appropriate mobile applications that are part of the mobile health (m-Health) systems [155]. Health sensors are embedded into real-time wireless systems that enable the continuous monitoring of the patient's physiological activity (e.g., heart rate, respiration rate, body temperature), physical activity (e.g., motor movement), and mental activity (e.g., sleep patterns) toward the improvement of the individual's life quality, the evaluation of a drug's progress, and the early diagnosis of symptoms in numerous medical conditions and chronic disorders, such as Parkinson's disease, diabetes, chronic obstructive pulmonary disease, and dementia, among many others.

2.4.3 Electronic health records

The electronic health records (EHRs) are digital, individual-level, health records for the management and analysis of health information [156]. EHRs enable the collection and structured management of individual-level, health-related data (e.g., laboratory tests, patient history, clinical claims, medical images) and the "true" analysis of the health information using computerized methods against the traditional PRs that are purpose-specific and thus focus more on the population characteristics, as well as on deriving the health information before the collection and analysis of the health information [156]. EHRs can enhance the clinician's decision process regarding the patient outcomes and make the communication between the clinician and the patient much easier, as well as ensure the safety of the stored medical records and avoid medical errors during the data entry in digital forms. In addition, the EHRs can enable the electronic prescription of drugs and promote medical data sharing through their interconnectivity with other EHR systems toward the establishment of a global EHRs system. The majority of the national healthcare systems have already adopted the EHR technology for transforming their traditional PR systems due to the former's (proven) positive effectiveness in patient care [156].

The majority of the EHR systems adopt the Health Level 7 (HL7) standard clinical document architecture (CDA) [157]. The latter is based on a referenced information model (RIM) that serves as a semantic model that consists of a set of structural components (e.g., classes with data types) and semantic relations (e.g., a healthcare provider "belongs to" an organization where the healthcare provider is an actor who is represented by a class and has a name, a surname, etc.) that are used to represent clinical notes in the form of an extensible markup language document [157]. The CDA is part of the HL7 version 3 family and uses the RIM as an object-oriented model to define the actors (e.g., clinicians, healthcare providers), the clinical document standard terms, and the targets (i.e., patients). CDA can also include multimedia content, such as medical images. All this information is organized into a human readable and standardized format that enables the interlinking of EHRs among different clinical centers.

2.4.4 Genome registries

The genome registries are collections (records) of genomic information (e.g., biochemical genetic tests) that are stored in large databases [158]. These databases include molecular-level health information regarding the genes, the chromosomes, the DNA and RNA sequences, the proteins, the haplotypes, the genotypes, and the phenotypes, which are related to a specific disease. In addition, these databases include demographic information that is related to the population characteristics of the patients so as to study for the effects of drugs in target groups during clinical trials. The genomic registries can help the clinicians to study the genotype–phenotype relationships, the genetic profiles of various diseases in a cellular level, and the SNPs associations with genetic variants and thus can shed light into the underlying genetic variations in chronic and rare diseases. The sharing of genomic data between genome registries can help geneticists to comprehend the mechanisms behind genetic disorders and yield clinical decisions with higher statistical power and thus enhance patient care [158]. Furthermore, similar types of genome registries can be also generated for other areas of omics studies, such as proteomics, lipidomics, transcriptomics, and metabolomics, among others. In addition, the individual omics registries can be combined together to establish multiomics registries [159].

2.4.5 Clinical trials

Large pharmaceutical and biotechnological companies conduct large-scale clinical trials with the purpose of examining and validating the effectiveness of a proposed drug in a specific group of patients, which is usually referred to as the target group [160]. This group consists of patients (usually referred to as subjects) who are drawn from the overall population, fulfilling specific criteria, i.e., under the same medical condition or disease. The patient selection process is usually based on stringent and eligibility criteria and is compliant with all the legal, ethical, and patient privacy requirements. The properly conducted large-scale clinical trials can yield significant outcomes and pharmaceutical claims regarding the underlying mechanisms of a disease's genetic profile, as well as the negative implications of a proposed drug's (e.g., toxicities) on the target group. Clinical trials have numerous applications in cancer and other chronic diseases [161]. Clinical trials can be also part of governmental health plans that envisage to promote patient care toward the development of new public health policies or the reinforcement of the existing ones. The outcomes from the small- or medium-scale clinical trials are gathered according to the case report form (CRF) designing architecture that is based on well-structured, high-quality, user-friendly questionnaires that summarize the outcomes of the clinical trial, as well as information regarding the specifications and actors of the clinical trial in the form of concise questions and prompts [162]. The actors of the CRF are the investigators that conducted the clinical study and the recipients are clinicians, programmers, researchers, etc. The CRF content provides population-related demographic information (e.g., age, sex), patient history (e.g., medical conditions, smoking history, therapies), personal details (e.g., body weight), and information regarding the presence of adverse events, among others. In the case of large-scale clinical studies, the electronic case report form reduces the time effort needed for data entry and processing.

2.4.6 Clinical claims

Clinical claims are diagnostic reports from clinicians, general practitioners, and healthcare stakeholders related to the diagnosis of a patient's medical condition/disease or the patient's outcomes under a specific administered drug. For example, a clinician can evaluate a patient's medical condition through the thorough examination of the patient's medical images (i.e., imaging reports), laboratory measures (i.e., laboratory examination reports), etc. These diagnostic reports can usually be found in EHR systems or in traditional PRs, as well as in the form of published articles in international medical journals. Similar to the medical claims, the pharmaceutical (and biotechnological) claims are clinical trial reports that are derived by pharmaceutical and biotechnology companies and health-related stakeholders during small-, medium-, and large-scale clinical trials, i.e., during drug research, where the effects (e.g., negative implications) of a proposed drug are extensively studied on the participating group of subjects, i.e., the target group (see Section 2.4.5 for more information about clinical trials).

2.4.7 Additional data sources

Additional sources where various medical data can be found are public or private clinical centers, urban clinics, diagnostic centers, and clinical laboratories, as well as in social media (e.g., Twitter, Facebook), private or public blogs, and web pages, through which an individual can post any health-related information. Furthermore, a variety of different types of medical data have already been made publicly available by large healthcare platforms and funded projects for promoting clinical research worldwide, such as the Human Connectome Project, which has made available fMRI and diffusion neuroimaging data from more than 1000 healthy subjects [163], and the Entrez Gene database at the National Center for Biotechnology Information [][164], which includes more than 7 million gene-specific records (e.g., from fungi, bacteria, eukaryotes, etc.) since 2010, among many others. Medical data can also be found in cohort studies that will be extensively discussed separately due to their clinical significance toward the investigation of a disease's onset and progress, against the interventional studies which are often high cost and involve treatment randomization [43–47].

2.5 Cohorts

2.5.1 Origins

The term "cohort," which has been extensively used in the field of clinical study design, has its roots back in the Roman Empire, where the word "cohors" was used to denote groups of soldiers who were part of a military Roman legion [43]. In synchronous epidemiology, the term "cohort" is used to describe a group of follow-up individuals who exhibit a common medical condition or a specific type of rare and/or chronic disease [43–47]. The cohort studies are part of a larger group of studies, which are widely known in the literature as analytical (or inferential) observational studies, along with the case–control and the cross-sectional studies [43]. An observational study is a specific type of medical study design that mainly aims to answer important questions regarding a disease's origins and examine the descriptive statistics of the population that is under examination. A well-designed cohort study can be of great clinical importance because it can introduce a temporal dimension (follow-up) that is able to provide clinical outcomes about a disease's prevalence and reveal the associations between the causes and the disease's outcomes [45].

The longitudinal cohort studies are of great importance in the vast domain of the observational cohort studies. A longitudinal cohort study is a special type of observational study that involves the repetitive collection of patient data over long (or short) time periods [165]. This is a key characteristic of the longitudinal cohort studies that enables the accurate examination of differences between the same patients over the time as the recall biases in patients are reduced along with the well-known "cohort effect," which introduces erroneous components in the cohort (e.g., erroneously parsed ages) [165]. The longitudinal cohort studies can provide deeper insight on a disease's progress over time, yielding more accurate risk predictors for the disease under investigation [165]. An example of a longitudinal cohort study is the prospective cohort study, where the same patients are examined over time although the longitudinal studies can also take the form of retrospective cohort studies where the data from patients who have experienced the same medical condition or disease have been already collected for a long time period in the past [165].

In general, the observational studies can either be descriptive or analytical [43]. The former focus more on the examination of a disease's characteristics either on a single patient (i.e., a case report study) or on a series of patients (i.e., a case series study), as well as the disease's prevalence (i.e., a cross-sectional study). The descriptive observational studies make use of descriptive statistics, such as mean, variance, etc., to study the population characteristics and identify the causes of a disease or medical condition. On the other hand, the inferential observational studies include (i) the cross-sectional studies, which quantify the associations between the causes and the disease's outcomes, on a specific time period (i.e., no follow-up information is recorded), (ii) the case–control studies, which are retrospective studies that compare the disease prevalence between a case group and a control group, where the latter is a group of healthy individuals and the former is a group of individuals that share common disease characteristics, and (iii) the cohort studies that are prospective or retrospective studies that investigate the associations between the causes and the disease's outcomes, as well as the effect of various prognostic factors (e.g., laboratory measures) on the clinical outcomes [45].

2.5.2 Cohort study design

In cohort studies, the participating individuals (subjects) are selected on their exposure to a specific disease (or medical condition) and according to specific exclusion criteria that are set during the design of the cohort study. These criteria are important and comprise the fundamental basis of a cohort study. The exclusion criteria pose restrictions to minimize the introduction of bias into the cohort study , for example, to avoid the inclusion of patients that exhibit adverse events due to drug usage [44]. Once all the necessary exclusion criteria have been set, the exposed and unexposed groups are defined, where both groups are drawn from the same population. The exposed group comprises the subjects that will be exposed to a particular type of drug, whereas the control group includes the healthy ones (in the absence of any clinical outcome) and serve as the reference group. The drug exposure time may vary from weeks or months, depending on the initial hypothesis of the cohort study. The fact that a cohort study is designed to investigate the disease versus the outcome (e.g., a symptom such as fever) associations, on a temporal basis, can reveal significant information regarding the disease's onset and progress over the time.

In the meantime, the association between the drug exposure and the outcome can be measured in terms of the RR [46]. The RR quantifies the chance that the outcome is increased or decreased by the drug exposure and is defined as the ratio between an incidence in the exposed group versus the incidence in the reference group. A cohort study is finalized either when the drug exposure timescale comes to an end or when the expected outcomes are fulfilled [43–45]. Other quite similar measures include the HR, the risk difference (RD), and the odds ratio (OR). HR corresponds to the hazard of death after treatment and is used to measure the frequency of death between the drug-exposed group and the reference group [47]. OR measures the odds that the outcome of interest will occur given the drug exposure against the opposite and is mostly used in cross-sectional and case–control studies where the RR cannot be estimated [166]. Finally, RD is defined as the difference between an incidence in the exposed group and an incidence in the reference group [167]. The absolute risk reduction is also used to measure whether the outcome of interest is increased by the drug exposure. In multivariate statistics, a $2x2$ cross-tabulation table (or contingency table) [168] is usually used to compute the risk measures through the examination of the associations between the outcomes and the groups.

A cohort study can either use prospective or retrospective data. The difference between the prospective and retrospective cohort studies lies on the fact that, on the former, the cohort data are expected to be updated across time, where a temporal dimension is used to keep track of the upcoming data (on a patient basis), whereas on the latter, the cohort data are not expected to be updated in the future. This means that in prospective (where "pro" denotes future) cohort studies, the clinical information can be collected both at baseline and during follow-up, where multiple reoccurrences of the same symptom(s) might exist. On the other hand, in retrospective (where "retro" denotes past) cohort studies, the clinical researchers analyze the association between the exposures and the disease's prevalence using past patient data. Thus, in a retrospective cohort study, the starting point begins from the present and "looks back" in time, whereas in a prospective cohort study, the starting point begins from the present and "looks ahead."

2.5.3 Comparison with other study designs

The cohort studies can address several limitations that are present in the randomized clinical trials, which are part of the interventional studies [160]. The cohort studies [43–45] (i) can involve a large number of patients who are often followed for a large timescale (i.e., longitudinal cohort studies) instead of the clinical trials where the tracking of new patient data is often lost and the sensitive population groups (e.g., the children) are often excluded from the experiments, (ii) can measure patient-specific outcomes instead of surrogate end points, (iii) do not use single-blind or double-blind procedures, where the tracking of the subjects in the exposure group is partially or completely lost, (iv) are less expensive than large-scale clinical trials, and (v) are easier for the subjects to be part of instead of the clinical trials where the subjects are often ambiguous. The cohort studies do not always involve any experimentation with the subjects in terms of treatment randomization and the mandatory use of placebos or other interventions for the control groups. The cohorts, however, could be used for conducting low-cost, realistic clinical trials.

In a cohort study, the different types of medical data can be either prospective or retrospective and thus can be collected both at baseline and during follow-up, including data collected in the past, as well. The case–control studies share a common basis with the retrospective cohort studies, where the analysis of the disease versus outcome associations involves the analysis of the data that have been already collected in the past (Fig. 2.1). In addition, in the cross-sectional studies, the investigation of the disease versus outcome associations occurs only in the presence, i.e., at a single time point (Fig. 2.1). The clinical significance of a cohort study can be met in the special case where the exposure is a rare condition, such as a chronic disease. In that case, a cohort study can be a powerful tool. Understanding a disease's onset and progress over time can address clinical unmet needs, such as

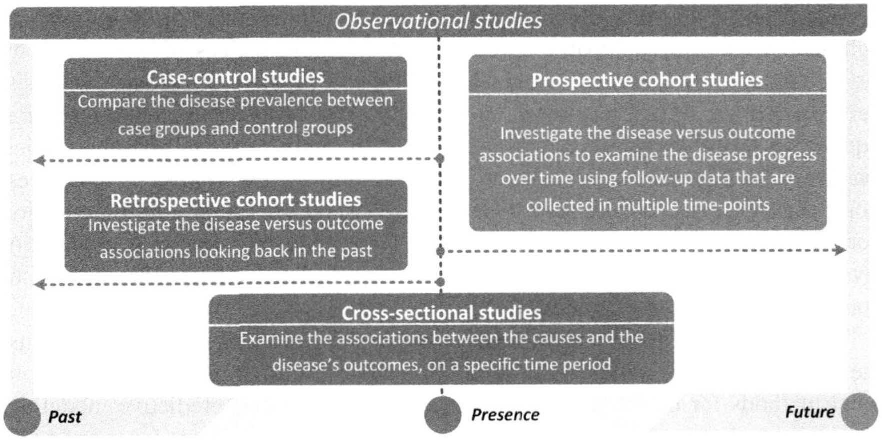

FIGURE 2.1

An illustration of the temporal dimension for each type of observational studies.

the early identification of high-risk individuals (patient stratification), the discovery of new targeted therapeutic treatments, and the development of new public health policies for promoting healthcare worldwide.

2.6 Big data in medicine

Under the ages of our rapidly advancing technological era, the vast amount of daily generated digital data has led to a scientific breakthrough with huge benefits in many fields of our everyday lives including finance, medicine, and industry [1–6]. The term "big data" has been extensively used to characterize these massively accumulated datasets, which are mainly characterized by the four V's [1–6]: (i) volume, (ii) velocity, (iii) veracity, and (iv) variety, where each dimension has a unique scope. The "volume" dimension refers to the massive amounts of collected data elements, whereas the "velocity" refers to the speed of the continuously generated data flows. The "veracity" dimension refers to the biases that are introduced during the data collection process, and the "variety" refers to the different types of data sources (e.g., different formats, structural differences). The definition of big data overcomes the classic definition of the ordinary datasets, which is only limited to their size (i.e., volume). Big data is a promising tool that provides broader and more comprehensive insight from large data elements, a fact that greatly enhances their impact in various scientific and research areas, especially in healthcare. However, the size of the collected data and the speed of the data generation process, combined with the different types and the complexity of the big data, are crucial challenges that still need to be addressed by the scientific community.

Currently, there are many types of big data in healthcare. More specifically, medical big data can be found in the medical imaging domain, where the thin-slice technology that has already been adopted by the modern diagnostic imaging scanners (e.g., CT, MRI, OCT) is able to capture thousands (>2000 slices) of high-quality (in terms of spatial and temporal resolution) slices of different body parts, in a very small amount of time. In addition, in the field of genomic analysis, the advances in HTS technology has led to the NGS or second-generation sequencing technology that is able to capture the entire human genome (which consists of 30,000–35,000 genes), producing millions of DNA and RNA sequences. Moreover, in the field of biomedical signal analysis, the continuous, high-resolution monitoring (e.g., for days or even weeks) of a patient's physiological, mental, or physical activity produces large amounts of recorded waveforms that consist of thousands of samples per second.

The application of big data in healthcare is promising and with many benefits. The vast amount of generated data can improve the statistical power of the conventional methods for data analytics including data mining and predictive modeling. Furthermore, the big data can boost the clinical decision-making process and yield clinical outcomes with higher statistical power (i.e., higher scientific impact due to the large number of participating subjects) and improved accuracy [1–6]. As a result, the patient care will be greatly improved as the patients will avoid the risk of unnecessary surgery operations, as well as the negative implications of unnecessary (or even false) drug administration. Big data can enhance the performance of traditional machine learning methods, giving rise to a field that is known in computer science as deep learning [1–6], a modern technique that makes use of multilayer neural networks that are able to capture valuable patterns and associations that are hidden between the large data elements, such as in multislice medical images, omics, biomedical signals, etc. For example, the high-resolution, multidimensional, four-dimensional medical images, such as the PET images that can capture additional information regarding the metabolic effects of a radioisotope apart from the anatomic structure, or the fMRI images that can depict the brain activations under a specific physical activity, can greatly enhance the accuracy of clinical diagnosis.

The application of big data in medicine can also enhance the patient stratification process according to which straightforward machine learning methods can be applied to identify and discriminate high-risk individuals from large populations [1–6], i.e., groups of patients having high risk for the development of a type of malignancy, such as lymphoma. This will also yield significant improvements in personalized medicine for the selection of appropriate therapies by taking into consideration molecular-level health information. Furthermore, the multivariate analysis of large population groups can also reveal significant statistical associations between the disease's manifestations and different demographic factors, such as age, gender, etc., and other medication related factors. Moreover, the outcomes from large-scale clinical trials and clinical research studies that make use of big data from large populations can enable the development of new, low-cost, targeted therapies for chronic and rare diseases, as well as the development of new public health policies toward a global and sustainable healthcare system.

The software advancements toward the development of methods for big data analytics are an emerging field. The current software advancements in neuroimaging have led to the voxel-wise analysis of hundreds of thousands of voxels (>100,000) within the human brain yielding large-scale similarity matrices, i.e., brain networks, which are able to simulate the brain activations across different ROIs, yielding millions of connections between the voxels [169]. These large-scale networks have been widely used to study the brain activation patterns during resting state or under specific physical and mental tasks [170]. Furthermore, the analysis of omics big data using high-computing resources can reveal important clinical information concerning the genetic variants and cellular functionalities in different types of diseases, as well as assist the development of effective drugs with reduced implications in the participating subjects. A great example can be found in the field of interactomics [171], where the PPI networks are constructed, on a cellular basis, to study the stable and transient interactions among proteins [107]. In addition, in signal analysis, the applications of deep learning methods for the prediction of disease outcomes have shown significant performance yielding high sensitivity and low specificity scores in numerous cases, such as the prediction of epileptic events [172], among others.

Understanding big data is a difficult and demanding task for researchers and data analysts. With the growing number of large volumes of daily generated data from health sensors, social media posts, medical images, laboratory tests, electronic patient records, blogs, and web pages, the estimated amount of data is expected to overcome the zettabyte (10^{21} gigabytes) and even the yottabyte (10^{24} gigabytes) [35]. Therefore, the development of straightforward software architectures along with hardware components and computer-aided tools and systems toward the efficient storage, management, quality assessment, high-performance computing analysis, and visualization of big data is a constant and increasing demand. For example, in medical imaging analysis, emphasis must be given to the development of methods for big data compression (e.g., image compression), registration and mapping of thousands of slices, and methods for segmentation of anatomical structures across these slices. A scientific researcher who is able to understand the nature (e.g., the data patterns) of big data can discover new opportunities for the development of new methods for big data analytics.

There is no doubt that the benefits of big data in healthcare are many. However, there are several technical and clinical challenges that still need to be addressed. The main challenge is the fact that the sources of big data are disparate, heterogeneous, and costly, a fact that increases the computational complexity of handling large volumes of data, as well as hampers the application of traditional statistical and machine learning methods for big data analytics. In addition, the big data are often incomplete with several discrepancies due to the lack of a global protocol for big data acquisition. As a result, data standardization methods need to be adopted to overcome this structural heterogeneity. Moreover, the big data are difficult to manage due to their size and structural complexity. Furthermore, the risk of data misuse is increased in big data with the data quality assessment process being a significant challenge along with the lack of the researcher's skills that might hamper the quality of the data yielding unreliable outcomes. The big data are often prone to the existence of missing values and measurement errors throughout their context, which pose significant obstacles toward their effective analysis. As a result, the irrational use of machine learning methods for predictive modeling in large datasets might lead to false outcomes, with no clinical importance at all.

There are also privacy issues that lurk behind the use of big data [35]. Ethical and legal issues must be carefully taken into consideration during the collection and processing of big medical data from multiple data sources. As the big data are large collections of patient data, it is difficult and even impossible to obtain signed informed consent forms from every single patient. In addition, the large volume of medical data shall not be stored in centralized databases as the risk for data abuse is greatly increased. Therefore, the data should be stored in cloud environments that are compliant with data protection regulations and should be collected under appropriate data protection agreements based on international privacy and protection standards. The researchers and data analysts must be fully aware of the data protection regulations during the collection and processing of the data. Furthermore, there is an increased necessity toward the development of machine learning methods for analyzing data that are distributed in multiple sites, a fact that remains a great challenge (see Chapter 7 for methods that deal with the analysis of distributed data).

2.7 Conclusions

The medical domain has been overwhelmed by the big data. The dramatic increase in the speed of the data collection process along with the large volumes of accumulated data from dispersed data sources has led to a scientific breakthrough, especially in healthcare. The types of big data in healthcare are many (Fig. 2.2), varying from medical images and laboratory tests to biosignals and omics. The volume size in each type of data varies from megabytes (e.g., the size of the data in the laboratory tests depends on the number of samples, and the size of the recorded biosignals depends on the sampling frequency and the time duration) to gigabytes (e.g., the size of medical images depends on the pixel depth and the number of slices) and even Terabytes (e.g., the size of omics data depends on the type of sequencing) of generated data. The mining of knowledge from big data can shed light into the unmet needs of various diseases and lead to the development of more effective treatments and public health policies. In addition, the rapid advances in volume-rendering methods have led to 3D medical image reconstruction, a method that has significantly improved the quality of image interpretation during the clinical diagnosis. Moreover, the coregistration of different types of medical images, such as PET-CT and SPECT-CT, can significantly enhance the diagnostic accuracy through the construction of images that combine both high spatial resolution and temporal information regarding the metabolic effects of the organs. Meanwhile, the current advances in thin slicing technology have enabled the acquisition of thousands of slices, in short time, yielding images with high spatial resolution from different parts of the human body. Powerful 10.5T MRI scanners have been also constructed (i.e., the one located at the Center for Magnetic Resonance Research [173]) to further enhance the image resolution. Apart from the fusion of medical images, the high temporal resolution of the biomedical signals can be also combined with medical imaging systems, such as MRI and CT, to provide both high spatial and temporal information.

The evolution of NGS has enabled the parallel generation of millions of sequences, yielding powerful RNA and DNA sequencers that are able to study the whole transcriptome and even generate the genetic profiles of the whole human microbiome. This breakthrough has shed light into the mechanisms of cancer cells, the genetic profiles of various microorganisms, the genetic variants in different diseases, and the gene regulation in different tissues and organs. In addition, the advances in lipidomics and proteomics have made the identification of lipids and proteins much easier and have provided great insight on their importance in gene regulation. Modern techniques, such as microarray analysis, MS, and TLC, have offered many capabilities toward the examination of proteins, lipids, and the metabolic changes in biological cells in inflammatory and neurodegenerative diseases.

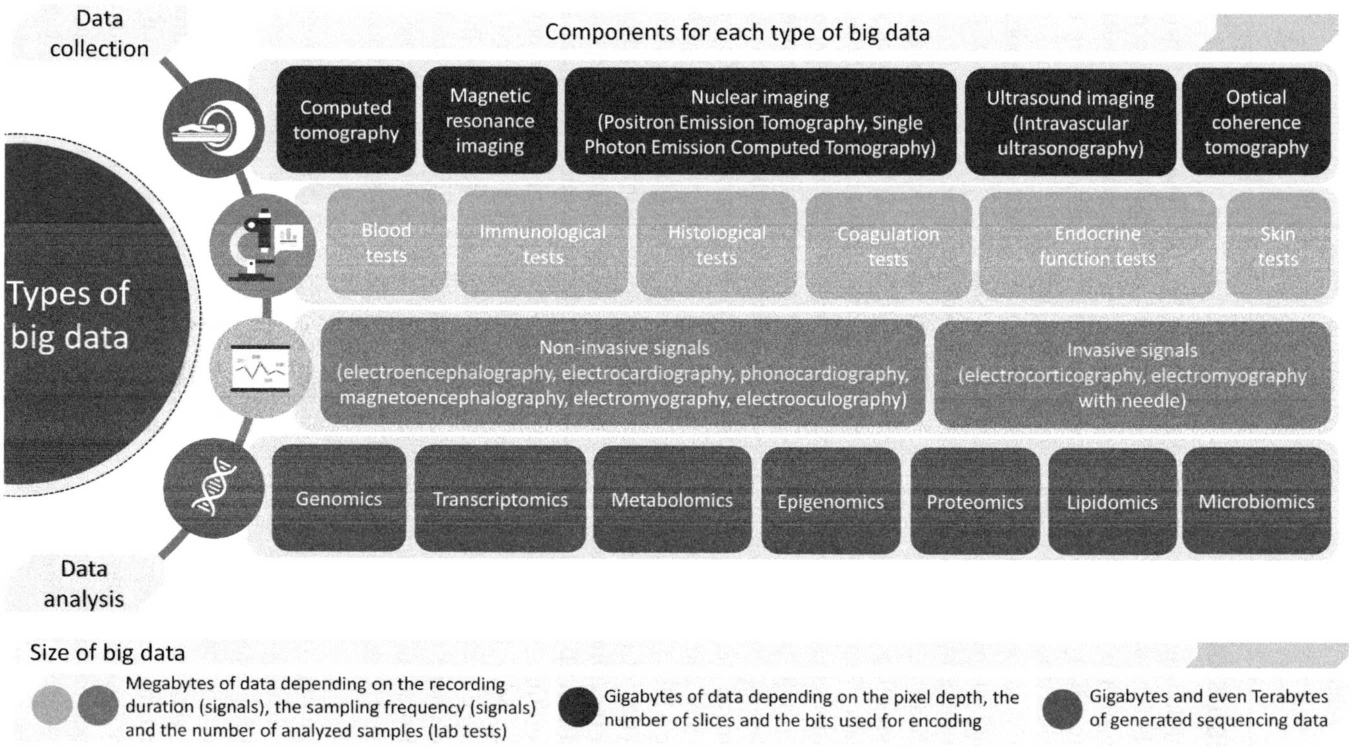

FIGURE 2.2

A summary of the types of big data in healthcare along with the volume dimension.

Furthermore, the application of Graph Theory has enabled the construction of PPI networks to understand the associations between the proteins and the identification of proteins with similar functional characteristics. Meanwhile, the NGS technology in the field of epigenomics has also enabled the examination of the implications of different pathological factors in the whole-genome expression.

The advances of the postgenomic era aim at providing straightforward tools toward the analysis and interpretation of the different omics data types. These advances are referred to as additional spans of the omics field, including [174] (i) the phylogenomics that involves the development of computational methods for the prediction of gene function using gene sequencing data, (ii) the physiomics that uses Graph Theory to construct networks that represent the interactions between genes so as to identify highly associated genes and predict the gene function, (iii) the pharmacogenomics that studies the effect of drugs on the gene function by computing the associations between SNPs and drug substances, (iv) the interactomics that uses Graph Theory to construct networks that represent the interactions between proteins (PPIs) and genes as well, and (v) the nutrigenomics that studies the implication of nutrition in gene function, among others. In addition, the construction of multiomics registries that are able to combine the uniqueness of every individual omics registry can provide great insight toward the analysis of the whole genome, the genetic profiling, and shed light into the underlying mechanisms of rare viruses and chronic disorders.

As far as the laboratory tests are concerned, the hematological analyzers that are widely used to analyze the chemical compounds of the blood cells can provide great insight on blood clotting disorders, inflammatory disorders, and blood cancer, as well as check-up information through the measurement of the HCT level, the WBC and RBCs, the cholesterol levels, and the glucose levels, among others. The coagulation analyzers can provide additional information regarding the thrombin and prothrombin time and blood clotting time, based on optical detection methods. Histological and immunological analyzers using powerful optical and electron microscopes along with the chemiluminescence effect (based on the IHC) can provide valuable information regarding the existence of pathological factors on different types of tissue specimens that are obtained using sectional approaches from (e.g., epithelial) and thus test for the existence of tumor cells (through color changes) or not. The traditional urine strips in urinalysis can reveal urinary deficiencies and the use of drug substances, as well as, prostate cancer and various kidney and liver disorders. Moreover, the direct fluorescent test that is used in serological analyzers can detect the presence of antigens on tissue antibodies and thus confirm the existence of specific compounds in the tissue sample for the detection of fungal and rubella infections and autoimmune disorders, among others. Finally, the endocrine tests can reveal important clinical information regarding the endocrine gland function by measuring the hormone levels, where high hormone levels denote the existence of endocrine gland dysfunction that can be expressed by endocrine tumors or other disorders.

The retrospective and prospective cohort studies are able to overpass several limitations that are introduced by case–control studies, clinical trial studies, and cross-sectional studies. The temporal dimension that is present in the prospective cohort studies can yield important information regarding the disease's onset and evolution over the time, as well as reveal valuable information that is related to the disease versus outcome associations. In addition, the fact that the cohort studies are patient-specific and composed of groups of individuals who share a common disease or a medical condition along with the ability to track follow-up data can address the unmet needs in rare and autoimmune diseases toward more effective therapeutic treatments and health policies, as well as better prediction models toward the identification of high-risk individuals who are more prone to the development of a malignancy or a pathological condition. Cohort studies have been used in a variety of diseases including heart failure [175], Alzheimer's disease [176], rheumatoid arthritis [177], diabetes [178], epilepsy [179], and breast cancer [180], among many others.

Several laboratories and clinical centers, however, use their own measurement units, a fact that hampers the coanalysis of their data with those from other laboratories and clinical centers, as well as the concept of data sharing. Moreover, the structural heterogeneity of the acquired medical data introduces crucial biases toward the effective analysis of medical data, and thus emphasis must be given on the development of new automated methods and international guidelines for data standardization and harmonization. The ethical and legal compliance of the data acquisition procedures are of great importance. All the necessary consent forms must be signed by the individuals, a fact that is difficult or even impossible in the case of big data as the data sources are often dispersed. Data protection agreements are also important toward the sharing of medical data. Security compromise is also a serious threat. Having all these large volumes of data stored in centralized databases poses significant threats for data abuse, and additional emphasis must be given on the compliance with data security standards. Thus, emphasis must be given on the development of distributed databases with security measures to ensure the confidentiality of the data.

The fact that the profile of the big data is complex, with multiple and dispersed data sources and different data formats, hampers the application of the traditional approaches for data management and analytics. The sources of medical data are many (Fig. 2.3), including PRs, EHRs, genome registries, clinical centers, social media posts, and clinical claims, to name but a few, and thus the variety of these disparate sources pose several issues regarding the structural heterogeneities that are introduced during the analysis of such data. This is a crucial barrier that highlights the need for the development of new software architectures along with high-performance computational resources that are able to deal with big data management, analysis, and visualization. In addition, before the application of cohort studies and clinical trials, the researchers and data analysts must be able to understand the structure of the big data so as to avoid any data misuse and biases that are introduced during the data preprocessing stage, as well as during the statistical analysis procedure. Furthermore, the researchers and data analysts must be well

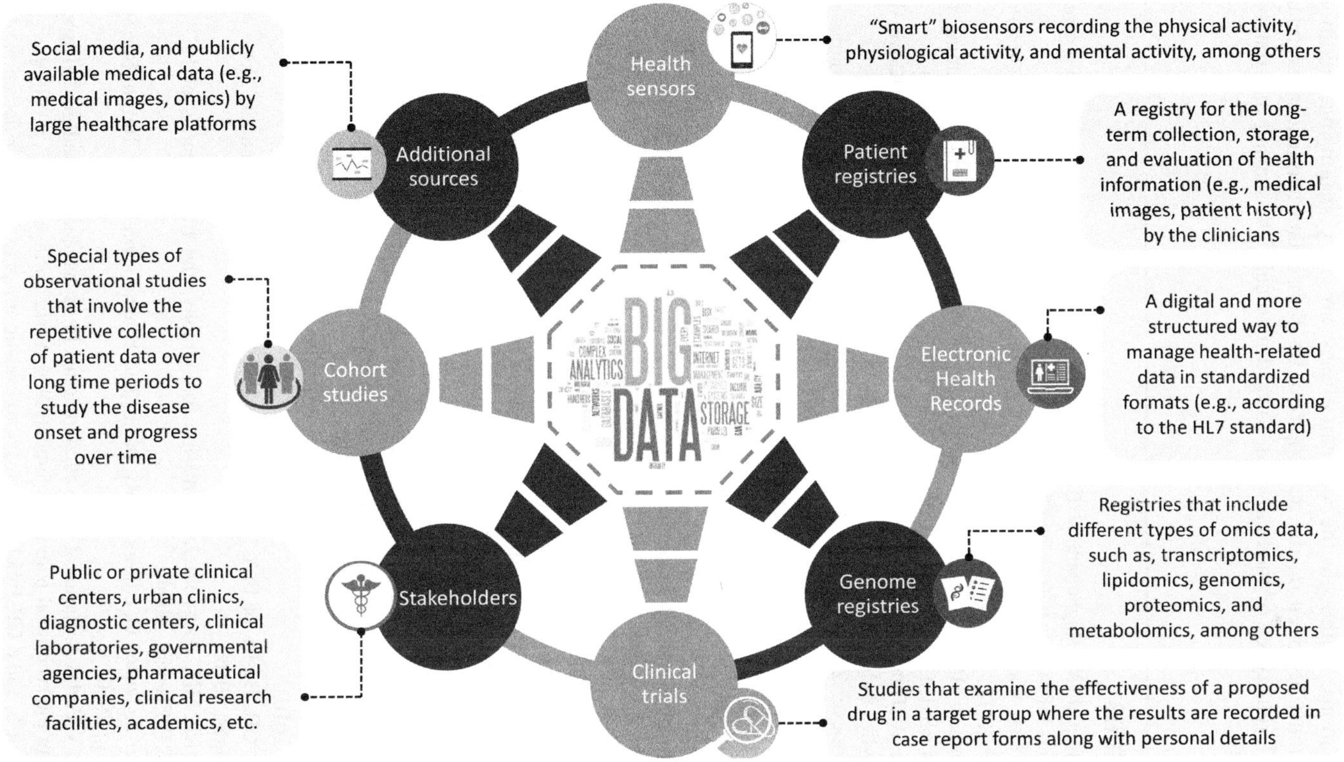

FIGURE 2.3

A summary of the sources of big data in healthcare.

aware of the ethical and legal issues that are posed by the international data protection regulations. Although the current efforts are very promising, additional emphasis must be given on the development of global data acquisition standards for the different types of medical data, toward medical data sharing and database federation.

References

[1] Raghupathi W, Raghupathi V. Big data analytics in healthcare: promise and potential. Health Inf Sci Syst 2014;2(1):3.

[2] Lee CH, Yoon HJ. Medical big data: promise and challenges. Kidn Res Clin Pract 2017;36(1):3.

[3] Krumholz HM. Big data and new knowledge in medicine: the thinking, training, and tools needed for a learning health system. Health Aff 2014;33(7):1163–70.

[4] Belle A, Thiagarajan R, Soroushmehr SM, Navidi F, Beard DA, Najarian K. Big data analytics in healthcare. BioMed Res Int 2015;2015,370194.

[5] Obermeyer Z, Emanuel EJ. Predicting the future—big data, machine learning, and clinical medicine. N Engl J Med 2016;375(13):1216.

[6] Bates DW, Saria S, Ohno-Machado L, Shah A, Escobar G. Big data in health care: using analytics to identify and manage high-risk and high-cost patients. Health Aff 2014; 33(7):1123–31.

[7] AlGhatrif M, Lindsay J. A brief review: history to understand fundamentals of electrocardiography. J Community Hosp Intern Med Perspect 2012;2(1):14383.

[8] Britton JW, Frey LC, Hopp JL, Korb P, Koubeissi MZ, Lievens WE, et al. Electroencephalography (EEG): an introductory text and atlas of normal and abnormal findings in adults, children, and infants. Chicago: American Epilepsy Society; 2016.

[9] Hari R, Salmelin R. Magnetoencephalography: from SQUIDs to neuroscience: neuroimage 20th anniversary special edition. Neuroimage 2012;61(2):386–96.

[10] Mills KR. The basics of electromyography. J Neurol Neurosurg Psychiatry 2005; 76(Suppl. 2):ii32–5.

[11] Usakli AB, Gurkan S, Aloise F, Vecchiato G, Babiloni F. On the use of electrooculogram for efficient human computer interfaces. Comput Intell Neurosci 2010;2010:1.

[12] Khadivi KO. Computed tomography: fundamentals, system technology, image quality, applications. Med Phys 2006;33(8):3076.

[13] Hashemi RH, Bradley WG, Lisanti CJ. MRI: the basics. Lippincott Williams & Wilkins; 2010.

[14] Saha GB. Basics of PET imaging: physics, chemistry, and regulations. New York: Springer; 2016.

[15] Wernick MN, Aarsvold JN. Emission tomography: the fundamentals of PET and SPECT. Elsevier; 2004.

[16] Ulmer S, Jansen O, editors. fMRI: basics and clinical applications. Berlin: Springer-Verlag; 2010.

[17] Chan V, Perlas A. Basics of ultrasound imaging. In: Atlas of ultrasound-guided procedures in interventional pain management. New York: Springer; 2011. p. 13–9.

[18] Mallidi S, Luke GP, Emelianov S. Photoacoustic imaging in cancer detection, diagnosis, and treatment guidance. Trends Biotechnol 2011;29(5):213–21.

[19] Ferrari M, Quaresima V. A brief review on the history of human functional near-infrared spectroscopy (fNIRS) development and fields of application. Neuroimage 2012;63(2):921–35.

[20] Griffiths AJF, Miller JH, Suzuki DT, Lewontin RC, Gelbart WM. An introduction to genetic analysis. 7th ed. New York: W. H. Freeman; 2000 Available from: http://www.ncbi.nlm.nih.gov/books/NBK21766/.

[21] Yang K, Han X. Lipidomics: techniques, applications, and outcomes related to biomedical sciences. Trends Biochem Sci 2016;41(11):954–69.

[22] Aslam B, Basit M, Nisar MA, Khurshid M, Rasool MH. Proteomics: technologies and their applications. J Chromatogr Sci 2017;55(2):182–96.

[23] Tzoulaki I, Ebbels TM, Valdes A, Elliott P, Ioannidis JP. Design and analysis of metabolomics studies in epidemiologic research: a primer on-omic technologies. Am J Epidemiol 2014;180(2):129–39.

[24] Shukla SK, Murali NS, Brilliant MH. Personalized medicine going precise: from genomics to microbiomics. Trends Mol Med 2015;21(8):461–2.

[25] Jones PA, Baylin SB. The epigenomics of cancer. Cell 2007;128(4):683–92.

[26] Lowe R, Shirley N, Bleackley M, Dolan S, Shafee T. Transcriptomics technologies. PLoS Comput Biol 2017;13(5):e1005457.

[27] Metzker ML. Sequencing technologies – the next generation. Nat Rev Genet 2010; 11(1):31.

[28] Ozsolak F, Milos PM. RNA sequencing: advances, challenges and opportunities. Nat Rev Genet 2011;12(2):87.

[29] Lavinder JJ, Horton AP, Georgiou G, Ippolito GC. Next-generation sequencing and protein mass spectrometry for the comprehensive analysis of human cellular and serum antibody repertoires. Curr Opin Chem Biol 2015;24:112–20.

[30] Fuchs B, Süß R, Teuber K, Eibisch M, Schiller J. Lipid analysis by thin-layer chromatography – a review of the current state. J Chromatogr A 2011;1218(19):2754–74.

[31] Ljosa V, Carpenter AE. Introduction to the quantitative analysis of two-dimensional fluorescence microscopy images for cell-based screening. PLoS Comput Biol 2009; 5(12):e1000603.

[32] Wang D, Bodovitz S. Single cell analysis: the new frontier in 'omics'. Trends Biotechnol 2010;28(6):281–90.

[33] Polak JM, Van Noorden S, editors. Immunocytochemistry: practical applications in pathology and biology. Butterworth-Heinemann; 2014.

[34] Dabbs DJ. Diagnostic immunohistochemistry E-book: theranostic and genomic applications. Elsevier Health Sciences; 2017.

[35] Greene CS, Tan J, Ung M, Moore JH, Cheng C. Big data bioinformatics. J Cell Physiol 2014;229(12):1896–900.

[36] Ashley EA, Niebauer J. Conquering the ECG. In: Cardiology explained. England: Remedica; 2004.

[37] Teplan M. Fundamentals of EEG measurement. Meas Sci Rev 2002;2(2):1–11.

[38] Pan X, Sidky EY, Vannier M. Why do commercial CT scanners still employ traditional, filtered back-projection for image reconstruction? Inverse Probl 2009;25(12):123009.

[39] Gordon R, Bender R, Herman GT. Algebraic reconstruction techniques (ART) for three-dimensional electron microscopy and X-ray photography. J Theor Biol 1970; 29(3):471–81.

[40] Abeida H, Zhang Q, Li J, Merabtine N. Iterative sparse asymptotic minimum variance based approaches for array processing. IEEE Trans Signal Process 2013;61(4): 933–44.

[41] Xu M, Wang LV. Universal back-projection algorithm for photoacoustic computed tomography. Phys Rev 2005;71(1):016706.

[42] Hegde P, Qi R, Abernathy K, Gay C, Dharap S, Gaspard R, et al. A concise guide to cDNA microarray analysis. Biotechniques 2000;29(3):548–63.

[43] Song JW, Chung KC. Observational studies: cohort and case-control studies. Plast Reconstr Surg 2010;126(6):2234.

[44] Gamble JM. An introduction to the fundamentals of cohort and case–control studies. Can J Hosp Pharm 2014;67(5):366.

[45] Süt N. Study designs in medicine. Balkan Med J 2014;31(4):273.

[46] Robbins AS, Chao SY, Fonseca VP. What's the relative risk? A method to directly estimate risk ratios in cohort studies of common outcomes. Ann Epidemiol 2002;12(7): 452–4.

[47] Kim HY. Statistical notes for clinical researchers: risk difference, risk ratio, and odds ratio. Restor Dent Endod 2017;42(1):72–6.

[48] McNutt LA, Wu C, Xue X, Hafner JP. Estimating the relative risk in cohort studies and clinical trials of common outcomes. Am J Epidemiol 2003;157(10):940–3.

[49] Buzsaki G. Rhythms of the brain. Oxford University Press; 2006.

[50] Kropotov JD. Quantitative EEG, event-related potentials and neurotherapy. Academic Press; 2010.

[51] Nishida K, Morishima Y, Yoshimura M, Isotani T, Irisawa S, Jann K, et al. EEG microstates associated with salience and frontoparietal networks in frontotemporal dementia, schizophrenia and Alzheimer's disease. Clin Neurophysiol 2013;124(6): 1106–14.

[52] Kindler J, Hubl D, Strik WK, Dierks T, König T. Resting-state EEG in schizophrenia: auditory verbal hallucinations are related to shortening of specific microstates. Clin Neurophysiol 2011;122(6):1179–82.

[53] Penolazzi B, Spironelli C, Vio C, Angrilli A. Brain plasticity in developmental dyslexia after phonological treatment: a beta EEG band study. Behav Brain Res 2010;209(1):179–82.

[54] Proudfoot M, Woolrich MW, Nobre AC, Turner MR. Magnetoencephalography. Pract Neurol 2014;14(5):336–43.

[55] Kharkar S, Knowlton R. Magnetoencephalography in the presurgical evaluation of epilepsy. Epilepsy Behav 2015;46:19–26.

[56] Stam CJ. Use of magnetoencephalography (MEG) to study functional brain networks in neurodegenerative disorders. J Neurol Sci 2010;289(1–2):128–34.

[57] Preston DC, Shapiro BE. Electromyography and neuromuscular disorders E-book: clinical-electrophysiologic correlations (expert consult-online). Elsevier Health Sciences; 2012.

[58] Sartori M, Reggiani M, Farina D, Lloyd DG. EMG-driven forward-dynamic estimation of muscle force and joint moment about multiple degrees of freedom in the human lower extremity. PLoS One 2012;7(12):e52618.

[59] Hsu WY, Lin CH, Hsu HJ, Chen PH, Chen IR. Wavelet-based envelope features with automatic EOG artifact removal: application to single-trial EEG data. Expert Syst Appl 2012;39(3):2743–9.

[60] Deng LY, Hsu CL, Lin TC, Tuan JS, Chang SM. EOG-based Human-Computer Interface system development. Expert Syst Appl 2010;37(4):3337–43.

[61] Szirmai A, Keller B. Electronystagmographic analysis of caloric test parameters in vestibular disorders. Eur Arch Oto-Rhino-Laryngol 2013;270(1):87–91.

[62] Holm H, Gudbjartsson DF, Arnar DO, Thorleifsson G, Thorgeirsson G, Stefansdottir H, et al. Several common variants modulate heart rate, PR interval and QRS duration. Nat Genet 2010;42(2):117.

[63] Khorrami H, Moavenian M. A comparative study of DWT, CWT and DCT transformations in ECG arrhythmias classification. Expert Syst Appl 2010;37(8):5751–7.

[64] Acharya UR, Fujita H, Lih OS, Adam M, Tan JH, Chua CK. Automated detection of coronary artery disease using different durations of ECG segments with convolutional neural network. Knowl Based Syst 2017;132:62–71.

[65] Emmanuel BS. A review of signal processing techniques for heart sound analysis in clinical diagnosis. J Med Eng Technol 2012;36(6):303–7.

[66] Konishi E, Kawasaki T, Shiraishi H, Yamano M, Kamitani T. Additional heart sounds during early diastole in a patient with hypertrophic cardiomyopathy and atrioventricular block. J Cardiol Cases 2015;11(6):171–4.

[67] Kumar SBB, Jagannath M. Analysis of phonocardiogram signal for biometric identification system. In: Proceedings of the 2015 IEEE International Conference on Pervasive Computing (ICPC); 2015. p. 1–4.

[68] Hill NJ, Gupta D, Brunner P, Gunduz A, Adamo MA, Ritaccio A, et al. Recording human electrocorticographic (ECoG) signals for neuroscientific research and real-time functional cortical mapping. J Vis Exp 2012;64.

[69] Tripathi M, Garg A, Gaikwad S, Bal CS, Chitra S, Prasad K, et al. Intra-operative electrocorticography in lesional epilepsy. Epilepsy Res 2010;89(1):133–41.

[70] Picht T, Schmidt S, Brandt S, Frey D, Hannula H, Neuvonen T, et al. Preoperative functional mapping for rolandic brain tumor surgery: comparison of navigated transcranial magnetic stimulation to direct cortical stimulation. Neurosurgery 2011;69(3):581–9.

[71] Lusic H, Grinstaff MW. X-ray-computed tomography contrast agents. Chem Rev 2012;113(3):1641–66.

[72] Auer M, Gasser TC. Reconstruction and finite element mesh generation of abdominal aortic aneurysms from computerized tomography angiography data with minimal user interactions. IEEE Trans Med Imaging 2010;29(4):1022–8.

[73] Herman GT, Kuba A, editors. Discrete tomography: foundations, algorithms, and applications. New York: Springer Science & Business Media; 2012.

[74] Miglioretti DL, Johnson E, Williams A, Greenlee RT, Weinmann S, Solberg LI, et al. The use of computed tomography in pediatrics and the associated radiation exposure and estimated cancer risk. JAMA Pediatr 2013;167(8):700–7.

[75] Poser BA, Koopmans PJ, Witzel T, Wald LL, Barth M. Three dimensional echo-planar imaging at 7 Tesla. Neuroimage 2010;51(1):261–6.

[76] Yang H, Zhuang Y, Sun Y, Dai A, Shi X, Wu D, et al. Targeted dual-contrast T1-and T2-weighted magnetic resonance imaging of tumors using multifunctional gadolinium-labeled superparamagnetic iron oxide nanoparticles. Biomaterials 2011; 32(20):4584–93.

[77] Gordillo N, Montseny E, Sobrevilla P. State of the art survey on MRI brain tumor segmentation. Magn Reson Imag 2013;31(8):1426–38.

[78] Tsili AC, Argyropoulou MI, Giannakis D, Sofikitis N, Tsampoulas K. MRI in the characterization and local staging of testicular neoplasms. Am J Roentgenol 2010;194(3):682–9.
[79] Del Grande F, Carrino JA, Del Grande M, Mammen AL, Stine LC. Magnetic resonance imaging of inflammatory myopathies. Top Magn Reson Imaging 2011;22(2):39–43.
[80] Khoo MM, Tyler PA, Saifuddin A, Padhani AR. Diffusion-weighted imaging (DWI) in musculoskeletal MRI: a critical review. Skelet Radiol 2011;40(6):665–81.
[81] Basser PJ, Pierpaoli C. Microstructural and physiological features of tissues elucidated by quantitative-diffusion-tensor MRI. J Magn Reson 2011;213(2):560–70.
[82] Barch DM, Burgess GC, Harms MP, Petersen SE, Schlaggar BL, Corbetta M, et al. Function in the human connectome: task-fMRI and individual differences in behavior. Neuroimage 2013;80:169–89.
[83] Formaggio E, Storti SF, Bertoldo A, Manganotti P, Fiaschi A, Toffolo GM. Integrating EEG and fMRI in epilepsy. Neuroimage 2011;54(4):2719–31.
[84] Han D, Yu J, Yu Y, Zhang G, Zhong X, Lu J, He W. Comparison of 18F-fluorothymidine and 18F-fluorodeoxyglucose PET/CT in delineating gross tumor volume by optimal threshold in patients with squamous cell carcinoma of thoracic esophagus. Int J Radiat Oncol Biol Phys 2010;76(4):1235–41.
[85] Boss A, Bisdas S, Kolb A, Hofmann M, Ernemann U, Claussen CD. Hybrid PET/MRI of intracranial masses: initial experiences and comparison to PET/CT. J Nucl Med 2010;51(8):1198.
[86] Hutton BF, Buvat I, Beekman FJ. Review and current status of SPECT scatter correction. Phys Med Biol 2011;56(14):R85.
[87] Wang Y, Qin L, Shi X, Zeng Y, Jing H, Schoepf UJ, et al. Adenosine-stress dynamic myocardial perfusion imaging with second-generation dual-source CT: comparison with conventional catheter coronary angiography and SPECT nuclear myocardial perfusion imaging. Am J Roentgenol 2012;198(3):521–9.
[88] Zhou Y, Chakraborty S, Liu S. Radiolabeled cyclic RGD peptides as radiotracers for imaging tumors and thrombosis by SPECT. Theranostics 2011;1:58.
[89] Willeumier KC, Taylor DV, Amen DG. Elevated BMI is associated with decreased blood flow in the prefrontal cortex using SPECT imaging in healthy adults. Obesity 2011;19(5):1095–7.
[90] Izatt JA, Choma MA, Dhalla AH. Theory of optical coherence tomography. Cham: Springer; 2015.
[91] Jia Y, Wei E, Wang X, Zhang X, Morrison JC, Parikh M, et al. Optical coherence tomography angiography of optic disc perfusion in glaucoma. Ophthalmology 2014;121(7):1322–32.
[92] Jia H, Abtahian F, Aguirre AD, Lee S, Chia S, Lowe H, et al. In vivo diagnosis of plaque erosion and calcified nodule in patients with acute coronary syndrome by intravascular optical coherence tomography. J Am Coll Cardiol 2013;62(19):1748–58.
[93] Kirtane TS, Wagh MS. Endoscopic optical coherence tomography (OCT): advances in gastrointestinal imaging. Gastroenterol Res Pract 2014;2014:376367.
[94] Sutoko S, Sato H, Maki A, Kiguchi M, Hirabayashi Y, Atsumori H, et al. Tutorial on platform for optical topography analysis tools. Neurophotonics 2016;3(1):010801.
[95] Eierud C, Craddock RC, Fletcher S, Aulakh M, King-Casas B, Kuehl D, et al. Neuroimaging after mild traumatic brain injury: review and meta-analysis. Neuroimage Clin 2014;4:283–94.

[96] Leff DR, Orihuela-Espina F, Elwell CE, Athanasiou T, Delpy DT, Darzi AW, et al. Assessment of the cerebral cortex during motor task behaviours in adults: a systematic review of functional near infrared spectroscopy (fNIRS) studies. Neuroimage 2011; 54(4):2922−36.
[97] Abramowicz JS. Benefits and risks of ultrasound in pregnancy. Semin Perinatol 2013; 37(5):295−300.
[98] Biswas M, Sudhakar S, Nanda NC, Buckberg G, Pradhan M, Roomi AU, et al. Two- and three-dimensional speckle tracking echocardiography: clinical applications and future directions. Echocardiography 2013;30(1):88−105.
[99] Räber L, Taniwaki M, Zaugg S, Kelbæk H, Roffi M, Holmvang L, et al. Effect of high-intensity statin therapy on atherosclerosis in non-infarct-related coronary arteries (IBIS-4): a serial intravascular ultrasonography study. Eur Heart J 2014;36(8): 490−500.
[100] Wong CS, McNicholas N, Healy D, Clarke-Moloney M, Coffey JC, Grace PA, et al. A systematic review of preoperative duplex ultrasonography and arteriovenous fistula formation. J Vasc Surg 2013;57(4):1129−33.
[101] Salonen JT, Salonen R. Ultrasound B-mode imaging in observational studies of athero-sclerotic progression. Circulation 1993;87(3 Suppl):II56−65.
[102] Beard P. Biomedical photoacoustic imaging. Interface Focus 2011;1(4):602−31.
[103] Wang X, Pang Y, Ku G, Xie X, Stoica G, Wang LV. Noninvasive laser-induced photo-acoustic tomography for structural and functional in vivo imaging of the brain. Nat Biotechnol 2003;21(7):803.
[104] Devlin TM. Textbook of biochemistry. John Wiley & Sons; 2011.
[105] Seeb JE, Carvalho G, Hauser L, Naish K, Roberts S, Seeb LW. Single-nucleotide poly-morphism (SNP) discovery and applications of SNP genotyping in nonmodel organisms. Mol Ecol Resour 2011;11:1−8.
[106] Sager M, Yeat NC, Pajaro-Van der Stadt S, Lin C, Ren Q, Lin J. Transcriptomics in cancer diagnostics: developments in technology, clinical research and commercialization. Expert Rev Mol Diagn 2015;15(12):1589−603.
[107] Nepusz T, Yu H, Paccanaro A. Detecting overlapping protein complexes in protein-protein interaction networks. Nat Methods 2012;9(5):471.
[108] Mazereeuw G, Herrmann N, Bennett SA, Swardfager W, Xu H, Valenzuela N, et al. Platelet activating factors in depression and coronary artery disease: a potential biomarker related to inflammatory mechanisms and neurodegeneration. Neurosci Bio-behav Rev 2013;37(8):1611−21.
[109] Wishart DS. Emerging applications of metabolomics in drug discovery and precision medicine. Nat Rev Drug Discov 2016;15(7):473.
[110] Buermans HPJ, Den Dunnen JT. Next generation sequencing technology: advances and applications. Biochim Biophys Acta 2014;1842(10):1932−41.
[111] Bernstein BE, Stamatoyannopoulos JA, Costello JF, Ren B, Milosavljevic A, Meissner A, et al. The NIH roadmap epigenomics mapping consortium. Nat Bio-technol 2010;28(10):1045.
[112] Sandoval J, Esteller M. Cancer epigenomics: beyond genomics. Curr Opin Genet Dev 2012;22(1):50−5.
[113] Hoffman R, Benz Jr EJ, Silberstein LE, Heslop H, Anastasi J, Weitz J. Hematology: basic principles and practice. Elsevier Health Sciences; 2013.
[114] Hillman RS, Ault KA, Rinder HM. Hematology in clinical practice, (LANGE clinical medicine). New York: McGraw-Hill; 2011.

[115] Wu X. Urinalysis: a review of methods and procedures. Crit Care Nurs Clin 2010; 22(1):121–8.
[116] Gratzke C, Bachmann A, Descazeaud A, Drake MJ, Madersbacher S, Mamoulakis C, et al. EAU guidelines on the assessment of non-neurogenic male lower urinary tract symptoms including benign prostatic obstruction. Eur Urol 2015;67(6):1099–109.
[117] Wine Y, Horton AP, Ippolito GC, Georgiou G. Serology in the 21st century: the molecular-level analysis of the serum antibody repertoire. Curr Opin Immunol 2015;35:89–97.
[118] Guarner J, Brandt ME. Histopathologic diagnosis of fungal infections in the 21st century. Clin Microbiol Rev 2011;24(2):247–80.
[119] Triplett DA. Coagulation and bleeding disorders: review and update. Clin Chem 2000; 46(8):1260–9.
[120] Peckham M. Histology at a glance, vol. 50. John Wiley & Sons; 2011.
[121] Öberg K, Castellano D. Current knowledge on diagnosis and staging of neuroendocrine tumors. Cancer Metastasis Rev 2011;30(1):3–7.
[122] Delves PJ, Martin SJ, Burton DR, Roitt IM. Essential immunology. John Wiley & Sons; 2017.
[123] Chandrashekara S. The treatment strategies of autoimmune disease may need a different approach from conventional protocol: a review. Indian J Pharmacol 2012; 44(6):665.
[124] Helfand M, Mahon SM, Eden KB, Frame PS, Orleans CT. Screening for skin cancer. Am J Prev Med 2001;20(3):47–58.
[125] Heinzerling L, Mari A, Bergmann KC, Bresciani M, Burbach G, Darsow U, et al. The skin prick test—European standards. Clin Transl Allergy 2013;3(1):3.
[126] Greenstein B, Wood DF. The endocrine system at a glance. John Wiley & Sons; 2011.
[127] Yu J. Endocrine disorders and the neurologic manifestations. Ann Pediatr Endocrinol Metab 2014;19(4):184.
[128] Spence N. Electrical impedance measurement as an endpoint detection method for routine coagulation tests. Br J Biomed Sci 2002;59(4):223–7.
[129] Adan A, Alizada G, Kiraz Y, Baran Y, Nalbant A. Flow cytometry: basic principles and applications. Crit Rev Biotechnol 2017;37(2):163–76.
[130] Rose HH. Optics of high-performance electron microscopes. Sci Technol Adv Mater 2008;9(1):014107.
[131] Parmigiani G, Garrett ES, Irizarry RA, Zeger SL. The analysis of gene expression data: an overview of methods and software. New York: Springer; 2003.
[132] Cinquanta L, Fontana DE, Bizzaro N. Chemiluminescent immunoassay technology: what does it change in autoantibody detection? Auto Immun Highlights 2017;8(1):9.
[133] Odell ID, Cook D. Optimizing direct immunofluorescence. Methods in molecular biology (methods and protocols), vol. 1180. New York: Humana Press; 2014.
[134] Aydin S. A short history, principles, and types of ELISA, and our laboratory experience with peptide/protein analyses using ELISA. Peptides 2015;72:4–15.
[135] Knopp T, Kunis S, Potts D. A note on the iterative MRI reconstruction from nonuniform k-space data. Int J Biomed Imaging 2007;2007:24727.
[136] Srinivasan S, Pogue BW, Carpenter C, Yalavarthy PK, Paulsen K. A boundary element approach for image-guided near-infrared absorption and scatter estimation. Med Phys 2007;34(11):4545–57.
[137] Larobina M, Murino L. Medical image file formats. J Digit Imaging 2014;27(2): 200–6.

[138] Gorgolewski KJ, Auer T, Calhoun VD, Craddock RC, Das S, Duff EP, et al. The brain imaging data structure, a format for organizing and describing outputs of neuroimaging experiments. Sci Data 2016;3:160044.

[139] Mazurenka M, Di Sieno L, Boso G, Contini D, Pifferi A, Dalla Mora A, et al. Non-contact in vivo diffuse optical imaging using a time-gated scanning system. Biomed Opt Express 2013;4(10):2257−68.

[140] Sanger F, Nicklen S, Coulson AR. DNA sequencing with chain-terminating inhibitors. Proc Natl Acad Sci USA 1977;74(12):5463−7.

[141] Maxam AM, Gilbert W. A new method for sequencing DNA. Proc Natl Acad Sci USA 1977;74(2):560−4.

[142] Harrington CT, Lin EI, Olson MT, Eshleman JR. Fundamentals of pyrosequencing. Arch Pathol Lab Med 2013;137(9):1296−303.

[143] Bronner IF, Quail MA, Turner DJ, Swerdlow H. Improved protocols for illumina sequencing. Curr Protoc Hum Genet 2014;21(80). 18.2.1-42.

[144] Cock PJ, Fields CJ, Goto N, Heuer ML, Rice PM. The Sanger FASTQ file format for sequences with quality scores, and the Solexa/Illumina FASTQ variants. Nucleic Acids Res 2009;38(6):1767−71.

[145] Leinonen R, Sugawara H, Shumway M. The sequence read archive. Nucleic Acids Res 2011;39:D19−21.

[146] Ishizu H, Siomi H, Siomi MC. Biology of PIWI-interacting RNAs: new insights into biogenesis and function inside and outside of germlines. Genes Dev 2012;26(21): 2361−73.

[147] Yamamoto M, Wakatsuki T, Hada A, Ryo A. Use of serial analysis of gene expression (SAGE) technology. J Immunol Methods 2001;250(1−2):45−66.

[148] Takahashi H, Kato S, Murata M, Carninci P. CAGE (cap analysis of gene expression): a protocol for the detection of promoter and transcriptional networks. Humana Press; 2012.

[149] Singh G, Pratt G, Yeo GW, Moore MJ. The clothes make the mRNA: past and present trends in mRNP fashion. Annu Rev Biochem 2015;84:325−54.

[150] Workman TA. Engaging patients in information sharing and data collection: the role of patient-powered registries and research networks. Rockville (MD): Agency for Healthcare Research and Quality (US); 2013.

[151] Fleurence RL, Beal AC, Sheridan SE, Johnson LB, Selby JV. Patient-powered research networks aim to improve patient care and health research. Health Aff 2014;33(7): 1212−9.

[152] Bashir R. BioMEMS: state-of-the-art in detection, opportunities and prospects. Adv Drug Deliv Rev 2004;56(11):1565−86.

[153] Holzinger M, Le Goff A, Cosnier S. Nanomaterials for biosensing applications: a review. Front Chem 2014;2:63.

[154] Black AD, Car J, Pagliari C, Anandan C, Cresswell K, Bokun T, et al. The impact of eHealth on the quality and safety of health care: a systematic overview. PLoS Med 2011;8(1):e1000387.

[155] Kay M, Santos J, Takane M. mHealth: new horizons for health through mobile technologies. World Health Organ 2011;64(7):66−71.

[156] Häyrinen K, Saranto K, Nykänen P. Definition, structure, content, use and impacts of electronic health records: a review of the research literature. Int J Med Inform 2008; 77(5):291−304.

[157] Dolin RH, Alschuler L, Beebe C, Biron PV, Boyer SL, Essin D, et al. The HL7 clinical document architecture. J Am Med Assoc 2001;8(6):552–69.
[158] Rubinstein WS, Maglott DR, Lee JM, Kattman BL, Malheiro AJ, Ovetsky M, et al. The NIH genetic testing registry: a new, centralized database of genetic tests to enable access to comprehensive information and improve transparency. Nucleic Acids Res 2012;41(D1):D925–35.
[159] Hasin Y, Seldin M, Lusis A. Multi-omics approaches to disease. Genome Biol 2017; 18(1):83.
[160] Umscheid CA, Margolis DJ, Grossman CE. Key concepts of clinical trials: a narrative review. Postgrad Med 2011;123(5):194–204.
[161] Unger JM, Cook E, Tai E, Bleyer A. The role of clinical trial participation in cancer research: barriers, evidence, and strategies. Am Soc Clin Oncol Educ Book 2016; 36:185–98.
[162] Bellary S, Krishnankutty B, Latha MS. Basics of case report form designing in clinical research. Perspect Clin Res 2014;5(4):159–66.
[163] Van Essen DC, Smith SM, Barch DM, Behrens TE, Yacoub E, Ugurbil K. The Wu-Minn human connectome project: an overview. Neuroimage 2013;80:62–79.
[164] Maglott D, Ostell J, Pruitt KD, Tatusova T. Entrez gene: gene-centered information at NCBI. Nucleic Acids Res 2005;33(Suppl. 1_1):D54–8.
[165] Caruana EJ, Roman M, Hernández-Sánchez J, Solli P. Longitudinal studies. J Thorac Dis 2015;7(11):E537–40.
[166] Szumilas M. Explaining odds ratios. J Can Acad Child Adolesc Psychiatry 2010;19(3): 227–9.
[167] Austin PC. The performance of different propensity-score methods for estimating differences in proportions (risk differences or absolute risk reductions) in observational studies. Stat Med 2010;29(20):2137–48.
[168] Olivier J, Bell ML. Effect sizes for 2× 2 contingency tables. PLoS One 2013;8(3): e58777.
[169] Rubinov M, Sporns O. Complex network measures of brain connectivity: uses and interpretations. Neuroimage 2010;52(3):1059–69.
[170] Glasser MF, Coalson TS, Robinson EC, Hacker CD, Harwell J, Yacoub E, et al. A multi-modal parcellation of human cerebral cortex. Nature 2016;536(7615):171–8.
[171] De Las Rivas J, Fontanillo C. Protein–protein interactions essentials: key concepts to building and analyzing interactome networks. PLoS Comput Biol 2010;6(6): e1000807.
[172] Tsiouris KM, Pezoulas VC, Zervakis M, Konitsiotis S, Koutsouris DD, Fotiadis DI. A long short-term memory deep learning network for the prediction of epileptic seizures using EEG signals. Comput Biol Med 2018;99:24–37.
[173] Eryaman Y, Zhang P, Utecht L, Kose K, Lagore RL, DelaBarre L, et al. Investigating the physiological effects of 10.5 Tesla static field exposure on anesthetized swine. Magn Reson Med 2018;79:511–4.
[174] Lindon JC, Nicholson JK, Holmes E, editors. The handbook of metabonomics and metabolomics. Elsevier; 2011.
[175] Christensen S, Mogelvang R, Heitmann M, Prescott E. Level of education and risk of heart failure: a prospective cohort study with echocardiography evaluation. Eur Heart J 2010;32(4):450–8.

[176] Vos SJ, Xiong C, Visser PJ, Jasielec MS, Hassenstab J, Grant EA, et al. Preclinical Alzheimer's disease and its outcome: a longitudinal cohort study. Lancet Neurol 2013;12(10):957–65.

[177] Hyldgaard C, Hilberg O, Pedersen AB, Ulrichsen SP, Løkke A, Bendstrup E, et al. A population-based cohort study of rheumatoid arthritis-associated interstitial lung disease: comorbidity and mortality. Ann Rheum Dis 2017;76(10):1700–6.

[178] Driver JA, Smith A, Buring JE, Gaziano JM, Kurth T, Logroscino G. Prospective cohort study of type 2 diabetes and the risk of Parkinson's disease. Diabetes Care 2008;31(10):2003–5.

[179] Shorvon SD, Goodridge DM. Longitudinal cohort studies of the prognosis of epilepsy: contribution of the National General Practice Study of Epilepsy and other studies. Brain 2013;136(11):3497–510.

[180] Kuehn T, Bauerfeind I, Fehm T, Fleige B, Hausschild M, Helms G, et al. Sentinel-lymph-node biopsy in patients with breast cancer before and after neoadjuvant chemotherapy (SENTINA): a prospective, multicentre cohort study. Lancet Oncol 2013;14(7):609–18.

CHAPTER

Medical data sharing

3

Chapter outline

3.1 Overview

Medical data sharing involves all those mechanisms concerning the protection of patient's rights and privacy. It comprises the core of a federated platform as it enables

Medical Data Sharing, Harmonization and Analytics. https://doi.org/10.1016/B978-0-12-816507-2.00003-7

the interlinking of medical cohorts worldwide [1]. A data sharing framework is responsible for two major functionalities: (i) the assessment of whether the data origin and acquisition, as well as the processes that are undertaken in a federated platform, fulfill the guidelines posed by the corresponding data protection regulations (i.e., the legal aspect) and (ii) the assessment of the quality and completeness of medical data (i.e., the data quality aspect) by taking into consideration existing clinical domain knowledge and related public health policies. The latter is usually referred to as data governance and is related to (i) the evaluation of data quality metrics, (ii) the inspection of the data organizational structure, and (iii) the overall information management [2].

The data sharing framework constitutes the primary stage before the development and application of the federated data analytics services. From a legal point of view, a clinical center that wishes to share clinical data to a federated platform must provide all the necessary ethical and legal documents, before any further data manipulation. These documents depend on the data protection regulations posed by each party (e.g., according to the General Data Protection Regulation [GDPR] guidelines in Europe or the Health Insurance Portability and Accountability [HIPAA] guidelines in the United States) and usually include (i) precise definition of legitimate interests, (ii) complete data protection impact assessments, (iii) exact purposes of processing, (iv) signed consent forms for the processing of personal data from the data subjects, (v) purposes of transferring to third parties, (vi) data protection guarantees, and (vii) notifications to the data subject about the processing, among many others. A federated platform that is responsible for data sharing must first provide complete definitions for the primary data collectors and the secondary analysts. Informed consent forms for pooled data analysis are also necessary for data analysis through a process that is currently referred to as "handshaking" [3]. Ethical issues for data collection introduced by different countries inside and outside the EU must also be taken into consideration. Moreover, the fear for data abuse and losing the control of the data is a crucial barrier toward data sharing. Secure data management and data deidentification is thus mandatory for privacy preserving so as to enable the sharing of sensitive data.

From the data quality point of view, under the data governance part of the data sharing framework lies a fundamental procedure, known as data quality assessment [4–7], which aims to improve the quality of the data in terms of consistency, accuracy, relevance, completeness, etc. Data cleansing [8–10], also referred to as data curation, is a multidisciplinary process that comprises the core of the data quality assessment procedure and deals with duplicate fields, outliers, compatibility issues, missing values, etc., within a raw clinical dataset. Nowadays, automated data curation is a crucial technical challenge for data analysts and researchers worldwide who wish to manage and clean huge amounts of data [7]. For this reason, emphasis must be given on the development of tools for realizing such a concept. In addition, it is important to define a common format for the clinical datasets, i.e., a template of predefined variables, data ranges, and types, for a specific domain, which serves as a model that can be used to develop rules for (i) matching variables across heterogeneous datasets and (ii) normalizing them where necessary. The former is an intermediate step of data harmonization [1,12,13], and the latter is known as data standardization [1,11] (Chapter 5).

Several data sharing initiatives have been launched toward the integrity of clinical research data [14–17]. These initiatives aim at providing frameworks and guidelines for sharing medical and other related data. They mainly focus on the transparency of the data collection protocols and the patients deidentification process to promote medical research worldwide. Most of these initiatives develop publicly available gateways in the form of data warehouses, which host data from thousands of highly qualified clinical studies worldwide, including prospective and retrospective data from clinical trials, case-studies, genome-wide association studies (GWAS), etc., with the purpose of providing access to large amounts of data for scientific purposes [15–17]. Powerful cloud-based systems have been launched, with all the processes (registration, deidentification, quality control) being conducted automatically through the web. Thus, the meaningful interpretation of the outcomes of studies that make use of such data is reassured due to the increased statistical power they offer. Centralized patient databases, however, are often prone to data breach and sometimes unable to comply with data protection regulations [14]. A promising solution to this can be accomplished using multisite databases that serve as remote data warehouses that communicate in a distributed manner, giving emphasis on the "sharing of information from the data instead of sharing the data themselves" [14,17]. This approach overcomes several data sharing barriers discussed previously as the fear for data abuse can be controlled through distributed firewalls and individual data monitoring mechanisms. Moreover, the need to transfer sensitive data is nullified as an individual researcher can work independently on each site through coordinating systems that distribute the commands per site [17].

A federated platform should take into consideration several technical challenges. Treating patients with respect is a key factor toward its establishment. Emphasis must also be given to the cost scalability over security that is a crucial trade-off, as well as on software and copyright licenses for all the tools that will be employed in the platform. Big data monitoring, validation, storage, and multidimensional interoperability (legal, regulatory) are a few examples of such challenges. In this chapter, emphasis will be given on the scope of data sharing, the data quality part of the data sharing framework, as well as on global initiatives. The legal part of the data sharing framework (data protection) will be presented in Chapter 4.

3.2 The rationale behind medical data sharing

Why is medical data sharing so important? Imagine what you could do in a medical area if you had access to almost all of the medical data in this area. To answer this question in a realistic way, we will focus on presenting four clinical needs that have been identified as of great importance in several cohort studies: (i) patient stratification, (ii) identification of new biomarkers and/or validation of existing ones, (iii) new therapy treatments, and (iv) development of new health policies. Each of these needs highlights the necessity of medical data sharing in promoting research worldwide. As it was already mentioned in Chapter 1, cohort studies are capable of resolving crucial scientific questions related to the predictive modeling of a disease's onset and progress, the clinical significance of genetic variants, and the adequate identification of high-risk individuals. Although data sharing is valuable for the public, the ignorance of knowing what the denominators and the requirements are in a study leads to contradictory findings.

3.2.1 Patient stratification

So, how can medical data sharing help the scientific community identify high-risk patients, i.e., a subgroup of patients that are more likely to develop a pathological condition (e.g., malignant lymphoma) than the rest of the population? Patient stratification involves not only the early identification of high-risk patients but also (i) the estimation of the risk of organ involvement, disease course, and comorbidities, (ii) the prescription of specific treatments to different subgroups of the population, and (iii) patient selection for clinical trials [18,19]. Patient stratification can also decrease the risk of producing unsatisfactory results in clinical trials employing expensive drugs and quantify the effectiveness of a treatment as it may vary among different subgroups [19]. According to the majority of the cohort studies worldwide, computational methods are usually recruited for dealing with this need that is considered to be a classification/clustering problem[1] [19–21]. Such methods involve the use of data mining for training models that are able to yield accurate disease prognosis. The complexity of such models varies according to the data structure and size.

Although cohort, case–control, and cross-sectional studies [18] are capable of resolving crucial scientific questions related to risk stratification—(i) the predictive modeling of a disease's onset and progress, (ii) the clinical significance of genetic variants, and (iii) the identification of high-risk individuals—the fact that these cohorts are dispersed withholds important clinical information and leads to small-scale studies with reduced statistical significance and thus poor clinical importance. In addition, the application of data mining algorithms is more or less trivial due to the large number of open source and proprietary commercial software. None of these data models yields meaningful clinical interpretations about the disease prognosis and/or occurrence, unless the population size that is used to construct these models is large enough to be considered as of high statistical power. So, the real question here is how can the population size of a cohort study be efficiently increased to yield accurate results toward effective patient stratification?

---

[1] A clinical dataset with laboratory measures is usually used as input to classify patients as low or high risk.

The answer to this question is data sharing. By taking into consideration the fact that a key factor for making accurate predictions is the population size, different cohorts can be interlinked to realize the concept of federated databases [17,22–24]. Federated databases are not only able to interlink heterogeneous cohorts but also may lead to more accurate studies of rare diseases, i.e., studies with high statistical power. However, this "gathering" of data is often obscured due to the heterogeneity of the data protection regulations of the country where each cohort belongs to. Another limitation is the infrastructure. For example, where and how will the data be stored? This is another question that poses significant technical challenges. A common practice is to create a centralized database where the data from different cohorts will be gathered. This type of storage, however, is prone to data breach and poses several data sharing issues (e.g., in the case of pseudonymized data under the GDPR in Europe), especially in the case of prospective studies where the patient data are updated. The distributed analysis concept, however, is less prone to data breach as the data never move out from the clinical centers and thus serves as a promising solution [17]. In any data sharing system, however, the patient is the leading actor. The patient's power is significantly higher than the researchers could ever have as the patients are the real owners of their clinical data. Without the patient's consent, data sharing is pointless.

3.2.2 Identification of new biomarkers and/or validation of existing ones

A second clinical need that highlights the clinical importance of medical data sharing is the identification of new biomarkers and/or validation of existing ones. The primary focus of the majority of cohort studies worldwide is to confirm the association and test the predictive ability of previously suggested clinical laboratory predictors for disease diagnosis, development, and response to new therapy treatments [25,26]. An additional goal is to identify novel molecular and genetic biomarkers for disease diagnosis and future target therapies [25,26]. The validation of existing biomarkers, as well as the discovery of novel ones, can be usually formulated as a feature selection and feature ranking problem where data mining algorithms are applied with the purpose of identifying features of high significance.[2] Feature selection methods quantify how well a subset of features can improve the performance of a prediction model [27]. For example, feature selection methods are used in simple decision trees for decision-making [27,28]. Assuming a clinical dataset with laboratory measures or DNA microarrays, one can make use of such methods to quantify the significance of each variable. Then, the significant features can be used to create more accurate prediction models and thus lead to either the validation of existing biomarkers or the discovery of new ones.

[2] Subsets of variables that are separated according to their usefulness to a given predictor [27].

This problem is formulated as another classification problem (e.g., separate healthy patients from cancer patients, based on their gene expression profile), and thus population size matters. So, once more, the real question is how can the population size of a cohort study be efficiently increased to yield new and accurate biomarkers or validate existing ones? The answer is, again, data sharing. Through cohort interlinking, the researcher can improve the predictive ability of previous biomarkers in a larger population, validate the accuracy of the existing biomarkers, and even lead to the identification of new biomarkers. Apart from the existing feature selection methods that are often applied on centralized databases and are more or less trivial, several distributed feature selection algorithms have been proposed in the literature for identifying significant features in distributed environments [29,30] (see Chapter 7).

3.2.3 New therapy treatments

Several targeted therapies have been developed so far, and many of them have been shown to confer benefits, some of them being spectacular, especially in cancer drug development [25]. Traditional therapeutic clinical trial designs involve the application of small (approximately 100 patients), large (approximately 1000 patients), and very large (thousands of patients) clinical trials for measuring the disease prevalence, validating biomarkers, and defining the tolerability and pharmacological properties of agents involved in tumors [25]. Precision medicine is a modern field pursuing research advances that will enable better assessment of disease risk, understanding of disease mechanisms, and prediction of optimal therapy for many more diseases, with the goal of expanding the benefits of precision medicine into myriad aspects of health and healthcare [31,32]. Precision medicine's more individualized, molecular approach to cancer can enrich and modify diagnostics and effective treatments while providing a strong framework for accelerating the adoption of precision medicine in other spheres [31]. The most obvious spheres are the inherited genetic disorders and infectious diseases, but there is promise for more diseases. The importance of precision medicine toward the improvement of the current healthcare systems has been also highlighted by USA President Barack Obama, in 2015 [31].

Therapeutic research has turned to precision medicine to develop new approaches for detecting, measuring, and analyzing a wide range of biomedical information, including molecular, genomic, cellular, clinical, behavioral, physiological, and environmental parameters [31,32]. It is clear now that the basis of precision medicine lies on the sharing of medical data. Small- and large-scale cohort studies may lead to inappropriate conclusions about the safety and effectiveness of therapies, which may in turn harm patients and yield inaccurate scientific outcomes. On the other hand, a very large-scale cohort study including millions of patients is not always feasible and not sustainable at all. To reassure the effectiveness and safety of therapies and diagnostic tests, responsible data sharing is necessary. Interlinking small- and/or large-scale cohorts is a sustainable solution that reduces the complications that are posed during data processing and the effort needed for data management, as well.

This facilitates the development of straightforward data sharing models for precision medicine that envisage to enhance the quality of healthcare.

3.2.4 Development of new health policies

Responsible sharing of medical data benefits the public and relevant stakeholders including public insurance and healthcare payers, at the state and local levels, private insurers, and employers who cover health insurance costs. Stakeholders including health policy makers assess whether the produced health policy scenarios are cost effective and actually provide positive impact in healthcare systems, financial figures, and the society. A health policy can be defined as "the decisions, plans, and actions that are undertaken to achieve specific healthcare goals within a society," which can serve as a means of evidence-based policy-making for improvement in health (early diagnosis, new therapies) [33]. Health impact assessment (HIA) is a multidisciplinary process within which a range of evidence about the health effects of a policy is considered [34]. HIA is a combination of methods whose aim is to assess the health consequences to a population of a policy, project, or program [33]. A well-established health policy must also consider the opinions and expectations of those who may be affected by the proposed health policy.

In general, health systems are characterized by five dimensions: (i) financing arrangements, (ii) public sector funding, (iii) patient cost sharing and insurance, (iv) physician remuneration methods, and (v) gatekeeping arrangements [35]. Each of these dimensions is determined via discretionary health policy actions and is able to formulate a healthcare system [35]. Health policies, however, must be public and transparent to improve interagency collaboration and public participation [34]. However, a major barrier introduced during health policy-making is the lack of scientific evidence on the effectiveness of interventions [36]. In addition, their effectiveness and sustainability shall be configured and evaluated on patients from different countries by means of a shared health policy. Through data sharing, potential determinants of health, which are involved in the development of new health policies (biological factors, psychosocial environment, physical environment, socioeconomics) [34], are enhanced. Furthermore, strategic leadership and collaborative working for health promotes and protects the population's health, reduces inequalities, and preserves the ethics of people and resources yielding effective policy-making [37].

3.3 Data curation

Data quality assessment has been characterized as a key factor for achieving sustainable data of great quality in various domains varying from finance to healthcare [4–7]. Lacking data quality results in bad data manipulation, which makes data useless and has numerous negative effects on further processing. Thus, emphasis must be given on the development of proper mechanisms for data quality assessment.

The latter lies under the well-known data governance part of a data sharing system. Data cleansing, also referred to as data curation [8–10], comprises the core of the data quality assessment procedure. It aims to transform a dataset into a new one that meets specific criteria according to predefined quality measures. Examples of data quality measures include (i) accuracy, (ii) completeness, (iii) consistency, (iv) interpretability, (v) relevancy, and (vi) ease of manipulation, among many others [4]. Data curation can be also used as a diagnostic tool for marking problematic attributes that exhibit incompatibilities (e.g., unknown data types, missing values, outliers). In this way, data curation can guide the clinician for fixing missing clinical misinterpretations that are not easy to be automatically detected, especially when fixing missing values.

Automated data curation overcomes the complexity of processing huge amounts of medical data and can be easily scalable in contrast with traditional manual data curation that is not feasible in the case of big data management. However, clinical evaluation is necessary to ensure the reliability and applicability of automation. According to Fig. 3.1, data curation can be seen as a sequential process, i.e., a series of methodological steps, which involves functionalities for curating both prospective and retrospective data. Mechanisms for curating retrospective data include (i) the detection and elimination of duplicate fields (i.e., deduplication), (ii) the characterization of data according to their context (i.e., data annotation), (iii) the identification of duplicate fields with highly similar distributions (i.e., similarity detection), (iv) the transformation of data into standardized formats (i.e., standardization), (v) dealing with missing values (i.e., data imputation), and (vi) outlier detection for detecting values that deviate from the standard data range. Mechanisms for curating prospective data can be incorporated in the form of check constraints.

3.3.1 Metadata extraction

The first step toward data curation is to get a first look into the dataset's structure and quality through (i) the extraction of structural (e.g., number of features and instances) and vocabulary information (e.g., types of features and range values), (ii) the computation of ordinary descriptive statistics (e.g., histogram), (iii) the categorization of attributes into numeric and categorical, and (iv) the detection of missing values per attribute. This process is known as metadata extraction and provides useful information that can be used to interlink cohorts that belong to the same clinical domain and thus enable data sharing. Metadata is a simple way to preserve the privacy of the patient data as no sensitive information is revealed.

3.3.2 Data annotation

Data annotation refers to the categorization of features into continuous/discrete and categorical/numeric according to their data type (e.g., integer, float, string, date) and range values. Features can also be classified according to their quality in terms of compatibility issues and missing values. For example, features with higher

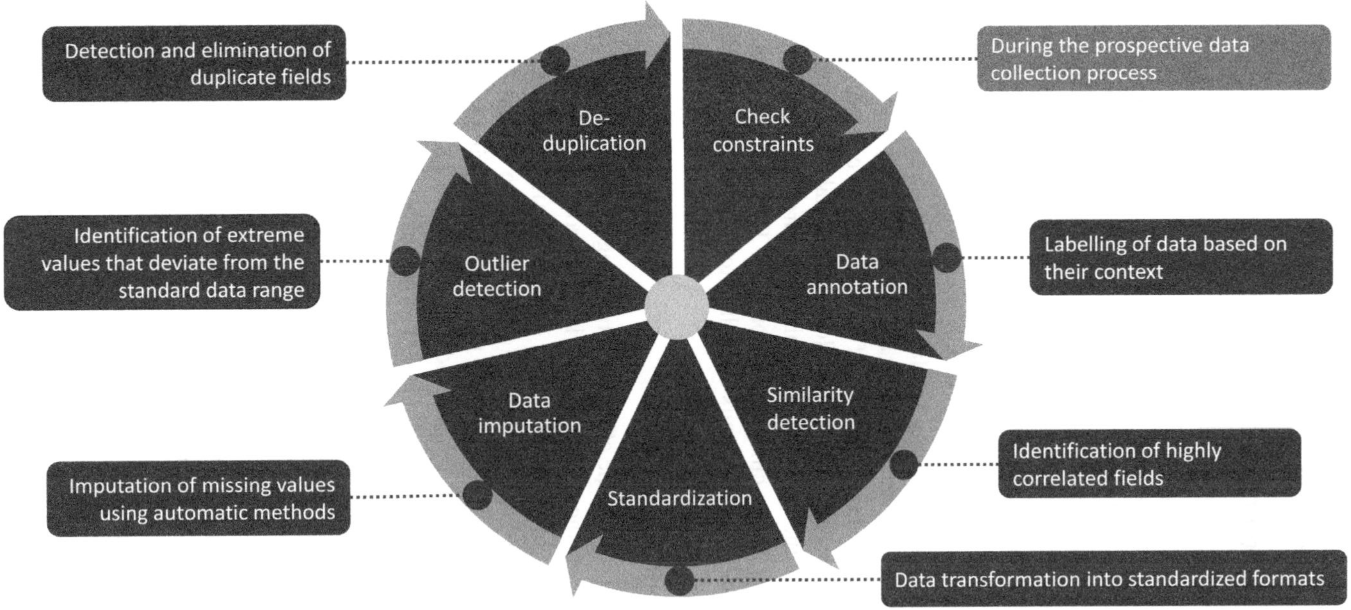

FIGURE 3.1

Steps toward prospective and retrospective data curation.

percentage of missing values and/or inconsistencies can be marked as "bad" for further removal.

3.3.3 Similarity detection and deduplication

A common problem that researchers face in large clinical datasets is the existence of duplicate attributes, i.e., features that exhibit similar characteristics, such as a common distribution or even a similar field name. To detect such features, one can compute a similarity or a distance measure for each pair of features. A widely used practice is to compute the correlation coefficient (e.g., Spearman's, Pearson's product moment) or the Euclidean distance between each pair of features across the dataset [38]. Assuming that the examined dataset contains m-features, the result of this procedure will be an mxm adjacency matrix, where the element (i,j) is the similarity between features i and j. In the case of correlation coefficient, highly correlated features signify a strong similarity, whereas features with high Euclidean distance values are considered nonsimilar as they are far away in the two-dimensional space. As a result, one can seek for pairs of features that are either highly correlated or close in the two-dimensional (2D) space. However, several biases might occur in both methods especially in the case where two binary features have common distributions but are not the same from a clinical point of view. Thus, the clinician's guidance is necessary to pick up the correct pairs according to the domain's knowledge.

3.3.4 Data imputation

A significant barrier toward the effective processing of medical data is the existence of missing values or unknown data types within the context of the data. The number of missing values can be outnumbered in case where (i) the data collection across different time points is not preserved (e.g., during the collection of follow-up data), (ii) biases are introduced into the measurements during the data acquisition process (e.g., unknown data types), and (iii) specific parameters within the data depend on the existence of other parameters (or groups of parameters) that can either be completely absent or partially missing due to the following two reasons. Toward this direction, various data imputation methods have been proposed in the literature, which deal with the presence of missing values extending from simple univariate methods [39–42] that replace the missing values using either the mean or the median, to supervised learning methods, such as regression trees, and nearest neighbors, which predict new values to replace the missing ones [39–42]. The univariate methods are computationally simple and can be performed by replacing the missing values (on a feature basis): (i) either with the feature's population mean or median/most frequent value, depending on whether the feature is a continuous or a discrete one, respectively, (ii) with preceding or succeeding values, and (iii) by selecting random values from the feature's distribution to replace each individual missing value [39–42]. Well-known supervised learning methods that are often used in practice for data imputation purposes include the support vector machines, the regression trees, the random forests, the k-nearest neighbors, and the fuzzy k-means, among others [39,42]. These methods are more suitable in the case of time series data, where multiple observations are acquired across different time points. The supervised learning models are individually trained on a subset of nonmissing values, per feature, to identify underlying data patterns that are used to predict new values for replacing the missing ones.

3.3.5 Outlier detection

Outlier detection, also referred to as anomaly detection, aims at separating a core of regular observations from some polluting ones, known as the outliers, which vary from the majority. According to the literature, a large variety of both univariate and multivariate methods have been proposed so far, some of which are discussed in the sequel. Most of these methods are standard approaches applied by clinical laboratories. The interquartile range (IQR) is a widely used approach that measures the statistical dispersion using the first and third quartiles of an attribute's range [43,44]. It is defined as the difference between the upper (Q3) and lower (Q1) quartiles of the data. Q1 is defined as the 25th percentile (lower quartile), whereas Q3 is the 75th percentile (upper quartile). Values lower than the first quartile or larger than the third quartile are considered to be outliers (Fig. 3.2) [43,44]. The IQR multiplied by

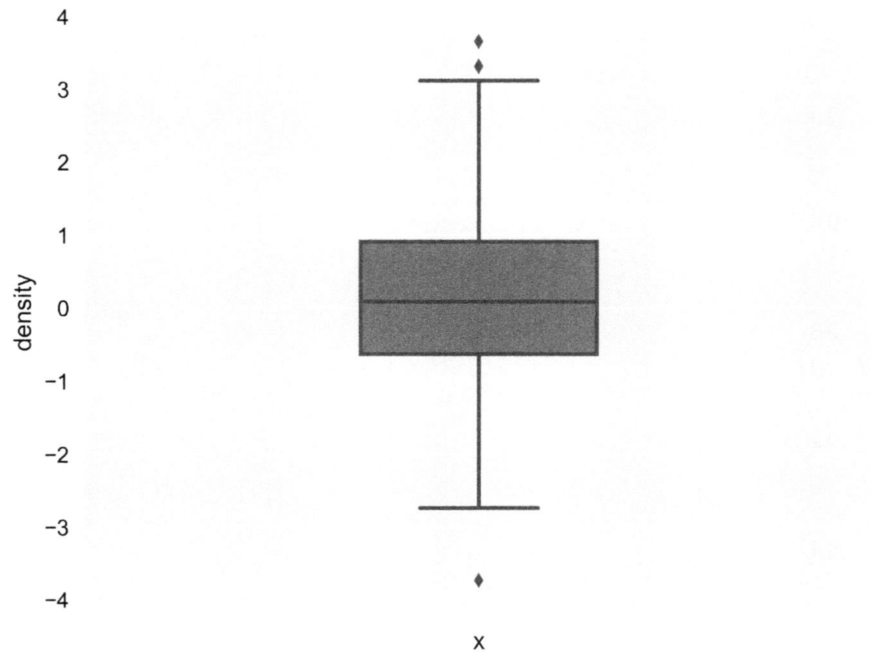

FIGURE 3.2

A typical boxplot for anomaly detection on a randomly generated and contaminated feature.

0.7413 yields the normalized IQR. The term 0.7413 comes from the inverse of the width of the standard normal distribution (1/1.3498). An example of a boxplot for outlier detection is displayed in Fig. 3.2, for a randomly generated and contaminated feature, $\boldsymbol{x} = (x_1, x_2, \ldots, x_N)$, where $x_i \in R$, and $N = 100$.

Another widely used statistical univariate measure for outlier detection is the z-score measure, which quantifies the distance between a feature's value and its population mean [43]. It is defined as

$$z = \frac{\boldsymbol{x} - \widehat{x}}{\sigma_x}, \tag{3.1}$$

where $\boldsymbol{x}$ is the feature vector, $\widehat{x}$ is its mean value, and σ_x is its standard deviation. In practice, features with z-values larger than 3 or smaller than −3 are considered as outliers (Fig. 3.3) [43]. However, the z-score might lead to misidentified outliers due to the fact that the maximum score is equal to $(n-1)/\sqrt{n}$, yielding small values due to the nonrobustness of the standard deviation that is used in the denominator, especially in small size data. For this purpose, a modified version has been proposed [44]:

$$z_{\text{mod}} = \frac{\boldsymbol{x} - \widetilde{x}}{\text{MAD}} = b\frac{\boldsymbol{x} - \widetilde{x}}{\text{median}(|\boldsymbol{x} - \widetilde{x}|)}, \tag{3.2}$$

where MAD stands for the median absolute deviation and $\widetilde{x}$ is the median. The constant 0.6745 comes from the fact that MAD is multiplied with the constant 1.483,

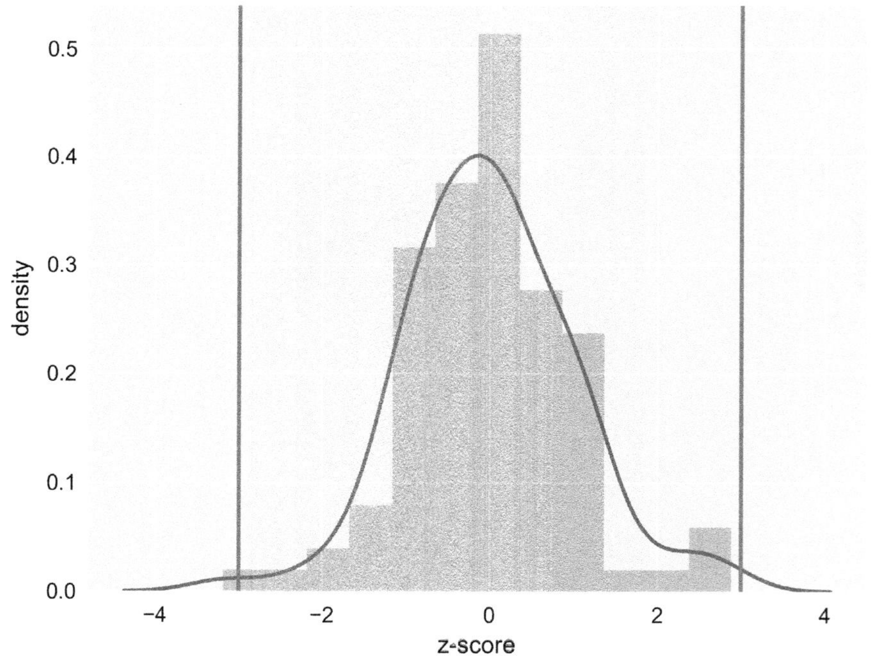

FIGURE 3.3

The z-score distribution of a randomly generated and contaminated feature.

which is a correction factor that makes the MAD unbiased at the normal distribution ($b = 1/1.483 = 0.6745$) [44]. The modified z-score yields more robust results due to the scale and location factors that are introduced by MAD in Eq. (3.2). An example of the z-score distribution of the previously generated and contaminated feature $\boldsymbol{x}$ is displayed in Fig. 3.3.

Outlier detection can be also performed using machine learning approaches. A more sophisticated approach is to use isolation forests. Isolation forest is a collection of isolation trees which (i) enables the exploitation of subsampling data to precisely detect outliers, (ii) does not make use of distance or density measures to detect anomalies, (iii) achieves linear time complexity, and (iv) is scalable [45,46]. The term "isolation" stands for the separation of an instance (a polluting one) from the rest of the instances (the inliers). Isolation trees are binary trees, where instances are recursively partitioned, and produce noticeable shorter paths for anomalies as (i) in the regions occupied by anomalies, less anomalies result in a smaller number of partitions—shorter paths in a tree structure and (ii) instances with distinguishable attribute values are more likely to be separated early in the partitioning process [45]. Thus, when a forest of random trees collectively produces shorter path lengths for some particular points, they are highly likely to be anomalies [45,46].

More specifically, the algorithm (Algorithm 3.1) initializes a set of N-isolation trees (an isolation forest), $\boldsymbol{T}$, where each tree has a standard tree height, h. At each iteration step, the algorithm partitions a randomly selected subsample $\boldsymbol{X}'$ with a subsampling size of M and adds an isolation tree, *iTree*, into the set $\boldsymbol{T}$.

Algorithm 3.1 ISOLATION FOREST [45,46]

Inputs: $X = \{x_1, x_2, \ldots, x_N\}$ (observations), N (number of trees), M (subsampling size)
Output: A set of N-isolation trees $\boldsymbol{T} = \{T_1, T_2, \ldots, T_N\}$

$\boldsymbol{T} = \{\}$;
$h = ceiling(\log_2 M)$;
for $i = 1{:}N$ **do**
    $\boldsymbol{X}' \leftarrow sample(\boldsymbol{X}, M)$;
    $\boldsymbol{T} \leftarrow \boldsymbol{T} \cup iTree(\boldsymbol{X}')$;
end for
return $\boldsymbol{T}$;

define ***iTree****(X)*:
    if X can not be divided **then**:
        return $exNode\{Size \leftarrow |\boldsymbol{X}|\}$;
    else:
        $\boldsymbol{x}' = rand(\boldsymbol{X}')$;
        $p = rand([\min(\boldsymbol{x}'), max(\boldsymbol{x}')])$;
        $\boldsymbol{X}_{left} \leftarrow filter(\boldsymbol{X}', \boldsymbol{x}' < p)$;
        $\boldsymbol{X}_{right} \leftarrow filter(\boldsymbol{X}', \boldsymbol{x}' \geq p)$;
        return $inNode\{Left \leftarrow iTree(\boldsymbol{X}_{left}), Right \leftarrow iTree(\boldsymbol{X}_{right}),$
                $SplitAtt \leftarrow \boldsymbol{x}', SplitValue \leftarrow p\}$;
    end if

The *iTree* is computed by (i) randomly selecting an attribute $x' \in X$ and a split point p between the range of x', (ii) partition $\boldsymbol{X}'$ into a left branch X_{left} and a right branch X_{right}, based on the whether x' is smaller or larger than (or equal to) the split point p, respectively, and (iii) repeating the procedure for the left and right partitions of X, until the path length converges or until the number of ensemble trees, N, is met. The path length is defined as the total number of edges starting from the root node down to the external node and is used as an anomaly indicator based on a predefined height limit (see Ref. [45] for more information). The functions *exNode* and *inNode* denote the external and internal nodes, respectively. The external node is formed when the partition can no longer be divided, and the internal node refers to a node that can be split into additional nodes. The space complexity of the model is $O(MN)$ and the worst case is $O(MN^2)$ [45].

The subsample size controls the training data size and affects the reliability of outlier detection, whereas the number of trees controls the size of the ensemble trees [45]. In practice, M is set to 2^8 and N is set to 100. The anomaly score is finally defined as follows:

$$s(x, M) = 2^{-\frac{E(h(x))}{c(M)}}, \tag{3.3}$$

where $c(M)$ the average path length of unsuccessful searches, $h(x)$ is a harmonic number that is defined as $\ln(x)$ plus the Euler's constant, and $E(h(x))$ is the average of $h(x)$ from a collection of isolation forests [45]. Scores very close to 1 indicate anomalies; scores much smaller than 0.5 are inliers, and scores close (or equal) to 0.5 are safe instances. An example of an application of the isolation forest for outlier detection is depicted in Fig. 3.4, which demonstrates the efficacy of the method. For illustration purposes, two randomly generated and contaminated features, assume $\boldsymbol{x}$ and $\boldsymbol{y}$, were generated, each one consisting of 1000 random observations that were drawn from a Gaussian distribution with zero mean and variance equal to 0.1. For testing purposes, 5% of the data were used for training, and 5% of the data were contaminated.

The Grubb's statistical test is a univariate statistical measure that tests for the hypothesis that there are outliers in the data [43]. The test statistics is given as

$$G = \frac{\max(|\boldsymbol{x} - \widehat{x}|)}{\sigma_x}. \tag{3.4}$$

In fact, the Grubb's test statistics is defined as the largest absolute deviation from the sample mean in units of the sample standard deviation [43]. Here, we are interested in testing whether the minimum value or the maximum value of $\boldsymbol{x}$ is an outlier, i.e., a two-sided test. A value is considered to be outlier if the null hypothesis is rejected at the 0.05 significance level. Another test statistics is the Hampel's test, which is defined as the difference of each sample from its population median value (median deviation). A sample is an outlier if its absolute Hampel value is 4.5 times larger than (or equal to) the median for deviation [43].

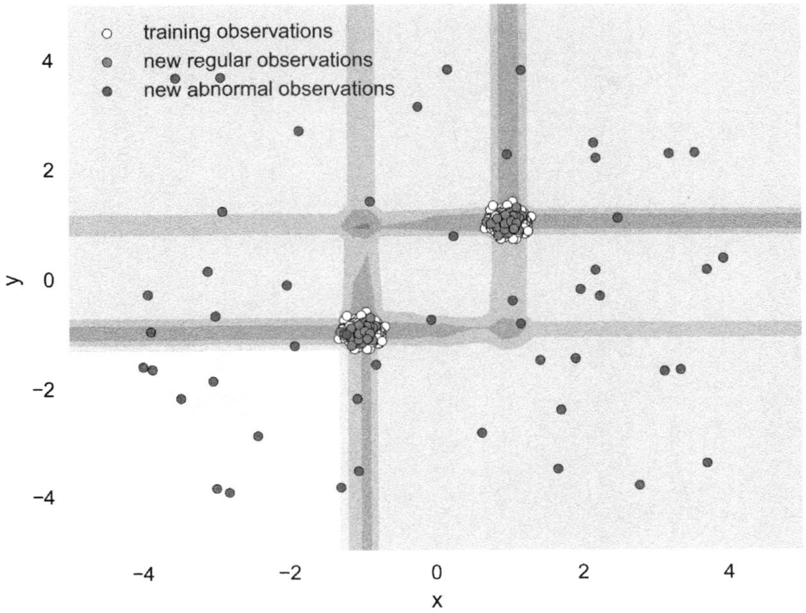

FIGURE 3.4

An example of isolation forest for outlier detection. The training observations are depicted in white, the new regular observations (inliers) are depicted in green (gray in print version), and the new abnormal observations (outliers) are depicted in red (dark gray in print version). Decision boundaries are depicted in a gray mesh color.

The local outlier factor (LOF) [47] is a density-based approach that measures the local density of a given data point with respect to its neighboring points, where the number of nearest neighbors determines the accuracy of the model. In fact, it uses the density of a point against its neighbors to determine the degree of the whether the point is an inlier or an outlier. For a point x, the local reachability density (*lrd*) of x, $lrd(x)$, is defined as [47]

$$rd(x) = \frac{N_k(x)}{\sum_{x' \in N_k(x)} r(x, x')}, \tag{3.5}$$

where $N_k(x)$ is the set of k-nearest neighbors for $\boldsymbol{x}$ and $r(x, x')$ is the reachability distance that is defined as the distance between x and its k-nearest neighbor. The reachability distance is the true distance between two points. The LOF is given by Ref. [47]

$$\mathrm{LOF}(x) = \frac{\sum_{x' \in N_k(x)} (lrd(x')/lrd(x))}{N_k(x)} = \sum_{x' \in N_k(x)} lrd(x') \sum_{x' \in N_k(x)} r(x, x'), \tag{3.6}$$

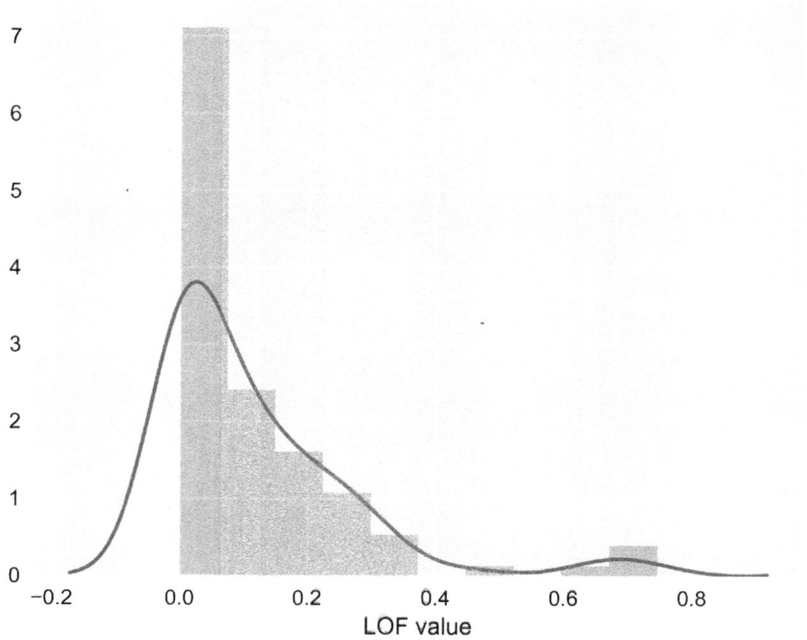

FIGURE 3.5

The local outlier factor (LOF) distribution between two randomly generated features.

which is equal to the average local reachability density of the neighbors divided by the point's own local reachability density. The lower the local reachability density of x, the higher the local reachability density of the kNN of x and thus the higher the LOF. The higher the LOF, the more likely the point is an outlier. An LOF distribution for two randomly generated and contaminated features, each one consisting of a set of 100 random observations following a normal distribution, can be seen in Fig. 3.5, where the neighboring size has been set to 2. The dense concentration around zero means that there are many inliers, while the extreme areas (around 0.7) include possible outliers.

A common distance measure, which is widely used for anomaly detection in properly scaled datasets, is the Euclidean distance. In multivariate datasets, however, the Euclidean distance suffers from the covariance that exists between the variables [48]. A distance measure that accounts for such effects, in multivariate datasets, is the Mahalanobis distance, which uses the eigenvalues to transform the original space into the eigenspace, so as to neglect the correlation among the variables of the dataset [49] and is defined as:

$$D(\boldsymbol{x}) = \sqrt{(\boldsymbol{x} - \boldsymbol{\mu})\boldsymbol{\Sigma}^{-1}(\boldsymbol{x} - \boldsymbol{\mu})}, \tag{3.7}$$

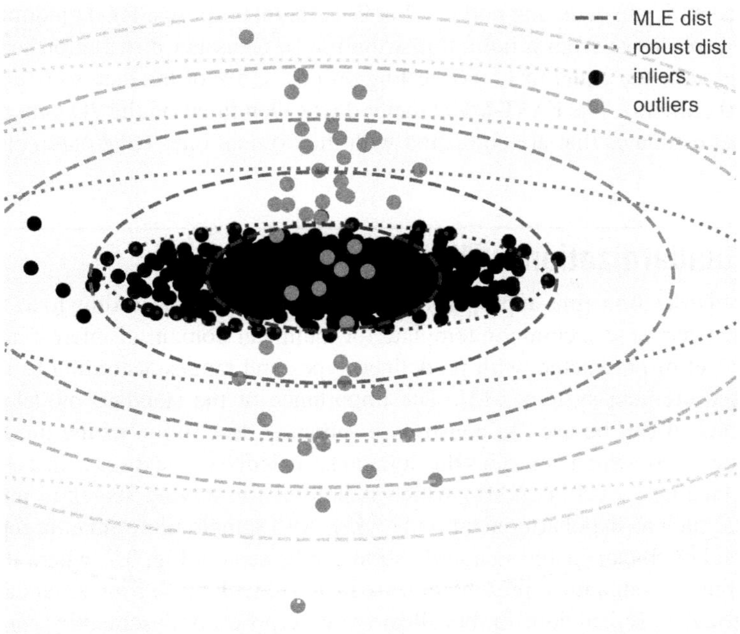

FIGURE 3.6

Illustration of the elliptic envelope approach for anomaly detection. The Mahalanobis distance (true parameters) is depicted in blue (light gray in print version), the robust distance estimates (FAST-MCD) are depicted in brown (gray in print version), the inliers in black, and the outliers in red (dark gray in print version).

where $\boldsymbol{x}$ is an n-dimensional feature vector where the observations are stacked in columns, $\boldsymbol{\mu}$ is the mean vector across the observations, and Σ^{-1} is the inverse covariance matrix. Note that if the covariance matrix is the identity matrix, Eq. (3.7) yields the Euclidean distance, whereas if the covariance matrix is diagonal, Eq. (3.7) yields the normalized Euclidean distance. A way to visualize the result of Eq. (3.7) is to use an Elliptic envelope as shown in Fig. 3.6. Data within the ellipse surface are inliers, whereas data outside of the ellipse are outliers. The Elliptic envelope (also referred to as elliptical envelope) models the data as high-dimensional Gaussian distributions that consider for the covariance between the observations. The FAST-Minimum Covariance Determinant (FAST-MCD) [49] is widely used to estimate the size and the shape of the ellipsis. The algorithm conducts initial estimations of the mean vector μ and the covariance matrix Σ (Eq. 3.7) using nonoverlapping subsamples of the feature vector. Then the algorithm proceeds with new subsamples until the determinant of the covariance matrix converges. An example of an application of the Elliptic envelope for outlier detection is depicted in Fig. 3.6, which demonstrates the efficacy of the method, as well as the importance of the parameter estimation that determines the envelope's size and as a matter of fact the accuracy in detecting outliers. For illustration purposes, 1000 random observations were drawn from a normal distribution (a Gaussian distribution with zero mean and variance equal to 1). For testing purposes, 5% of the data were contaminated (50 outliers). The FAST-MCD method was then fitted on the 2D data producing robust estimates that are compared with the original ones (true parameters).

3.4 Standardization

Data standardization aims at properly transforming a dataset according to a standard model that serves as a common template for a clinical domain of interest and consists of a set of parameters with predefined types and ranges according to international measurement systems [13]. The importance of the standard model lies on the fact that it can be used to construct a semantic description of the model, i.e., an ontology [13], which describes the domain knowledge of the disease in a common markup language, such as XML, which is helpful for solving semantic matching problems, such as data harmonization [13,55]. An example of a semantic representation and how it can enable standardization can be seen in Fig. 3.7, where the variables of the raw semantic representation are matched with those from a standard one, for a particular clinical domain. For illustration purposes, both semantic representations consist of indicative classes, subclasses, and/or variables that can be instances of larger semantic representations of a typical clinical domain.

The standard semantic representation (left side of Fig. 3.7) consists of the parent class "Patient," which includes four subclasses, i.e., "Demographics," "Laboratory examinations," "Treatments," and "Biopsies." Each subclass consists of further classes and/or different types of variables (e.g., the class "Laboratory examinations" consists of the class "Blood tests" that includes the variable "Hemoglobin"). In a similar way, the raw semantic representation (right side of Fig. 3.7) includes three subclasses under the parent class "Patient" (i.e., "Therapies," "Demographics," and "Clinical tests"), as well as additional subclasses and/or variables (e.g., the class "Demographics" includes the variable "Gender"). The standardization process involves (i) the identification of similar variables among the two schemas and (ii) the normalization of the matched terms of the raw semantic representation according to the predefined ranges of the standard semantic representation. In this example, the variables "Sex," "HGB," and "number of foci" of the raw schema are matched with the variables "Gender," "Hemoglobin," and "focus score" of the standard schema. Furthermore, standardization considers for the normalization of the former variables according to the ranges of the standard semantic representation, i.e., {0, 1} for "Sex," {low, normal, high} for "HGB," and {normal, abnormal} for "number of foci."

One way to match the terms between two semantic representations is to use string similarity measures in the light of a procedure known as lexical matching [50–54]. A widely used string matching method is the sequence matching algorithm [51,52], which seeks for "matching" blocks between two strings. The sequence

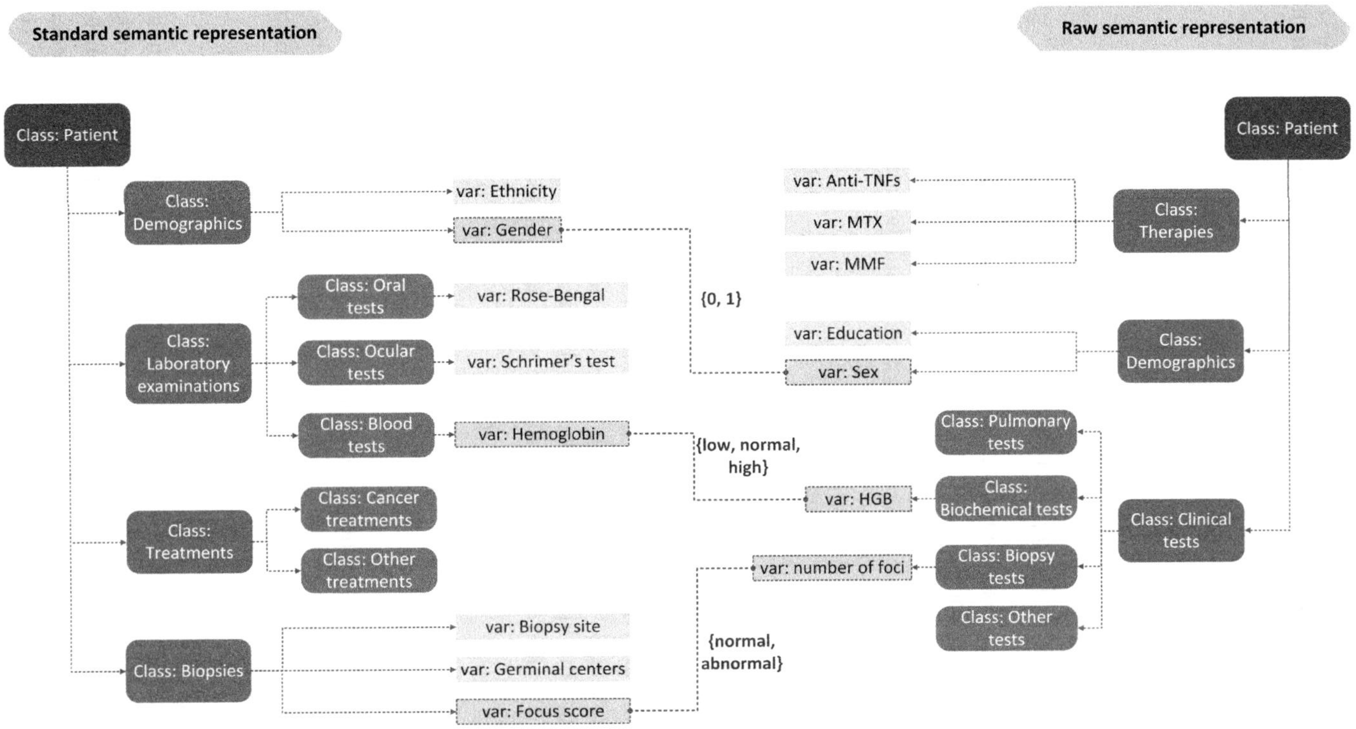

FIGURE 3.7

Illustration of a typical data standardization procedure.

matching algorithm calculates the edit distance between two sequences, assume a and b. The edit distance, $d_{a,b}$, is defined as the minimum number of operations, i.e., deletions, insertions, and replacements, that are required to transform a into b:

$$d_{a,b}(i,j) = \begin{cases} i & , \quad i = 0 \\ j & , \quad j = 0 \\ d[i-1, i-1] & , \quad i,j > 0 \text{ and } a_i = b_j \\ \min \begin{Bmatrix} d[i-1, j-1] + 1 \\ d[i-1, j] + 1 \\ d[i, j-1+1] \end{Bmatrix} & , \quad o.w. \end{cases} , \tag{3.8}$$

where $d_{a,b}(i,j)$ is the distance between the first i-characters of a and the first j-characters of b. Another popular metric for sequence matching is the Levenshtein distance [51,52] that measures the similarity between two strings, assume a and b, in terms of the number of deletions, insertions, or substitutions that are required to transform a into b:

$$lev_{a,b}(i,j) = \begin{cases} \max(i,j) & , \quad \min(i,j) = 0 \\ \min \begin{cases} lev_{a,b}(i-1, j) + 1 \\ lev_{a,b}(i, j-1) + 1 \\ lev_{a,b}(i-1, j-1) + 1_{(a_i \neq b_j)} \end{cases} & , \quad o.w. \end{cases} , \tag{3.9}$$

where a Levenshtein distance of zero denotes identical strings (i.e., zero operations to match the strings). The Jaro distance [50–52] is another widely used string similarity measure that quantifies the similarity between two strings. For two given strings, a and b, the Jaro string similarity measure, sim_J, is defined as

$$sim_J = \begin{cases} 0 & , \quad x = 0 \\ \frac{1}{3} \cdot \left(\frac{x}{|a|} + \frac{x}{|b|} + \frac{x-y}{x} \right) & , \quad o.w. \end{cases} , \tag{3.10}$$

where x is the number of coincident characters and y is half the number of transpositions [50–52]. The Jaro–Winkler distance measure is a modification of the Jaro distance measure that uses an additional prefix scale c to give more weight to strings with common prefix of a specific length. It is defined as follows:

$$sim_{JW} = sim_J + (lc(1 - sim_J)), \tag{3.11}$$

where l is the length of common prefix at the start of the string up to a maximum of four characters. The prefix weight is the inverse of the lenght l that is needed to consider both strings as identical. For example, the Jaro–Winkler distance between the terms "lymphocyte number" and "lymphoma score" is equal to 0.89, whereas the Jaro distance is equal to 0.73. In the same example, the Levenshtein distance is equal to 9, which denotes the number of the operations that are needed to match the two strings. Lexical matching does not consider for semantic relations but instead focuses more on matching variables with identical patterns, whereas semantic matching further seeks for semantic relations [12]. More emphasis on semantic matching methods for data harmonization will be given in Chapter 5.

3.5 Data sharing frameworks

Let us assume now that the cohort data (or any type of data) are ready to be shared. How will these data be shared? How is cohort interlinking going to be established? The answer is through a federated data sharing framework where all operations within it are executed with absolute privacy and high security levels. In this section, we discuss two major data sharing frameworks that enable the coanalysis of shared data with respect to data protection regulations and the subjects' privacy. In general, a simple data sharing framework shall be capable of providing (i) details about the structure of the data, i.e., metadata, (ii) the ability to run queries over and receive results from aggregated, perturbed, or obfuscated data without moving the data at all, (iii) properly defined user authentication mechanisms, and (iv) the ability to request and gain secure access to the deidentified data so as to be able to execute queries for data analytics.

3.5.1 Framework for responsible sharing for genomic and health-related data

The EU BioSHARE Project has developed, under the aegis of the Global Alliance for Genomics and Health, the Framework for Responsible Sharing of Genomic and Health-Related Data[3] [56–59]. The ensuing Framework has established a set of foundational principles for responsible sharing of genomic and health-related data: (i) respect individuals, families, and communities, (ii) advance research and scientific knowledge, (iii) enhance the existing healthcare systems with equal patient rights, and (iv) foster trust, integrity, and reciprocity. In addition, it has set out 10 core elements complementing the interpretation of the aforementioned principles: (i) transparency, (ii) accountability, (iii) engagement, (iv) data quality and security, (v) privacy, data protection, and confidentiality, (vi) risk–benefit analysis, (vii) recognition and attribution, (viii) sustainability, (ix) education and training, and (x) accessibility and dissemination (Fig. 3.8). Each element highlights the importance behind the sharing of genomic and health-related data, and all the elements together can be combined to formulate a set of guidelines for reliable data sharing.

[3] An initiative that is comprised by more than 100 healthcare and research organizations that cooperate toward the realization of three main goals: (i) to enable open standards for interoperability of technology platforms for managing and sharing genomic and clinical data, (ii) to provide guidelines and harmonized procedures for privacy and ethics internationally, and (iii) to engage stakeholders to encourage responsible sharing of data and of methods [57].

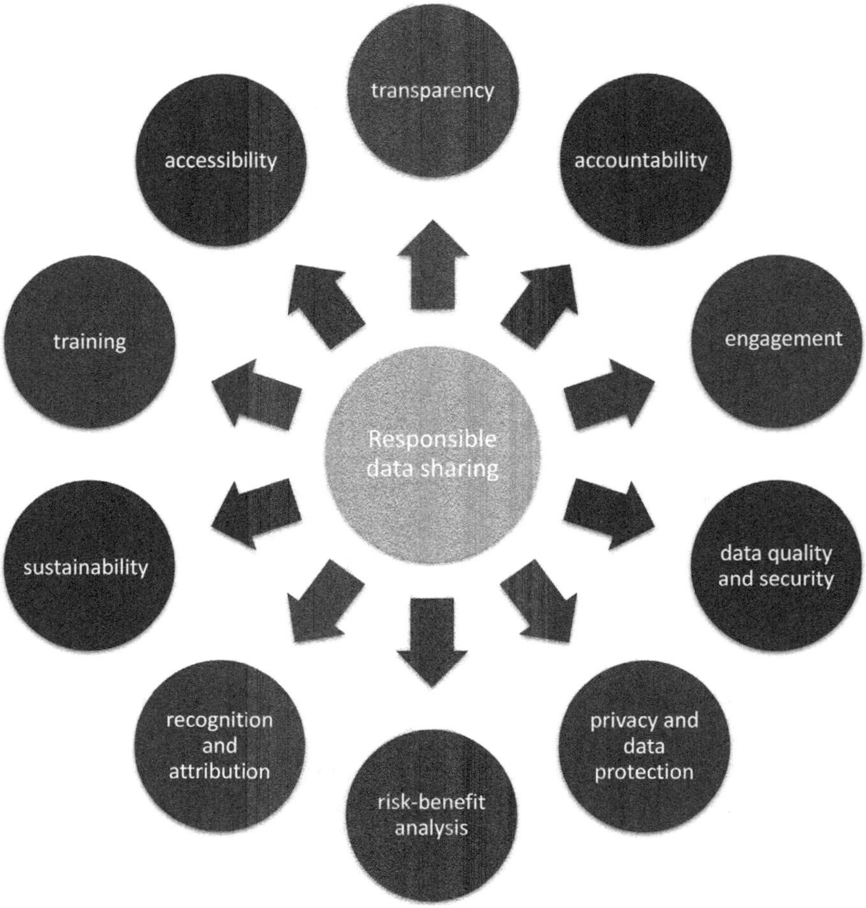

FIGURE 3.8

The 10 core elements for achieving responsible data sharing [57].

Transparency can be quantified by (i) defining the real purpose of data sharing and (ii) describing all the underlying procedures that are involved during data sharing, such as the data collection process, the potential involvement of third parties, data expiration dates, etc. Accountability has to do mainly with the development of mechanisms that are able to manipulate potential conflicts of interest, misuse of data, etc. Engagement involves the participation of the citizens (i.e., social involvement) in the evaluation of the Framework's efficacy. Data quality exhibits common characteristics with those from the data quality assessment procedure in Section 3.1. This core element aims at improving specific data quality measures including the (i) accuracy, (ii) consistency, (iii) interoperability, and (iv) completeness, among many others. Privacy and data protection refers to the process of preserving the patients' privacy and integrity during the (i) collection, (ii) sharing, (iii) storage, and (iv) processing of sensitive data (data aggregation methods must be taken into consideration before genomic and health-related data sharing).

Risk–benefit analysis entails the identification and management of potential risks that might appear during data sharing, such as data breach, invalid conclusions about the integrity of the data, and confidentiality rupture, whereas the term "benefits" refers to the extent to which the impact of data sharing is meaningful among different population groups. Any system that supports data sharing shall take into consideration two additional factors: (i) recognition to the contributors of the system's establishment and (ii) attribution to the primary and secondary purposes of the data sharing system. Furthermore, emphasis must be given to the coherent development of the mechanisms for data sharing, i.e., the system's specifications, response time, error handling, etc., so as to yield a sustainable system. To advance data sharing, it is also crucial to dedicate education and training sources for improving the data quality and integrity. Finally, the data sharing system must be easily accessible with respect to the ethical and legal issues, and its contents must be disseminated in terms of public good, i.e., minimizing risks and maximizing benefits (trade-off).

3.5.2 The DataSHIELD framework

An exemplar technological solution to preventing reidentification of an individual has been proposed within the DataSHIELD initiative by "*taking the analysis to the data, not the data to the analysis*" concept, which confines that the control researchers retain over the data. In particular, DataSHIELD "*enables the co-analysis of individual-level data from multiple studies or sources without physically transferring or sharing the data and without providing any direct access to individual-level data*" [60,61]. The latter feature contributes significantly to properly addressing several ethics-related concerns pertaining to the privacy and confidentiality of the data, the protection of the research participant's rights, and the post data sharing concerns. In addition to standard technical and administrative data protection mechanisms, DataSHIELD includes (i) a systematic three-level validation process of each DataSHIELD command for risks of disclosure, (ii) the definition of obligations to avoid the potential identification of an object, (iii) automatic generation of new subject's identifiers by Opal; original subject's identifiers are stored securely in a distinct database in Opal, (iv) protection mechanisms from potential external network attacks, and (v) encrypted and secured REST (Representational State Transfer) communications (Chapter 6).

DataSHIELD supports both single-site and multisite analysis (Fig. 3.9). Single-site analysis involves the analysis of a single data provider's data, i.e., application of statistical analysis, computation of descriptive statistics, etc., in R environment. Multisite analysis, also referred to as pooled analysis, requires that the data are first harmonized, i.e., have a common structural format [62]. Each data provider manages a DataSHIELD server and is responsible for harmonizing his/her data using either before the final submission of their data to the opal data warehouse or afterward.

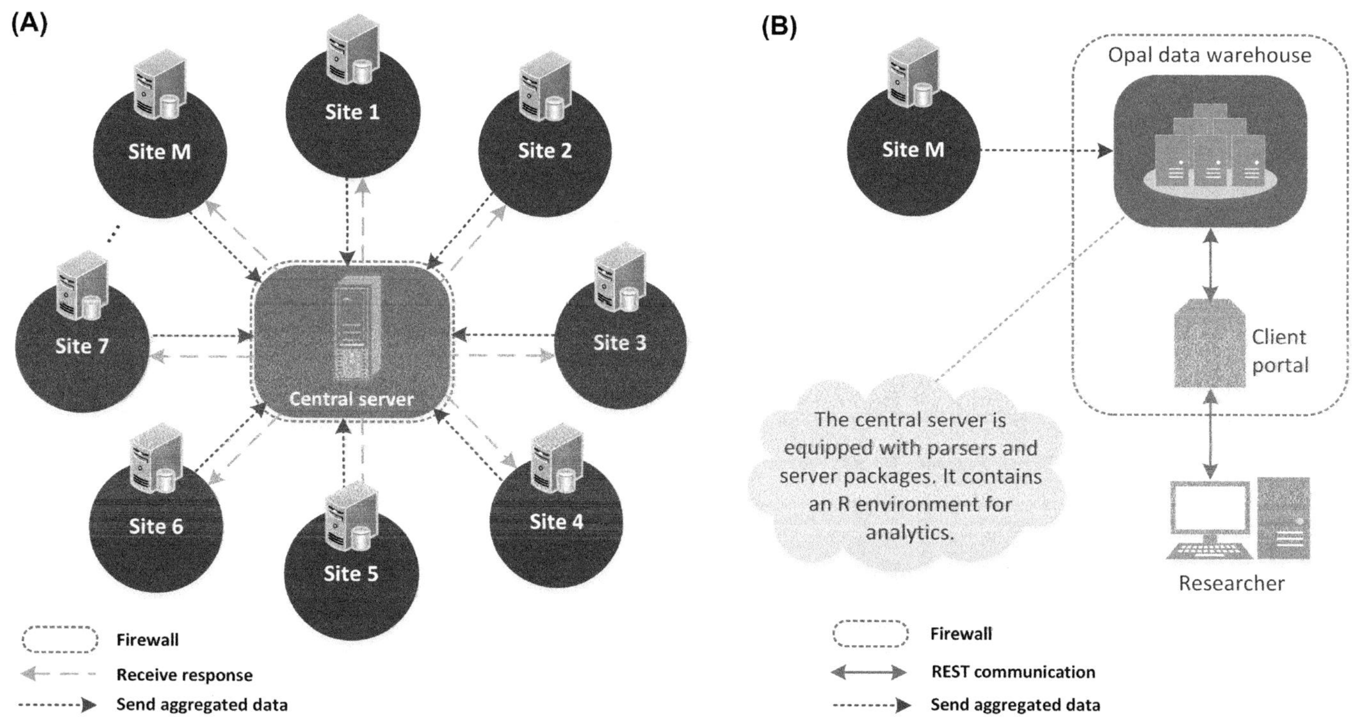

FIGURE 3.9

An illustration of the DataSHIELD's infrastructure across *M*-data computer sites. (A) A central server receives and handles the requests from each data provider's site. Data providers can send aggregated data, as well as receive responses from the central server regarding their submitted requests. (B) From the data processor's point of view, the researcher can log into the platform through a client portal and gain access to the data that lie in the opal data warehouse. The opal data warehouse is also equipped with a compute engine with an incorporated R environment for data analytics and data parsers [62].

The control of the datasets is protected by the DataSHIELD's firewall system that has been designed to prevent any potential data breach. At this point, it should be noted that (i) ethical, legal, (ii) data access approvals, and (iii) anonymization of the data constitute three necessary preliminary setup steps required for DataSHIELD-based analysis.

3.6 Global initiatives in the clinical domain

In this section, we discuss major data sharing initiatives whose aim is to enable out-of-border data sharing for promoting science. We further highlight their advantages and weaknesses toward this vision.

3.6.1 ClinicalTrials.gov

In 2000, a web-based clinical registry platform, namely the ClinicalTrials.gov, was created as a result of the efforts from the National Institute of Health (NIH) and the Food and Drug Administration (FDA) [15]. ClinicalTrials.gov is a remarkable example of a data sharing initiative whose aim is to make public and private clinical trials and studies available to the scientific community for promoting multidisciplinary science. It focuses only on clinical trials, also referred to as interventional studies, which involve medical data from human volunteers. According to ClinicalTrials.gov, a clinical trial is defined as a research study where the effects (e.g., health outcomes) of an intervention (e.g., a drug) are examined on human subjects who are assigned to this intervention and fulfill a set of predefined eligibility criteria (e.g., requirements for participation) [15]. The origin of the data from clinical trials can take many forms varying from uncoded patient-level data to analyzed summary data (metadata), although only the latter are published in the platform for data protection purposes. The medical information of each clinical study is updated by the corresponding principal investigator (PI) who leads the study and can be continuously updated throughout the study's lifetime. For each study, the user can find information regarding (i) the disease or condition under investigation, (ii) the title, description, and location(s) of each study, (iii) the eligibility criteria that were used to recruit human subjects, (iv) the contact information of the PI, and (v) links to supplementary material (external sources). For only selected studies, the user can gain further access to (i) population characteristics (e.g., age, gender), (ii) study outcomes, and (iii) adverse events.

ClinicalTrials.gov was launched in response to the FDA Amendments Act (FDAAA) of 1997. A few years later, the FDAAA 801 posed an extra requirement regarding the statement of the results of the clinical trials that were included in the platform. As a result, the ClinicalTrials.gov results database was launched in 2008 as an online repository that contains summary information on study participants and study outcomes. More specifically, the results database (also referred to as registry and results database) includes tabular summaries of (i) participant related information, such as periodical involvement in studies, etc., (ii) baseline characteristics, i.e., data summaries, (iii) primary and secondary outcomes from statistical analyses, and (iv) summaries of anticipated and unanticipated adverse events including affected organ system, number of participants at risk, and number of participants affected, by study arm or comparison group [14]. It is important to note that the information in the results database is considered summary information and does not include any individual patient data at all. The results of each individual study are first assessed by the National Library of Medicine, as part of the NIH, in terms of quality and scientific impact, and finally displayed to the user in an integrated way. Till 2013, the web platform included summary results of more than 7000 trials. Almost 2 years ago, in September 2016, an additional requirement that was issued by the US Department of Health and Human Services regarding the expandability of the platform so as to include more clinical studies and related outcomes took place. The regulation took effect 4 months later, and since then, the web platform contains more than 280,000 clinical studies in the United States (all 50 states) and in 204 countries all over the world (September 2018).

There is no doubt that ClinicalTrials.gov is a major initiative toward the establishment of a federated and transparent database of clinical registries worldwide. However, there are several key issues that are of great importance. Structural heterogeneity is a big problem because not all participating centers follow a common protocol for conducting clinical trials, a fact that introduces discrepancies during the analysis of such data. For example, measurement units for the same terms often vary across the studies. Moreover, the summary data are not always in a readable form. There are cases where no one can explain the structure of the trial or the analysis of the data even for trials that have already been published in the web [14]. Furthermore, the fact that the data can be continuously updated introduces structural changes to the trials and hampers the work of the data analysts who are not able to figure out important information from the prospective data (e.g., keep in track with previous data), which leads to information loss. There are also reported cases where several values make no sense (e.g., age ranges that vary from the normal). As a result, the summary data may not always be valid, a fact that obscures the consistency of the clinical trials and their reproducibility. Although the platform has already incorporated a variety of check constraints during data entry, ensuring the accuracy of the reported data is difficult and involves great effort.

3.6.2 The database for Genotypes and Phenotypes

The database for Genotypes and Phenotypes (dbGaP) is a public repository that was established by the National Center for Biotechnology Information (NCBI) with the purpose of archiving the scientific results of genotype and phenotype studies, including valuable information related to individual-level phenotype, exposure, genotype, and sequence data, as well as the associations between them [16]. The majority of these studies are (i) GWAS, (ii) studies involving sequencing experiments, and (iii) studies involving the association between genotype and nonclinical traits, resequencing traces. dbGaP ensures that before the data collection process (i.e., data sharing), the clinical data are deidentified. Furthermore, the database offers mechanisms for interlinking clinical studies according to the similarity of their protocols for providing quick access to the users. The metadata of each clinical study are publicly available (open access), whereas the access to sensitive data is controlled by an authorization system (private access). The private access management is undergone by a Data Access Committee that consists of technical and ethical experts who assess the user requests (i.e., signed documents, completed forms, etc.) through an application that is referred to as Data Use Certification [16].

PIs can upload their clinical study to the dbGaP platform only if the study fulfills specific data protection requirements. More specifically, the data submitted to the dbGaP must be compliant with the regulations posed by HIPAA of 1996 and the Title 45 of the Code of Federal Regulations [63,64]. The NIH is also involved in the data submission process. Only studies that are sponsored from the NIH are accepted, otherwise an agreement must be reached before the submission process. The deidentified raw data are distributed only to the private access users in an encrypted form through a secure server as it has been long stated that even when the data are deidentified, they can be linked to a patient if they are combined with data from other databases. On the other hand, dbGaP does not take into consideration the accuracy and the quality of the archived data. Meanwhile, the dbGaP documentation states that NCBI might cooperate with the primary investigators to identify and fix any related discrepancies within the data but are not responsible for any inconsistencies in the dbGaP data. It should be also noted that dbGaP offers a variety of statistical tools that can be used in cooperation with the PIs for the detection of statistical inflations within the data, such as false-positive errors. Examples of quality control metrics include (i) the Mendelian error rate per marker, (ii) the Hardy–Weinberg equilibrium test, (iii) the duplicate concordance rate, and (iv) the vendor-supplied genotype quality score, among many others [14,16].

In general, dbGaP contains four different types of data, namely (i) metadata, i.e., information regarding the data collection protocols, and related assays, (ii) phenotypic data, (iii) genetic data (e.g., genotypes, mapping results, resequencing traces), and (iv) statistical analysis results, i.e., results from genotype–phenotype analyses. Private access users have access to (i) deidentified phenotypes and genotypes for each study, (ii) pedigrees, and (iii) precomputed univariate associations between genotype and phenotype. In addition, dbGaP includes statistical tools for analyzing genotype and phenotype data, such as association statistics between phenotype(s) and genetic data, case–control associations, etc. Statistical analyses are currently performed only by the dbGaP's employees although the user's involvement is under consideration. The controlled access users are able to evaluate the analysis reports and view genome tracks and genetic markers that have been associated with the selected phenotype(s), as well as download the results for local use.

According to the final NIH official scientific-sharing summary of 2014, the number of registered studies in dbGaP was 483, with more than 700,000 involved individuals who are represented by clinical data from 169 institutions and organizations from 9 countries. The majority of these studies (341) come from institutes in the United States with only a few reported cases from Europe (Germany, United Kingdom), Australia, and Asia. All in all, dbGaP is another remarkable data sharing initiative, which is limited however among the US states.

3.6.3 The DataSphere Project

The DataSphere Project [65] is an example of an independent, nonprofit data sharing initiative that has been established by the CEO Roundtable on Cancer and aims at providing publicly accessible data sharing system for (Phase III) cancer clinical trials to reduce cancer mortality rates and advance scientific research in the cancer community. The CEO Roundtable on Cancer [66] is comprised by a board of CEOs from international pharmaceutical and technical companies who are involved in cancer research and treatment worldwide, including Johnson and Johnson, Novartis, and Bayer, among many others. Its main goal is to establish initiatives whose primary aim is to develop new methods for early cancer diagnosis and novel anticancer treatments for promoting public health. DataSphere is an example of such an initiative. Other programs that have been launched till now include (i) the CEO Cancer Gold Standard that was established in 2006, an accredited part of which is the National Cancer Institute, and (ii) the Life Sciences Consortium established in 2005 with the purpose of improving oncology treatments, among others.

The overall vision of Datasphere lies on the fact that oncologists shall make use of larger population groups to test for particular molecular drivers yielding outcomes with high statistical power. The key to this lies on the cooperation among the CEOs of the participating companies. Toward this direction, a web platform has been implemented by one of the technical companies that is currently comprised by 171 clinical datasets from more than 135,000 patients collected by 29 data providers. The platform consists of deidentified datasets with different cancer cases, such as colorectal, lung, head, prostate, and breast. The platform's database is scalable and secure with the overall deidentification process being HIPAA compliant. Guidelines for data aggregation and tools for advocacy have been made publicly available. The platform's reliability has been proven high since in 2015, Sanofi, a large pharmaceutical company, provided Phase III clinical studies to the platform. Every day, hundreds of cancer drugs are being developed and thousands of studies are registered on ClinicalTrials.gov. DataSphere is promising but needs to cooperate with third-party initiatives to keep in track with the advances and produce meaningful research.

3.6.4 Biogrid Australia

Biogrid is an advanced data sharing platform that currently operates across different sites in Australia [67]. Although its range is only limited to one continent, its federated design is promising and can be adopted by existing initiatives. Biogrid envisages to interlink and integrate data across various clinical domains apart from the nature of the involved research infrastructures, which makes its scope very interesting. This means that the platform is multidisciplinary (supports different types of data). In addition, it involves different domain experts, such as clinical researchers, epidemiologists, and IT engineers. The type of data that are supported by Biogrid includes thousands of records from (i) genomic data, (ii) imaging data (MRI, PET), (iii) clinical outcomes (diabetes, epilepsy, stroke, melanoma), and (iv) cancer data (breast, colorectal, blood, prostate, pancreatic), which are combined to formulate a federated database that serves as a virtual data store. This enables researchers to (i) identify genetic factors, (ii) create genetic profiles for each patient and combine them with the profiles of others, (iii) design and evaluate disease surveillance and intervention programs for early diagnosis, and (iv) promote precise personalized medicine. The platform also provides tools for data analytics, a fact that increases the platform's scalability and interoperability.

The authorization system complies with all the privacy legislations with rigorous attention to ethical and legal requirements. Each local research repository must first gain ethical approvals for participating in the initiative. The access to the platform is then provided by (i) data custodians, (ii) the Scientific Advisory Committee, and (iii) the BioGrid Australia Management Committee. The data are deidentified and converted to a coded form through probabilistic matching for record linkage per individual [67]. In fact, the clinical data from each site are linked to a federator that enables access across physical sites for query processing without storing any data at all. Only authorized researchers are able to access and analyze data through a secure virtual private network and secure web services in deidentified form with a record linkage key. All the queries to the federator are tracked and monitored to be stored into audit tables. Biogrid currently contains 32 data types from 33 research and clinical institutes and 75 data custodians.

3.6.5 The NEWMEDS consortium

The Novel Methods leading to New Medications in Depression and Schizophrenia (NEWMEDS) [68] is an international consortium that aims to bring together clinical researchers to discover new therapeutic treatments for schizophrenia and depression. Toward this goal, the NEWMEDS consortium comprises well-known partners from the academic and biopharmaceutical domains including the Karolinska Institutet, the University of Cambridge, the Novartis Pharma AG, and the Pfizer Limited, among many others, as well as partners from 19 institutes across 9 different EU countries, as well as from Israel, Switzerland, and Iceland. NEWMEDS currently comprises one of the major research academic–industry collaborations all over the world. It consists of two databases: the Antipsychotic database and the Antidepressant database with more than 35,000 patients. These databases include mainly fMRI and PET imaging scans for (i) the application of drug development methods and (ii) investigating how genetic findings influence the response to various drugs and how this can be used to select an appropriate drug for an individual patient.

A scientific advisory board that consists of four clinical members deals with the sharing of the clinical data. NEWMEDS strategy is comprised by three steps: (i) the development of animal models that truly translate the phenotype of a patient, (ii) the development of human tools that enable decision-making, and (iii) the discovery phase for validation. Toward this way, NEWMEDS has launched three tools [68]: (i) the DupCheck, which serves as a patient deduplication tool that excludes patients who are concurrently involved in other trial(s), (ii) the Pharmacological Imaging and Pattern Recognition, which can be used to analyze brain images in the context of drug development, and (iii) the Clinical Significant Calculator, which quantifies the significance of a newly identified biomarker. In addition, the consortium has published a large number of scientific discoveries that strongly suggest the discovery of common phenotypes in choroidal neovascularization animals and schizophrenia patients that might lead to the development of a platform for novel drug discovery. Moreover, their discoveries state the development of new approaches for shorter and more efficient clinical trials and effective patient stratification models [68].

3.6.6 The Query Health initiative

The Office of the National Coordinator for Health Information Technology Query Health initiative [17] is a novel example of a secure, distributed system for data sharing and distributed query processing, which aims to interlink diverse heterogeneous clinical systems, toward a learning health system. The distributed type of architecture that Query Health initiative has adopted is straightforward as the data never leave the individual sites. The initiative has developed a standard, open-source, reference methodology for distributed query processing, which has been successfully validated in two public health departments in New York City (NYC Department of Health and Mental Hygiene) and in one pilot study under the FDA's Mini-Sentinel project, which aims to provide medical product safety surveillance using electronic health data [69]. Undoubtedly, the Query Health initiative provides a federated architecture that can be used for cohort interlinking, and population health surveillance, to promote data sharing although several technical limitations exist concerning the structural heterogeneity among the cohorts.

Through the Query Health initiative, the users can develop public health queries that are then distributed across the sites of interest so that the data never leave the individual sites. This eliminates the need of a centralized data warehouse. In addition, the queries return only the minimum required information from each site, which overcomes several patient privacy obstacles. The queries have been developed based on a standard database schema, i.e., an ontology, according to the Consolidated Clinical Document Architecture (C-CDA) [69] and are converted into a standard format, namely the Health Quality Measures Format (HQMF), before the query distribution process. The aggregated results from the applied queries across multiple sites are finally combined and returned back to the user in the form of a report. Popular packages such as the hQuery and the PopMedNet have been used for data analytics and query distribution, respectively [17]. In addition, the i2b2 temporal query tool has been used as a query builder [70]. Interlinking cross-platform databases is indeed difficult due to the structural heterogeneity among the centers and the different data collection protocols. This is a crucial problem that the Query Health initiative faces, and thus, emphasis must be given to the development of data normalization procedures in terms of data harmonization.

3.7 Solutions against the misuse of clinical data

So far, it is clear that data sharing is indeed a benefit for the public good as it enables the interlinking of out-of-border medical and other related data, as well as the reuse of these data and thus promotes scientific research worldwide. The strong demand for biomedical research and innovation, as well as the existence of a smart healthcare system for disease surveillance and prevention, is a few of the clinical unmet needs that data sharing has been proven to fulfill. However, apart from the fear for data abuse and the privacy laws, which constitute the two main significant barriers toward data sharing, there is still one significant concern that can make data sharing harmful; and that is data misuse [71,72]. The misuse of shared data has bad consequences and is many-sided. In this section, we will discuss the reasons behind the misuse of shared data, as well as propose solutions for overcoming the fear regarding the misuse of shared data.

- *Absence of real evidence*: The researcher must make clear the reason behind data sharing, as well as state the ensuing opportunities. The absence of real evidence hampers the data sharing process and produces the exact opposite outcomes. Thus, emphasis must be given to the purpose of data sharing.
- *Lack of data quality control*: Before the analysis of the data, it is of primary concern to assess the quality of the data, i.e., to curate the data. However, the misuse of methods for data curation introduces biases during the analysis, which yields false outcomes. As we already mentioned during the description of the specifications for the data curator in Section 3.3, two of its important functionalities are the outlier detection and the data imputation. If a researcher performs data imputation before outlier detection, the dataset is very likely to be contaminated with false values (outliers) and thus will become useless. On the other hand, the outlier detection methods might identify mathematically correct extreme values but without any clinical interpretation. Therefore, the clinician's guidance is necessary not only to validate these findings but also to deal with missing values so as to avoid data contamination.
- *Lack of the researcher's skills*: The lack of knowledge regarding the hypothesis of a study makes the study pointless. A researcher must first state the hypothesis under examination and then develop tools toward this direction. In addition, the researcher must be well aware of the scientific advances in the domain of interest, as well as the software and tools that meet the specifications set by the hypothesis. Only high-quality researchers who are well aware of data quality problems and causal inference methodologies are more likely to produce reliable outcomes [71]. In addition, the public health policy makers and decision

makers might be too credulous sometimes, especially when the outcomes of a study involve large databases. As a matter of fact, the availability of big data does not always guarantee correct study outcomes, which yields another question here: Is bigger data always better?

- *Ignorance of the data collection protocols*: Not knowing the population characteristics of a study introduces many biases during the analysis procedure and produces false outcomes. In general, there are three types of biases that affect observational studies: (i) the selection bias, (ii) the confounding bias, and (iii) the measurement bias. The selection bias appears when the selected group of individuals for a particular study is not representative of the overall patient population [71]. Another appearance of this bias can be met in causal-effect studies, i.e., studies that involve the validation of a drug's treatment (benefit or harm effect on individuals). In this type of study, if a variable has a common effect on both the treatment/exposure factor and the outcome factor, it is considered as a collider bias, which is also known as "M-bias" [71]. An example of this type of bias occurs when a patient's follow-up data is lost either because the patient's treatment is harmful (treatment factor) or because the patient's treatment is good (outcome factor). The lack of such information yields false statistical associations between these two factors and introduces distortions on the true causal effect [71]. Selection bias introduces distortions (e.g., false positives) in the outcome measures, which hampers the disease prevalence and the risk exposure yielding false data models for patient stratification. It is, thus, important to appropriately adjust these type of variables during statistical analyses for obtaining true causal estimations.

Confounding bias, which is also met in causal-effect studies, is even worse than selection bias. A confounding variable is a variable that has a common cause on both the treatment/exposure and the outcome rather than a common effect [71]. A typical example of confounding occurs when a clinician's decision is affected by a patient's disease severity or duration, which in turn affects the treatment's outcome. Patients at an earlier stage of a disease receive different treatment than those in a later stage of the same disease, whereas sicker patients may have worse treatment outcomes than the healthy ones. In this example, the confounding variable is the degree of sickness that is exhibited by the patients who receive different treatments. Such types of variables must be identified and properly adjusted.

Finally, the measurement bias is a widely known bias that arises from errors during the data measurement and collection process. The main reasons behind measurement bias are the following: (i) improper calibration of the measurement systems, (ii) lack of the measurement system's sensitivity, (iii) lack of the physician's expertise during the data measurement process, (iv) lack of a patient's trust and confidence during a questionnaire competence, and (v) patient's medical state (e.g., dementia).

- *Ignorance of the privacy laws and ethics policies*: The lack of knowledge regarding the data protection legislations has severe consequences concerning the patients' privacy and obscures data sharing. This factor has nothing to do with the biases in the outcomes of a study or the strategy used for data analytics rather than the privacy legislations breached by the study. The patient data must be first deidentified and qualified by appropriate scientific advisory boards. The deidentified data must be maintained in secure databases with private networks undergoing strict authorization procedures.

- *Poor use of the available data*: This has to do again with the skills and expertise of the researcher. The lack of data management and domain knowledge from the researcher's point of view results to misconceived analyses with extremely harmful results for the public.
- *Different interpretations of the same outcome*: This is a common mistake that underestimates the findings of a study. Clinical centers and laboratories worldwide make use of different measurement systems and units for characterizing a patient's laboratory test. For example, a typical hemoglobin test may be recorded by a clinical center A in "mg/mL," whereas a clinical center B might record it in "g/dL." Moreover, the thresholds for characterizing the test's outcome might vary, e.g., the clinical center A may consider a hemoglobin value of 15.5 "g/dL" as the threshold above which the hemoglobin levels are abnormal, whereas clinical center B may consider a value of 17.5 "g/dL." A solution to this is to include a new variable that states whether the hemoglobin levels are normal or abnormal. Standardization is thus important for the normalization of common terms across heterogeneous data.

3.8 Conclusions

Data sharing envisages to promote the collaboration and cooperation among the members of the global research community beyond the boundaries of global organizations, companies, and government agencies. There is no doubt that notable progress has been made toward the adoption of the overall concept of data sharing from a large number of clinical studies involving big data, with the clinical outcomes demonstrating the value of data sharing in dealing with crucial clinical unmet needs, such as the early identification of high-risk individuals for effective patient stratification, the discovery of new therapeutic treatments, and precision medicine. In addition, the large number of existing and ongoing global data sharing initiatives in the clinical domain is noteworthy, a fact that confirms that the collaboration of pioneer technical and pharmaceutical companies can yield promising and well-established scientific outcomes regarding the identification of new biomarkers in chronic and autoimmune diseases, the development of targeted therapies for cancer, and the concise prediction of the disease progress, toward a "smart" healthcare system.

Beyond its worthiness, data sharing comes with a range of ethical and legal obligations, data protection requirements, and several concerns regarding the quality and misuse of shared data. In an attempt to address these challenges, the value of a framework for data curation and standardization is highlighted in terms of data quality control. Then, existing data sharing frameworks are stated along with ongoing global data sharing initiatives and proposed solutions against the misuse of shared data. Data quality assessment lies in the heart of a federated platform and is responsible for several functionalities and requirements regarding the data evaluation and quality control, including functions for anomaly (outlier) detection, data imputation, detection of inconsistencies, and further discrepancies. Emphasis was given on outlier detection mechanisms along with examples to demonstrate the importance of the data curator to the reader. Inevitably, the absence of data curative treatments can lead to falsified clinical studies with small statistical power. Although data quality has been recognized as a key factor in all operating processes both in the public and private sectors, curation methods shall be used with caution as it is more likely to make things worse.

Emphasis must be also given to the development of methods and tools that are able to deal with the heterogeneity of the interlinked cohorts in a federated environment. A promising solution to this challenge is data standardization, which involves the identification and normalization of the variables of a raw dataset according to a standard model. This model usually serves as a gold standard, i.e., a template, according to which a candidate (raw) dataset must be transformed to. Data standardization has been presented in terms of lexically matching similar variables between a raw dataset and a standard one, using different strategies. However, data standardization can be extended so as to consider for semantic relations among the variables. This is equivalent to solving a semantic matching problem, which enables data harmonization. A detailed description of this approach is presented and further discussed later in Chapter 5.

Every data sharing system that deals with federated databases must be well aware of the patient privacy and design an architecture that complies with the data protection regulations. Traditional centralized databases and data warehouses might be easy to work with but are prone to data breach and often noncompliant with existing data protection legislations. Currently, there is a strong need to the scientific research community concerning the definition of a standard protocol to address the technical challenges and requirements of patient privacy. For example, the term "anonymized" is often misinterpreted, with each clinical center appending a different meaning to this term, which has significant consequences on the processing of data. A standard protocol is thus needed to determine what the term "anonymized" means in a quantitative manner. In addition, a set of specifications that a secure system for data sharing shall meet must be defined. This will help the existing initiatives to develop data sharing platforms with respect to the legal frameworks.

Regarding the misuse of clinical data, the researcher must be well aware of the data collection protocols that each study uses, to avoid biases that are introduced during postanalysis. In addition, the researcher must be well aware of the state of the art regarding the data analytics methods that are going to be used, to avoid the poor use of the available data, which in turn yields falsified results. Controlled access is a solution against the poor-quality analysis of the shared data, which limits the amount of shared data instead of the registered access, which provides access to large groups of data. On the other hand, it raises new significant challenges regarding the development of improved privacy-enhancing technologies.

Data sharing needs to increase its social impact. Most patients have not thought much about data sharing, and this limits their participation in research studies. NIH has been working toward the development of a Common Rule [73], which reinforces the patient privacy and engagement during the clinical research process so as to empower people to participate in clinical research trials. Once this is achieved, someone can better comprehend the disease onset and progress and thus develop new therapies and public health policies. Besides, the patients are the real owners of their data and must have access to them anytime. In addition, the results from the published clinical trials must be made available to the scientific community worldwide and be used to their fullest potential for the public.

References

[1] Doiron D, Burton P, Marcon Y, Gaye A, Wolffenbuttel BHR, Perola M, et al. Data harmonization and federated analysis of population-based studies: the BioSHaRE project. Emerg Themes Epidemiol 2013;10(1):12.

[2] Khatri V, Brown CV. Designing data governance. Commun ACM 2010;53(1):148–52.

[3] Krishnan K. Data warehousing in the age of big data. Morgan Kaufmann Publishers Inc.; 2013.

[4] Pipino LL, Lee YW, Wang RY. Data quality assessment. Commun ACM 2002;45:4.

[5] Arts DGT, De Keizer NF, Scheffer GJ. Defining and improving data quality in medical registries: a literature review, case study, and framework. J Am Med Inform Assoc 2002; 9(6):600–11.

[6] Cappiello C, Francalani C, Pernici B. Data quality assessment from the user's perspective. In: Proceedings of the 2004 international workshop on information quality in information systems. IQIS; 2004. p. 68–73.

[7] Maydanchik A. Data quality assessment, Chapter 1 – causes of data quality problems. Technics Publications; 2007.

[8] Lord P, Macdonald A, Lyon L, Giaretta D. From data deluge to data curation. In: Proceedings of the UK e-science All Hands meeting; 2004. p. 371–5.

[9] Batini C, Cappiello C, Francalanci C, Maurino A. Methodologies for data quality assessment and improvement. ACM Comput Surv 2009;41:3.

[10] Stonebraker M, Bruckner D, Ilyas IF, Pagan A, Xu S. Data curation at scale: the Data Tamer system. In: Conference on innovative data systems research. CIDR; 2013.

[11] Nerenz DR, McFadden B, Ulmer C, editors. Race, ethnicity, and language data: standardization for health care quality improvement. Washington (DC): National Academies Press (US); 2009.

[12] Pang C, Sollie A, Sijtsma D, Hendriksen B, Charbon M, de Haan T, et al. SORTA: a system for ontology-based re-coding and technical annotation of biomedical phenotype data, vol. 18. Oxford): Database; 2015.

[13] Kourou K, Pezoulas VC, Georga EI, Exarchos T, Tsanakas P, Tsiknakis M, et al. Cohort harmonization and integrative analysis from a biomedical engineering perspective. IEEE Rev. Biomed. Eng.; 2018.

[14] Downey AS, Olson S, editors. Sharing clinical research data: workshop summary. Washington (DC): National Academies Press (US); 2013.
[15] US National Institutes of Health. ClinicalTrials.Gov. 2012. Link: https://clinicaltrials.gov/.
[16] Mailman MD, Feolo M, Jin Y, Kimura M, Tryka K, Bagoutdinov R, et al. The NCBI dbGaP database of genotypes and phenotypes. Nat Genet 2007;39(10):1181.
[17] Klann JG, Buck MD, Brown J, Hadley M, Elmore R, Weber GM, Murphy SN. Query Health: standards-based, cross-platform population health surveillance. J Am Med Inform Assoc 2014;21(4):650–6.
[18] Song JW, Chung KC. Observational studies: cohort and case-control studies. Plast Reconstr Surg 2010;126(6):2234–42.
[19] Jensen PB, Jensen LJ, Brunak S. Mining electronic health records: towards better research applications and clinical care. Nat Rev Genet 2012;13(6):395.
[20] Haas LR, Takahashi PY, Shah ND, Stroebel RJ, Bernard ME, Finnie DM, et al. Risk-stratification methods for identifying patients for care coordination. Am J Manag Care 2013;19(9):725–32.
[21] Fonarow GC, Adams KF, Abraham WT, Yancy CW, Boscardin WJ, ADHERE Scientific Advisory Committee. Risk stratification for in-hospital mortality in acutely decompensated heart failure: classification and regression tree analysis. JAMA 2005;293(5):572–80.
[22] Berger S, Schrefl M. From federated databases to a federated data warehouse system. In: Proceedings of the 41st annual Hawaii international conference on system Sciences. HICSS; 2008. 394-394.
[23] Saltor F, Castellanos M, García-Solaco M. Suitability of data models as canonical models for federated databases. ACM SIGMOD Rec. 1991;20(4):44–8.
[24] Friedman CP, Wong AK, Blumenthal D. Achieving a nationwide learning health system. Sci Transl Med 2010;2(57). 57cm29.
[25] de Bono JS, Ashworth A. Translating cancer research into targeted therapeutics. Nature 2010;467(7315):543.
[26] Bonassi S, Ugolini D, Kirsch-Volders M, Strömberg U, Vermeulen R, Tucker JD. Human population studies with cytogenetic biomarkers: review of the literature and future prospectives. Environ Mol Mutagen 2005;45(2-3):258–70.
[27] Guyon I, Elisseeff A. An introduction to variable and feature selection. J Mach Learn Res 2003:1157–82.
[28] Hall MA. Correlation-based feature selection of discrete and numeric class machine learning. In: ICML '00 Proceedings of the Seventeenth International Conference on Machine Learning; 2000. p. 359–66.
[29] Bolón-Canedo V, Sánchez-Maroño N, Alonso-Betanzos A. Distributed feature selection: an application to microarray data classification. Appl Soft Comput 2015;30:136–50.
[30] Morán-Fernández L, Bolón-Canedo V, Alonso-Betanzos A. Centralized vs. distributed feature selection methods based on data complexity measures. Knowl Based Syst 2017;117:27–45.
[31] Collins FS, Varmus H. A new initiative on precision medicine. N Engl J Med 2015;372(9):793–5.
[32] Mirnezami R, Nicholson J, Darzi A. Preparing for precision medicine. N Engl J Med 2012;366(6):489–91.
[33] Alliance for Health Policy and Systems Research, World Health Organization (WHO). Link: http://www.who.int/alliance-hpsr/en/.

[34] Lock K. Health impact assessment. Br Med J 2000;320(7246):1395–8.
[35] Paul MJ, Dredze M. A model for mining public health topics from twitter. Association for the Advancement of Artificial Intelligence; 2011.
[36] Brownson RC, Chriqui JF, Stamatakis KA. Understanding evidence-based public health policy. Am J Public Health 2009;99(9):1576–83.
[37] Abbott S, Chapman J, Shaw S, Carter YH, Petchey R, Taylor S. Flattening the national health service hierarchy. The case of public health; 2006. p. 133–48.
[38] Mukaka MM. A guide to appropriate use of correlation coefficient in medical research. Malawi Med J 2012;24(3):69–71.
[39] Schmitt P, Mandel J, Guedj M. A comparison of six methods for missing data imputation. Biom Biostat Int J 2015;6(1):1.
[40] Engels JM, Diehr P. Imputation of missing longitudinal data: a comparison of methods. J Clin Epidemiol 2003;56(10):968–76.
[41] Shrive FM, Stuart H, Quan H, Ghali WA. Dealing with missing data in a multi-question depression scale: a comparison of imputation methods. BMC Med Res Methodol 2006; 6(1):57.
[42] Bertsimas D, Pawlowski C, Zhuo YD. From predictive methods to missing data imputation: an optimization approach. J Mach Learn Res 2017;18(1):7133–71.
[43] Swarupa Tripathy S, Saxena RK, Gupta PK. Comparison of statistical methods for outlier detection in proficiency testing data on analysis of lead in aqueous solution. Am J Theor Appl Stat 2013;2(6):233–42.
[44] Rousseeuw PJ, Hubert M. Robust statistics for outlier detection. Wiley Interdisciply Rev: Data Min Knowl Discov 2011;1(1):73–9.
[45] Liu FT, Ting KM, Zhou ZH. Isolation-based anomaly detection. ACM Trans Knowl Discov Data 2012;6(1):3.
[46] Ding Z, Fei M. An anomaly detection approach based on isolation forest algorithm for streaming data using sliding window. IFAC Proc Vol 2013;46(20):12–7.
[47] Schubert E, Zimek A, Kriegel HP. Local outlier detection reconsidered: a generalized view on locality with applications to spatial, video, and network outlier detection. Data Min Knowl Discov 2014;28(1):190–237.
[48] Hoyle B, Rau MM, Paech K, Bonnett C, Seitz S, Weller J. Anomaly detection for machine learning redshifts applied to SDSS galaxies. Mon Not R Astron Soc 2015;452(4): 4183–94.
[49] Rousseeuw PJ, Driessen KV. A fast algorithm for the minimum covariance determinant estimator. Technometrics 1999;41(3):212–23.
[50] Cohen W, Ravikumar P, Fienberg S. A comparison of string metrics for matching names and records. In: Kdd workshop on data cleaning and object consolidation, vol. 3; 2003. p. 73–8.
[51] del Pilar Angeles M, Espino-Gamez A. Comparison of methods Hamming distance, Jaro, and Monge-Elkan. In: Proceedings of the 7th international conference on advances in databases, knowledge, and data applications. DBKDA; 2015.
[52] Jaro MA. Advances in record-linkage methodology as applied to matching the 1985 census of Tampa, Florida. J Am Stat Assoc 1989;84(406):414–20.
[53] Bozkaya T, Yazdani N, Özsoyoğlu M. Matching and indexing sequences of different lengths. In: Proceedings of the 27th ACM international conference on information and knowledge management. CIKM; 1997. p. 128–35.
[54] Rao GA, Srinivas G, Rao KV, Reddy PP. Characteristic mining of mathematical formulas from document-A comparative study on sequence matcher and Levenshtein distance procedure. J Comp Sci Eng 2018;6(4):400–4.

[55] Euzenat J, Shvaiko P. Ontology matching. Heidelberg: Springer-Verlag; 2007.
[56] Knoppers BM. International ethics harmonization and the global alliance for genomics and health. Genome Med 2014;6:13.
[57] Knoppers BM. Framework for responsible sharing of genomic and health-related data. HUGO J 2014;8(1):3.
[58] Knoppers BM, Harris JR, Budin-Ljøsne I, Dove ES. A human rights approach to an international code of conduct for genomic and clinical data sharing. Hum Genet 2014; 133:895–903.
[59] Rahimzadeh V, Dyke SO, Knoppers BM. An international framework for data sharing: moving forward with the global alliance for genomics and health. Biopreserv Biobanking 2016;14:256–9.
[60] Budin-Ljosne I, Burton P, Isaeva J, Gaye A, Turner A, Murtagh MJ, et al. DataSHIELD: an ethically robust solution to multiple-site individual-level data analysis. Public Health Genom 2015;18:87–96.
[61] Wallace SE, Gaye A, Shoush O, Burton PR. Protecting personal data in epidemiological research: DataSHIELD and UK law. Public Health Genom 2014;17:149–57.
[62] Wilson RC, Butters OW, Avraam D, Baker J, Tedds JA, Turner A, et al. DataSHIELD–new directions and dimensions. Data Sci J 2017;16(21):1–21.
[63] Health insurance portability and accountability act of 1996. Public Law 1996;104:191. Link: https://www.hhs.gov/hipaa/for-professionals/index.html.
[64] Public Welfare, Department of Health and Human Services, 45 CFR §§ 46. 2016. Link: https://www.gpo.gov/fdsys/pkg/CFR-2016-title45-vol1/pdf/CFR-2016-title45-vol1-part46.pdf.
[65] Green AK, Reeder-Hayes KE, Corty RW, Basch E, Milowsky MI, Dusetzina, et al. The project data sphere initiative: accelerating cancer research by sharing data. Oncologist 2015;20(5). 464-e20.
[66] CEO Roundtable on Cancer. Link: https://www.ceoroundtableoncancer.org/.
[67] Merriel RB, Gibbs P, O'brien TJ, Hibbert M. BioGrid Australia facilitates collaborative medical and bioinformatics research across hospitals and medical research institutes by linking data from diverse disease and data types. Hum Mutat 2011;32(5):517–25.
[68] Tansey KE, Guipponi M, Perroud N, Bondolfi G, Domenici E, Evans D. Genetic predictors of response to serotonergic and noradrenergic antidepressants in major depressive disorder: a genome-wide analysis of individual-level data and a meta-analysis. PLoS Med 2012;9(10):e1001326.
[69] Behrman RE, Benner JS, Brown JS, McClellan M, Woodcock J, Platt R. Developing the Sentinel System—a national resource for evidence development. N Engl J Med 2011; 364(6):498–9.
[70] Murphy SN, Weber G, Mendis M, Gainer V, Chueh HC, Churchill S. Serving the enterprise and beyond with informatics for integrating biology and the bedside (i2b2). J Am Med Inform Assoc 2010;17(2):124–30.
[71] McNeil BJ. Hidden barriers to improvement in the quality of care. N Engl J Med 2001; 345(22):1612–20.
[72] Hoffman S, Podgurski A. The use and misuse of biomedical data: is bigger really better? Am J Law Med 2013;39(4):497–538.
[73] Hudson KL, Collins FS. Bringing the common rule into the 21st century. N Engl J Med 2015;373(24):2293–6.

CHAPTER

Data protection

4

Chapter outline

4.1 Overview

The terms privacy and data protection have their roots several decades ago, when the digital technology was rapidly advancing and the necessity of a legal protection framework was emerging to prevent the breach, embezzlement, and misuse of sensitive personal data [1]. This legal framework was initially designed as a set of general guidelines that focused on how the processing of sensitive data should be conducted instead of a mechanism that would prevent the distribution of personal data. According to the former EU's Data Protection Supervisor, Peter J. Hustinx,

Medical Data Sharing, Harmonization and Analytics. https://doi.org/10.1016/B978-0-12-816507-2.00004-9

privacy and data protection can be defined as "*the right to respect for private life and the right to protection of one's personal data*" [1]. The term personal data refers to any kind of physical information that can be used to identify an individual. Examples of personal data include an individual's name, address, date of birth, ethnicity, etc. The term personal sensitive data is used to refer to any kind of physical information that might lead to the direct identification of an individual, such the social security number, the credit card number, the phone number, etc. Apart from the physical information though, there are other modern direct and indirect means as well, which should be seriously taken into consideration. Examples of direct means include an individual's Internet Protocol (IP) address that can be often used to identify the individual's physical location, public and private health databases, and governmental databases that contain civilians' sensitive data, credit card numbers in private bank databases, etc. An example of an indirect means can be often met by the online advertising companies that develop ways to monitor an individual's internet preferences so as to propose more precise and related offers based on the individual's profile [2].

It is obvious that information technology plays a prominent role both in an individual's public and private life. Early data protection laws envisaged at preserving the rights and interests of a single individual without considering the impact that information technology would have had in the individual's life. The Internet of things (IoT) [3], the social media networks, the online advertising companies, the smart systems, and the cloud computing infrastructures are only but a few examples of technological advancements that have an effect on an individual's rights and pose significant barriers in existing data protection regulations. In addition, the recent reports regarding intelligence agencies (see Refs. [4,5] for the US National Security Agency [NSA] case) concerning the unauthorized civilian online surveillance combined with several reports regarding social media network linkages [5] pose serious threats to well-known ethical and legal frameworks exemplars, such as the Directive 95/46/EC [6] in Europe, and its successor, the General Data Protection Regulation (GDPR) [7], and the Health Insurance Portability and Accountability Act (HIPAA) [8] in the United States.

The backbone of these exemplars lies on two major data protection Articles. The first one is the Article 12 of the Universal Declaration of Human Rights (as it was declared by the United Nations General Assembly on December 1948 in Paris) [9], according to which no one has the right to interfere with an individual's protection, family, and home (or correspondence), as well as with the individual's honor and reputation. The second one is the Article 8 of the European Convention on Human Rights [10], according to which every individual has the right to respect for his privacy, family, and home (or correspondence). As a matter of fact, the legal basis of every data protection law lies on the right to respect the privacy and home of each individual. Toward this direction, the Directive 95/46/EC [6] was adopted in 1995 with the purpose to introduce a common legal basis among the Member States of the European Union (EU). The backbone of the Directive was the Convention 108 of the Council of Europe [11], and its goal was to enable the free flow of personal information from the European citizens among the Member States. In May 2018, the Directive was replaced by GDPR (EU) 2016/679 [7] of the European Parliament and Council of EU, with the purpose of restructuring the Directive's legal basis to consider the radical advances and effects of information technology to the processing of personal data. The GDPR poses rules on data deidentification (data protection by design principle) giving emphasis on protection safeguards, such as data minimization (data protection by default principle), consent forms, and employment of data protection officers (DPO) for ensuring the GDPR compliance, all of them with respect to the individual rights who has the right to be forgotten and the right to revoke his/her consents anytime.

In the United States, HIPAA of 1996, Public Law 104−191, which is part of the Social Security Act, aims to embrace the sharing of certain patient administrative data and provide security provisions for promoting the healthcare industry [8]. The US Department of Health and Human Services (HHS), considering the fact that the technological advances could potentially erode the privacy of health information, developed two fundamental Rules under the HIPAA [8], namely the HIPAA Privacy Rule and the HIPAA Security Rule. The first sets national standards for the protection of healthcare electronic transactions including medical records and other personal health information that are conducted by healthcare providers, whereas the second sets national standards for the protection of personal health information that is created, received, used, or maintained by a covered entity[1] [8]. Both of the HIPAA Rules have been posed according to guidelines from the Food and Drug Administration [12], which serves as the federal US agency, responsible for the protection and the promotion of public health. The HIPAA presents many similarities with the Federal Trade Commission (FTC) Act of 1914 [13], which was composed to provide a legal basis against deceit actions and practices in the business area.

Data protection affects different fields of our everyday lives. Healthcare is just one aspect of the data protection regulations, and thus, great emphasis will be given on describing the impact of the existing data protection regulations and how they can be more effective toward a complete data protection framework. Additional emphasis will be given to describe the legal basis of the European and US data protection legislations, as well as their common characteristics in terms of out-of-border data sharing and privacy preservation. Steps toward the realization of GDPR and HIPAA compliant federated platforms will be further discussed as part of the corresponding legislations, and major international data protection initiatives will be presented as exemplars that effectively deal and comply with national data protection standards. Data protection compliance is a legal, ethical, and technical challenge, as well, and lies in the core of a federated platform as it enables data sharing. It is, therefore, important to comprehend the legal basis of data sharing before the establishment of a federated platform that deals with human rights and freedom.

[1] An entity that provides healthcare services, such as a doctor, a healthcare insurance company, a hospital, etc.

4.2 The fundamental basis of data governance

Data governance is a conceptual framework that refers to the specifications that must be met to effectively manage a federated healthcare platform[2] and serves as a safeguard for establishing the mechanisms of such a platform [14]. The data governance framework can be visualized as a multilayer data management framework that consists of a series of layers (see Fig. 4.1) including (i) the data protection layer, (ii) the risk management layer, (iii) the data sharing layer, (iv) the data quality control layer, and (v) the data analytics layer. Data protection lies in the basis of data governance (level 1) and involves the application of all the necessary regulatory compliance measures, such as compliance with laws, rules, standards, audits, policies, and transparency. The risk management layer follows, which includes mechanisms for (i) risk identification, (ii) risk assessment, (iii) risk analysis, and (iv) risk minimization. The risk analysis procedure is a fundamental procedure that is described as a basic principle for data protection both in the EU data protection law (under the GDPR [7]) and the US data protection law (under the FTC Act [13] and the HIPAA [8]).

The data sharing layer enables the sharing of the data with respect to secure data management standards and data access rights according to the physical actors of the platform (e.g., the data processor, the data provider, etc.). The data quality control layer lies right above the data sharing layer and is responsible for the assessment of different data quality measures, such as the: (i) accuracy, (ii) relevance, (iii) completeness, (iv) clarity, and (v) adequacy of the involved data. The data quality layer has been extensively described in Chapter 3 in the light of a procedure known as data curation. Furthermore, data quality comprises the backbone of several data protection principles (such as the risk management layer). Finally, the data analytics layer includes all those processes that are related to (i) data exploration, (ii) descriptive analytics, and (iii) machine learning applications on the qualified data.

In fact, the framework that is depicted in Fig. 4.1 is a complete series of operations that characterizes a federated platform. It is a multilayer framework that consists of five layers that are placed in a bottom-up manner, where each layer corresponds to a specific functionality (e.g., regulatory compliance, data security, data sharing, data quality assessment, data analytics) that is part of the overall data governance plan. The compliance with the data protection regulations combined with the development of efficient mechanisms for risk management to identify and reduce the underlying risks (e.g., security threats) can enable the sharing of data. The nonexistence of well-qualified data is a crucial technical challenge that has negative effects on the processing of the data. This highlights the need for the development of automated methods for data quality assessment. There is no doubt that the application of data analytics methods on well-qualified data might lead to results with high statistical power.

---

[2] An interlinked healthcare platform consisting of a set of interconnected individual healthcare platforms.

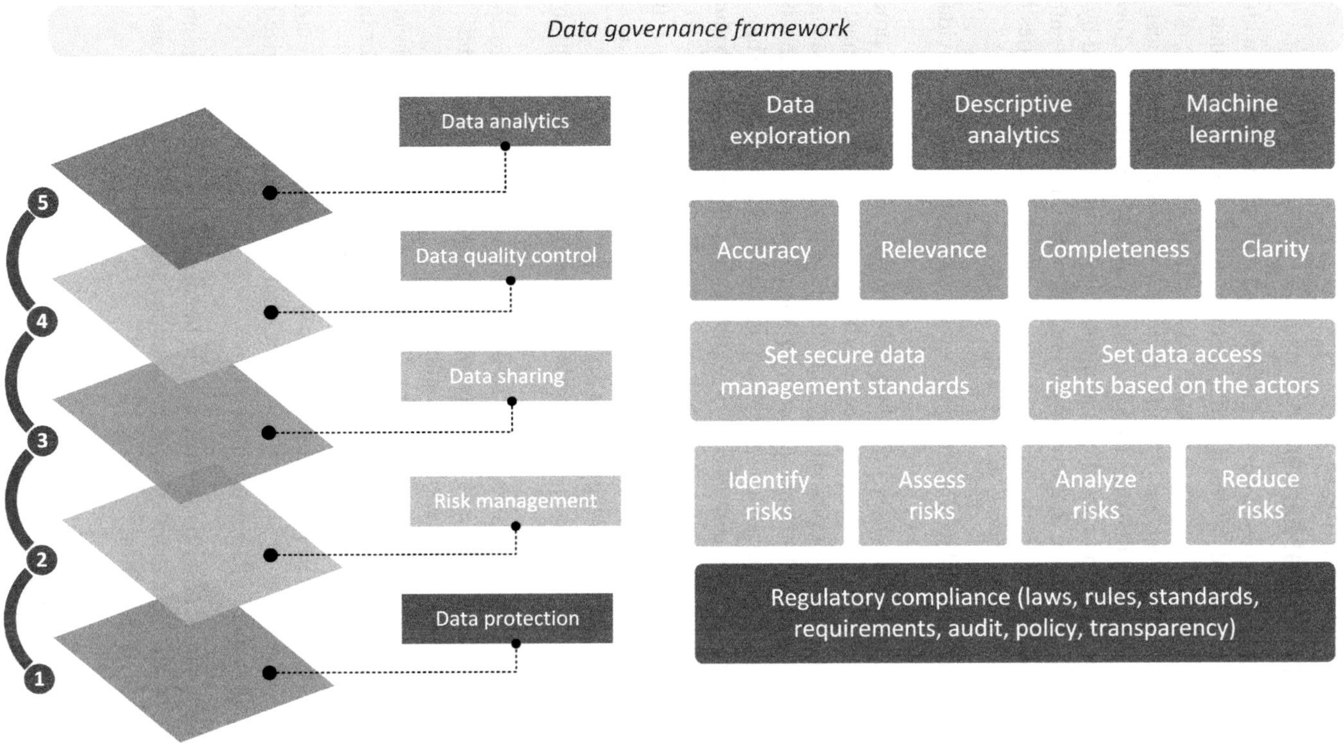

FIGURE 4.1

The conceptual architecture of a multilayer data governance framework.

4.3 Significant challenges

4.3.1 Legal and ethical barriers

The backbone of data governance lies on the legal and ethical compliance (with the requirements that are posed by the existing data protection legislation) that any entity (e.g., an organization), which wishes to get involved in the processing of personal data, must meet. A federated platform, however, faces significant ethical and legal compliance challenges that directly affect the privacy of personal data. Toward this direction, strict legal and ethical requirements [7–11] must be adopted by the data protection laws to ensure the privacy during the inner (i.e., within the entities of a country) and outer (i.e., between the entities of a country and the entities of a third country) personal data flows, including the following:

- The individuals' personal data must be processed with respect to the individual's rights and freedoms.
- Individual consent forms must be obtained by anyone who wishes to process personal data according to the purposes of processing. The individual must be informed about all types of processing, which involve his/her personal data and provide his/her informed consent according to the consequences (i.e., the risks) that might arise as a result of the processing of his/her personal data. This comprises a strict requirement that involves the participation of the individuals in the processing of their data.
- The individuals must be given the right to (i) access, rectify, and erase their personal data, (ii) object and restrict the processing of their data, and (iii) request to obtain their data when they wish to do so.
- The risks behind the processing of the individual data (i.e., risk assessment) must be clearly stated.
- Any cross-border data flows involving sensitive data must be subject to international legal requirements and data protection principles that require the cooperation of international supervised authorities.
- The sensitive data must not be transferred to third countries (parties) without the fulfillment of adequate data protection requirements and principles under the international data protection regulations.
- The existence of any third parties must be clearly defined in the related contracts and rules of conduct, along with any natural or legal person who is authorized to collect the data and further manipulate them.

All the entities must describe any measures to be taken to comply with the above requirements. In any other case, strict legal sanctions and ethical ramifications will be issued against such entities.

4.3.2 Patient privacy issues

The term personal data includes a variety of personal identifiers that can either directly or indirectly lead to the identification of the individual by any processing entity. Nowadays, the number of personal identifiers has been greatly increased due to the rapid digital advancements. These personal identifies include not only names, telephone numbers, license numbers, and social security numbers but also email addresses, biometric identifiers, bank account numbers, IP addresses, and any other unique digital identifiers [7,8]. The following privacy issues [7–11] shall be taken into consideration to protect the individual's identity:

- Personal data must be deidentified by either pseudonymizing or anonymizing them. Anonymization involves the complete removal of any information that can lead to the identification of the individual, whereas pseudonymization involves the partial removal of the individual data with an additional storage of information that can indirectly lead to the identification of the individual (e.g., an identifier).
- Only a small portion of the individuals' data must be processed according to the purposes of processing, which were already described during the legal and ethical compliance procedure (see Section 4.3.1).
- Common international standards and definitions must be introduced for the terms data anonymization and data pseudonymization to avoid any confusion during data collection and data processing.
- It must clearly define who is responsible for data collection (i.e., the primary data collectors) and data processing (i.e., the secondary analysts), as well as the existence of any involved third parties.
- Researchers and analysts must be well-qualified with appropriate expertise in data sharing, data protection to avoid data embezzlement, and further data misuse that might harm the individual.
- Audit trails are necessary so that the individuals can see who accessed their medical records. In the case of a patient privacy breach, the involved patients must be directly informed by the related authorities.
- All the data processing operations must be transparent and fair according to the individuals' rights.
- Strict data protection protocols are needed to avoid unauthorized surveillance and prevent data breach.
- Law enforcement agencies need to be involved in the out-of-border tracking of personal data flows.

4.3.3 Technical limitations

The technical limitations are similar to the technical challenges of a federated platform in the basis of data sharing and data protection [15]. Those technical limitations include the following:

- Secure mechanisms for user access management and multiple-factor user authentication services.
- Secure and encrypted communication mechanisms for the collection and transmission of personal data.

- Secure private data layers within the cloud for the storage of personal data (in remote private spaces).
- Effective deidentification mechanisms through the construction of unique identifiers per patient.
- Efficient methods for reducing the information that is needed during the processing of personal data.
- The "bring the analysis to the data" design where the sensitive data are stored in remote private spaces.
- Batch-based processing mechanisms (i.e., distributed methods for data analytics), especially when the personal data are stored in decentralized databases where secure communication is necessary.
- The data shall be made always available, which is a fundamental principle of data sharing in federated platforms. Data availability, i.e., the reuse of personal data, promotes scientific research worldwide.
- Automated error recovery mechanisms when the operating system fails to respond to any kind of functionality and especially when the operating system loses the control of the data, e.g., in the case of a data breach. In the latter case, supervised government authorities must be properly informed.
- Automated mechanisms for assessing and cross-checking the quality of the data (i.e., data curation). The data must be accurate, up-to-date, relevant, adequate, complete, and in a readable form.
- Digital forms to upload personal consent forms through highly remote secure systems.
- Continuous monitoring of the data input and export processes along with the logging and processing.
- Scalability and interoperability of the federated platform that accounts for data sharing and international data protection regulations. The interoperability factor includes legal, regulatory, and application issues. The scalability factor involves efficient resource management (e.g., IT infrastructure).
- Pooled (centralized) analysis must be supported only when informed consent forms are employed.

4.3.4 Other aspects

The significant challenges that a federated platform can face toward its compliance with the data protection regulations involve additional multidimensional aspects, including

- The heterogeneity of the data protection laws across different countries, i.e., the existence of legal and ethical inequalities across developed and developing countries, as well as ethical issues during the data collection process that is introduced by different countries.

- The heterogeneity of the data protection protocols across international laboratories and institutions. Different entities have different legal and ethical regulations regarding the data collection process.
- Additional bioethical regulations in the case of genome-wide studies must be taken into consideration. The health policies regarding the processing of genetic data are usually more strict and hard to follow.
- The negative implications of big data in privacy protection (e.g., the use of big data for the identification of individuals using information from the social media or any other information from the internet).
- The negative effect of centralized data warehouses in the case of data breach. It is easier for the hackers to breach centralized data repositories instead of distributed data repositories where the access to the rest of the repositories can be blocked in the case of data breach in a specific repository. On the other hand, distributed data repositories pose significant computational challenges (see Section 1.3).
- The existence of potential data obscuration/aggregation mechanisms in the form of malicious software. These mechanisms can be uploaded in the form of an ordinary software and cause serious leaks.
- The early detection and prevention of personal data information leaks in large-scale platforms. Large-scale platforms might be hard to breach, but a successful attempt can have serious consequences.
- Ineffective tracking of entities that falsely claim to be compliant with the data protection regulations. This is a serious issue that led to the repeal of the Safe Harbor [16] data protection agreement between the EU–US, in 2016, regarding the existence of secret unauthorized surveillance programs and the lack of data protection during the transatlantic data flows from the EU to the entities that lie within the United States.

4.4 Data protection regulations across Europe

4.4.1 The Directive 95/46/EC of the European Parliament and of the Council

The Directive 95/46/EC of the European Parliament and of the Council (Data Protection Directive) [6] was initially introduced in October 1995 and took effect 3 years later on October 1998. Since May 25, 2015, the Directive was replaced by GDPR [7]. As the GDPR includes common legal basis and descriptions related to the main actors of the law, which were already included in the Directive, it is important to take a first look into the Directive's scope. The Directive was composed to protect the information flow of individuals within the Member States of the EU. It is based on two primary objectives: (i) to protect the rights and freedoms of individuals with respect to the processing of personal data and (ii) to enable the free flow of information within the Member States. The Directive was the first one to introduce precise definitions [6] for (i) the data controller, (ii) the data processor, (iii) the third party, (iv) the data recipient, and (v) the data subject. In addition, the Directive adopts the term "personal data," which is defined as the physical information that can be used to identify an individual (either directly or indirectly), such as the individual's name, address, date of birth, and identification number. The Directive 95/46/EC, however, did not mention any kind of digital information that would be able to identify an individual perhaps due to the immaturity of the Information Technology at that time.

The Directive posed that all Member States, under the Article 6, Section I [6], shall ensure that the personal data are (i) processed fairly with respect to the existing laws, (ii) collected for certain purposes with respect to the safeguards provided by the data controllers, (iii) adequate and relevant to the purposes of processing, (iv) accurate and kept up to date, and (v) maintained in such a form that it enables the identification of the individuals and only for the specified time period that was described in the purposes of processing. Furthermore, according to Article 7, Section II [6], the Member States shall ensure that the personal data must be processed only if the data subject has given his informed consent to the processing of his data or if the processing is necessary for (i) protecting the subject's vital interests, (ii) preserving the validity of the contract to which the data subject is subjected to, (iii) the performance of a public study, which is conducted by a public authority, and (iv) the purposes of the processing (i.e., legitimate interests) that were set by the corresponding data controller who is responsible for the legitimate interest of his data subject(s).

The Directive highlighted five major actors. All these actors were defined in terms of automated means for data processing (either completely or partially automated). These actors can be physical or legal entities, public authorities, agencies, or any other bodies other than the actor itself. The individual is referred to as data subject and is defined as an actor who has the right to approve the processing of his personal data. According to Article 12, Section V [6], the data subjects have the right to obtain confirmation from their corresponding data controllers (who are responsible for their data) to (i) confirm or reject the processing of their data, (ii) be aware of the kind of processing actions on their data, (iii) erase their data in case they are not compliant with the Directive's regulations, and (iv) receive notifications from third parties concerning the processing of their data. In addition, according to Article 14, Section VII [6], the data subject has the right to object to the processing of his/her data, at any time.

The data controller is an important actor of the Directive. The data controller can define the purposes and means of processing personal data according to the existing ethical and legal obligations. Moreover, the controller can cooperate with other controllers, known as joint controllers, and provide information to the supervised authorities about (i) the purposes of processing, (ii) the involvement of third parties (transfer to other countries), (iii) the data recipients or categories of recipients, (iv) legal documents, and (v) personal information (e.g., the identity of the controller). A third party is another actor who is authorized to process the data other than the controller and the processor. Third parties must be well-defined and are able to process the data only under the supervision of the data controller or the data processor. The data recipient is an actor who has the right to disclose the personal data, with the public authorities being excluded.

According to Article 17, Section VII [6], the data controller shall employ strict measures, conduct risk management, and define appropriate security levels to protect the personal data against unauthorized data and privacy breach. Risk management is also part of the Member States responsibilities, which, according to Article 20, Section IX, shall provide the processing operations that are prone to exhibit risks that threaten the subject rights and carry out all necessary checks through the supervised authorities. Moreover, the controller shall carefully choose those data processors that fulfill sufficient requirements for processing the personal data. Furthermore, the controllers must also provide information to the data subjects regarding the collection process and the purposes of processing. In addition, the data controller must notify the supervised authorities of the Member States about the purposes of data processing before carrying out any complete or partially automated data processing operation (Article 18, Section IX).

The data processor grants the right from the data controller to process the subjects' data. The data processor shall fulfill all the restrictions that are posed by the data controller and transfer these restrictions to the rest of the data processors that he might be responsible for. In fact, the processor acts on behalf of the data controller under his instructions and of course under the obligations posed by the law (Article 16, Confidentiality of processing, Section VIII). The data processor shall be well-skilled (both technical and ethically aware) and compliant with the codes of conduct that are posed by the supervised authorities. In addition, the processor shall inform the controller about any information that is necessary during the latter's obligation to inform the supervised authorities. Both the processor and the controller are bound by a contract or a legal act to comply with the Directive's requirements. The contract (or legal act) is usually in a written form so as to keep proofs as defined by the law of the corresponding Member State.

The Member States shall provide information related to the data controller, the purposes of processing, the recipients of the data, and the third parties. In special cases, a personal data protection official can also be appointed for preserving the subjects' rights, which are possibly prone to be affected by the data processing operations [1,6]. In addition, an independent Working Party is established regarding the data processing operations. The Working Party acts independently and has the right to advise the Commission concerning the level of protection, the codes of conduct, and the impact to the rights and freedoms of the subjects.

4.4.2 The General Data Protection Regulation

In May 25, 2018, GDPR 2016/679 superseded the Directive 95/46/EC of the European Parliament and of the Council. The GDPR [7,17] gives emphasis on the data pseudonymization process and imposes rules on the use of big data, on the nature of biobanks and research infrastructures in general, as well as on the purposes of data processing, on legitimate and liability exemptions, and on data manipulation. According to Article 4 of the GDPR, the main actors that are highlighted by the regulation are (i) the data subject, (ii) the data controller (and the joint controllers), (iii) the data processor, (iv) the third party, (v) the data recipient, (vi) the representative who represents the controller or processor and is designated by them, and (vii) the enterprise who involves partnerships or associations engaged in economic activities. The GDPR introduces the term data pseudonymization [7], which refers to the processing of the personal data in such a way that the data can no longer be linked to the subject through any additional information that shall be kept separately in any case and be protected.

The data controller is authorized to manage the subject's data, whereas the data processor can apply any kind of (authorized) processing on the data. Emphasis is given on the data subject (i.e., patient) who is the core of the data protection regulation and lies at the intersection of these two roles with respect to a final role, the one of the DPO who is responsible for supervising the compliance of all the operations that involve the sharing and processing of personal data [1,7,17]. In fact, according to the Articles 16–22 of GDPR, the data subject has the right to (i) be informed about the actions and participate in any automated decision that involves his data (ii) access, rectify, and erase (i.e., the right to be forgotten) his data, (iii) object and restrict the processing of his data, and (iv) request his data (i.e., the right for data portability). Furthermore, according to Article 5 of the GDPR, the personal data shall be (i) processed with respect to the legal regulations (posed by Article 6) and in a secure way, (ii) adequate, (iii) accurate, and (iv) kept in such a form that they can be identified. As a matter of fact, the contents of Article 5 are similar with those from Article 12 in the former Data Protection Directive.

Chapter 4 of GDPR [7] poses crucial obligations on the actions of the data controller and the data processor. More specifically, the data controllers and the data processors shall prepare codes of conduct with respect to (i) the collection of personal data, (ii) the pseudonymization of personal data (data protection by design), (iii) the legitimate interests pursued by the data controllers, (iv) the transparency and fairness in data processing, and (v) data minimization (data protection by default), among many others [1,7,17]. According to Article 14 of the GDPR, the data controller is responsible for the implementation of technical and organizational measures to ensure that the processing of data is conducted with respect to the obligations posed by the GDPR. The same stands for the joint controller, i.e., one or more controllers that jointly determine the purposes and means of processing [7,17]. To conduct medical research, the data controllers must collect a variety of documents, varying from signed consent forms, the purpose of processing, and the contact details of the corresponding DPOs to data protection impact assessment (DPIA) reports and data protection guarantees. In addition, according to Article 25, the controllers shall take not only appropriate technical and organizational measures, e.g., pseudonymization, but also appropriate measures for ensuring that only the personal data that are necessary for each specific purpose will be processed, toward the establishment of safeguards for the individuals' rights. These fundamental principles of the GDPR are referred to as data protection by design and by default, respectively.

The data processor acts on behalf of the data controller regarding the processing of personal data with respect of course to the obligations posed by the GDPR. The processor is not allowed to engage additional processors without written consents and must sign a binding contract, with regard to the corresponding data controller, which states that the processor must [7] (i) process the data according to the instructions he receives from the data controller, (ii) ensure that other data processors are compliant, (iii) be compliant with Article 32, (iv) engage another processor under written consents (also in digital form), (v) assist the data controller to be compliant with the obligations according to Articles 32–36, (vi) make available to the controller any kind of information that is required to demonstrate the GDPR compliance of the processing operations in audits, and (vii) delete or return personal data at the choice of the controller or the Member State law. Moreover, a data processor can be considered as a data controller in case he is the one who determines the purposes and means of processing, with respect to that data processing.

DPO can be designated by a data controller or a data processor when the processing actions of the controller or the processor require regular and systematic monitoring of the data subjects, especially in large-scale studies where the rights of the subjects need to be preserved. The DPO is a physical or legal entity who is well qualified with expertise on data protection laws and practices. According to Article 39, the DPO is responsible for (i) informing and advising the data controller or the data processor, as well as the employees who participate in any processing of personal data, (ii) monitoring the compliance of the processing operations according to the obligations posed by GDPR, (iii) supporting the DPIA process, and (iv) cooperating with supervised authorities on issues related to the processing of personal data and consult them when necessary. The contact details of the DPO are published from the data controller or the processor. The data subjects can contact the DPO in case they want to be informed about the processing of their personal data and whether the processing fulfills their rights under the GDPR. The DPO reports only to the supervised authorities that are above the level of the data controller or the data processor. The latter support the DPO with any information that is referred to Article 39 but are not authorized to dismiss or penalize the DPO's actions in any way (Article 38). On the other hand, the DPO shall take into account the nature, scope, context, and purposes of processing so as to assess any risks that are associated with the processing of personal data.

GDPR repealed the Data Protection Directive 95/46/EC, highlighting the impact of digital technology in the rights of the data subjects and extending current laws to deal with such issues. The GDPR also applies to all matters regarding the protection of rights and freedoms of data subjects according to those set under the same objective with the Directive 2002/58/EC, which concerns the processing of personal data in the electronic communication sector [1,7]. Several data protection principles including the data protection by design and by default are included in the binding of corporate rules set by supervised authorities. The data subjects are now the real owners of their data and have the right for data portability and the right to be forgotten. Each federated platform shall adopt and integrate data protection principles and safeguard rights for individuals according to the data protection by design and by default.

4.4.3 The Lisbon Treaty and its impact in data protection law development

A treaty is defined as a binding agreement between the EU Member States. Under the treaties, the European Commission can propose legislations that can be applied only to those Member States whose policy area is cited in a treaty. Until now, the EU has developed eight main Treaties. The first and foremost Treaty is the Lisbon Treaty, which was initially introduced in 2007 and came into force 2 years later, in December 2009. The Lisbon Treaty is a well-known Treaty, which directly affects the development of all the EU data protection laws and policies. The most important aspect of the Lisbon Treaty is the fact that it enhances the citizens' involvement in all the decision-making processes with the purpose of increasing the latter's transparency and efficiency toward a more democratic EU. The Lisbon Treaty introduces the proven Community approach, as the basis for the development of any protection legislation within the EU.

The Lisbon Treaty emphasizes on the foundations of Democracy and toward this way promotes three fundamental principles, namely [18] (i) the democratic equality, (ii) the representative democracy, and (iii) the participatory democracy. The first two principles are well known by the authorities of any democratic society, whereas the third is not often taken into consideration by the existing EU policies. The third principle is the core of the Treaty, which allows more than 1 million EU citizens to participate in law development by inviting the Commission to submit a legislative proposal for a particular legal act of interest, according to the European's Citizen's Initiative [19]. Any act of privacy and data protection shall fulfill the guidelines posed by the EU Treaties and the EU Charter of Fundamental Rights [20]. It is notable though that the entry of Lisbon Treaty into force in 2009 legally equalizes both the EU Treaties and the EU Charter of Fundamental Rights [20]. This legal equalization is a crucial step toward the development of new data protection regulations, which consider the citizens' rights during the processing of data.

In an attempt to better comprehend this legal equation, great emphasis shall be given on the Treaty on the Functioning of the European Union (TFEU) of 2007 [21], also known as the Treaty of Rome, which is one of the two primary EU Treaties along with the Treaty on European Union (TEU) [22]. In the TFEU, Article 16 states that every citizen has the right to the protection of his or her personal data. Most importantly, it states that the European Parliament and the Council can enable the sharing of an individual's data only for those activities that are in line with the scope of the Union law and the requirements that permit the sharing of such data. Thus, the development of laws regarding the protection of personal data during processing is linked through the Lisbon Treaty to the EU Charter of Fundamental Rights. Through this relation, the Directive 95/46/EC, for example, has reached the level of EU primary law. In addition, the GDPR has already defined the right to data protection, which is in accordance with these rules. Through the Lisbon Treaty, the Commission has the role of introducing new legislations to be adopted by the Parliament and the Council and executed by the Court of Justice with respect to the proven Community approach [18].

In brief, the EU Charter of Fundamental Rights of 2000 states that every EU citizen has [20] (i) the right to freedom and security, (ii) the right for his/her privacy and family life, (iii) the right to the protection of his or her personal data, (iv) the freedom of assembly and of association at all levels, in particular in political matters, and (v) the right to education, among many others. Moreover, every citizen has (i) the right to vote and stand as candidate at elections to the European Parliament, (ii) the right to vote and stand as candidate at municipal elections, (iii) the right of access to documents of the institutions, bodies, offices, and agencies of the Union, (iv) the freedom of movement and of residence freely within the territory of the Member States, and (v) the right for Diplomatic and consular protection in third countries where the Member State of which he or she is a citizen is not represented. The Lisbon Treaty binds the Member States to propose legislations related to the rights, freedoms, and principles, on the basis of the EU Charter of Fundamental Rights, which has been given, according to Article 6(1) of the TEU, the same legal value as the Treaties. This is a tremendous impact of the Lisbon Treaty toward law development.

4.5 Data protection regulations across the United States

4.5.1 The Federal Trade Commission Act

The FTC Act was initiated in 1914 with the purpose of preventing unfair acts or practices that affect commerce [13]. The FTC Act empowers the US Congress to (i) prevent unfair acts and unfair methods of competition affecting commerce, (ii) establish requirements and develop safeguards to prevent such acts, (iii) gather information and conduct further investigations regarding the bodies that are engaged in commerce, and (v) submit related reports and recommendations to the US Congress. The FTC Act was inspired by the Sherman Antitrust Act of 1890, which subsequently aimed at protecting the public from the failure of the market [23]. In brief, the Sherman Antitrust Act posed significant barriers in artificial raising of prices and legal prosecution against (i) any contracts or conspiracies against trade or commerce and (ii) persons who attempt to monopolize any part of the trade or commerce across the US States. Similar to the Sherman Act, the FTC Act is much more consumer-oriented and poses significant data protection regulations regarding the protection of the consumers' personal data, which in turn constitute the fundamental basis of the US data protection law development.

In 1914, the FTC Act established FTC [13] as an independent US law enforcement agency whose main goal is to protect the consumers from unfair acts (e.g., deceits) according to Section 5 of the FTC Act, as well as enhance the competition across out-of-border economic fields. The FTC uses privacy protection measures and security programs to hamper law violations and penalize companies with unlawful behavior. These privacy protection measures include the tracking and deletion of any illegally obtained consumer information, the hosting of public workshops for privacy protection, and the cooperation with international partners for providing robust transparency to the consumers. Regarding the medical sector, the FTC has established safeguards providing (i) security guidelines and interactive tools for health application developers to check for federal law compliance, (ii) the FTC's Health Breach Notification Rule under which companies with security breach shall inform (a) the subjects whose information has been breached, (b) the media, and (c) the FTC, (iii) additional guidelines for companies in case of data breach, (iv) guidelines in case of medical identity theft, (v) advice for businesses that are to be involved with the IoT, and (vi) guides for peer-to-peer file sharing in businesses involving sensitive personal data, among others.

In 2010, the FTC issued the Health Breach Notification Rule [24], which provides guidelines and safeguards that are related to companies affected by a security breach. The FTC periodically posts a list of breaches to inform the companies regarding potential information leaks along with a related brochure that serves as a template for dealing with such issues. For breaches affecting health information of 500 or more individuals, FTC must be informed and in that case the media is involved as well. In cases where the number of affected individuals is less than 500, FTC tracks the breach events and reports them in an annual report. It is notable though the Health Breach Notification Rule applies to only those companies or organizations whose health information is not secured through technologies.[3] In addition, the Rule does not apply to those companies or organizations that are not subject to the HIPAA. In other case, the HIPAA-covered entities and business associates must comply with the HIPAA Breach Notification Rule [24].

The FTC has also launched two primary international privacy frameworks: (i) the EU—US Privacy Shield Framework and (ii) the Asia—Pacific Economic Cooperation (APEC) Cross-Border Privacy Rules (CPBR) System. The EU—US Privacy Shield [4,25] bridges the gap between the privacy protection regulations of the EU and the United States to enhance the protection of personal data transfers between the two continents. The Privacy Shield Principles were administered by the US Department of Commerce (DOC) in cooperation with the European Commission, as well as with industry experts and stakeholders, to provide a trustworthy mechanism for personal data transfers to the United States from the EU while ensuring that the EU citizens are protected by the safeguards posed by the EU when the processing of their transferred personal data takes place to non-EU

[3] These technologies are specified by the US Department of Health and Human Services.

countries (see Section 4.6 for the intersection with the U.S. law). Every US organization wishing to receive personal data from the EU must be compliant with the Privacy Shield Principles. In any case, the Principles affect neither the application of provisions under the Directive 96/46/EC (superseded now by GDPR) nor the regulations that apply under the US privacy law. Since 2016, FTC has an active role in privacy law enforcement under the EU–US Privacy Shield by reporting three enforcement actions and issuing the Swiss–US Privacy Shield Framework [26], in 2017, as part of the EU–US Privacy Shield, to enhance the protection of Swiss consumers in transatlantic commerce.

On the other hand, the APEC CPBR [27] introduces a privacy framework for secure information flow in the Asia Pacific region where the FTC serves as the privacy enforcement authority that monitors the privacy protection violations. The Framework involves the 21 APEC Economies and poses a set of 9 guidelines (i.e., Principles) for assisting the APEC Economies to develop safeguards for privacy protection during the information flow of sensitive data between the APEC members and the US [27]. Its main purpose is to (i) protect personal data from fraud and further misuse (enforce information privacy), (ii) enable the collection and processing of personal information of the APEC members from organizations, and (iii) assist privacy enforcement agencies. Any company or organization fulfilling the set of nine guidelines is considered as APEC CPBR system compliant.

Till 2017, FTC has brought more than 500 enforcement actions against well-established international companies, including Google, Microsoft, Facebook, and Twitter, among many others. Moreover, FTC has reported over 60 companies since 2002 for deceptive actions and practices against consumers regarding insecure storage in cloud databases and safety holes, which pose serious security issues [28,29]. The range of privacy issues that FTC has currently brought in the light includes (i) spam, (ii) social networking, (iii) advertising, (iv) spyware, (v) children's privacy issues, and (vi) file sharing. As far as the international privacy frameworks are concerned, the FTC has brought 4 actions under the APEC CPBR, 3 actions under the Privacy Shield (39 under the former US–EU Safe Harbor Program) [28]. The FTC's international impact in privacy law enforcement confirms its characterization as the "primary statue" of the US law.

4.5.2 The Health Insurance Portability and Accountability Act

HIPAA of 1996, Public Law 104–191, aims to [8] (i) improve the portability and accountability of health insurance coverage, (ii) protect the healthcare coverage of individuals, (iii) embrace the sharing of certain patient administrative data for promoting the healthcare industry, and (iv) prevent fraud and abuse in the healthcare industry. Right after the HIPAA took effect in August 1996, the US Department of HHS considering the great impact that the technological advances have in the violation of personal health information issued the first two HIPAA Rules, namely the HIPAA Privacy Rule (December 2000), which was modified in August 2002, and the HIPAA Security Rule (February 2003). The former poses national standards for the protection of individually identifiable health information to three types of covered entities: health plans (those that provide or pay the cost of medical care), healthcare providers (providers of medical or health services), and healthcare clearinghouses (e.g., health information systems), whereas the latter poses national standards for the protection of the (i) confidentiality, (ii) integrity, and (iii) availability of protected health information (PHI).

PHI includes information related to [8] (i) the individual's medical history (past, current, or future), (ii) individual healthcare provisions, and (iii) any kind of payment related to these provisions. In addition, the PHI includes a detailed list of 18 personal identifiers that can be used to match an individual either directly or indirectly. In brief, the list includes various personal identifiers, such as names, email addresses, telephone numbers, bank account numbers, social security numbers, license numbers, individual photos, IP addresses, biometric identifiers, and any other unique numbers of identifiers, among others [8]. This list is similar with the definition of the term "personal data" that was initially described in the Directive 95/46/EC and exists in GDPR as well. All the covered entities are responsible for the protection of the PHI under the HIPAA Privacy Rule.

The HIPAA has also published two additional but crucial Rules: (i) the Breach Notification Rule and (ii) the Enforcement Rule. The HIPAA Breach Notification Rule is similar with the FTC's Health Breach Notification Rule in the case of companies or organizations that are subject to the HIPAA (i.e., one of the HIPAA-covered entities) [24]. The Enforcement Rule includes provisions regarding the HIPAA compliance and poses audits and financial penalties in the case of the violation of the HIPAA Rules. More specifically, in October 2009, the HIPAA launched the Health Information Technology for Economic and Clinical Health (HITECH) Act [30], which establishes four categories of violations according to the impact and level of the HIPAA violation. In 2013, the US Department of HHS, under the HIPAA, published the Final Rule (Omnibus Rule) that implements a number of provisions mentioned in the HITECH Act to strengthen the privacy and security protection of health-related information flow.

In fact, the Omnibus Rule implements amendments under the HITECH Act to enhance the safeguards for PHI. In addition, the Final Rule modifies the HIPAA Privacy Rule to consider for the protection of genetic information by implementing Section 105 of Title I of the Genetic Information Nondiscrimination Act of 2008 [31]. The modifications apply also the HIPAA Security, Breach Notification, and Enforcement Rules to improve their workability and effectiveness. The HIPAA Rules require that all the covered entities, companies, and organizations shall [28]: (i) develop policies and security measures for ensuring the protection of sensitive data, (ii) conduct risk analysis for compliance accountability, and (iii) minimize access and develop safeguards for PHI. A deeper look into the four HIPAA Security Rules is presented in the next section.

4.5.3 Legislation and HIPAA security rules

The HIPAA Privacy Rule [32] is a fundamental legislative regulation that establishes national standards for the protection of healthcare electronic transactions including medical records and other personal health information that are conducted by healthcare providers. The Privacy Rule that came into force in 2003 focuses on the development of standards for the protection of individually identifiable health information, which is referred to, in the law, as PHI. Organizations that are subject to the Privacy Rule are referred to as covered entities. These covered entities include [32] (i) small- and large-scale health plans, (ii) healthcare providers who electronically transmit sensitive health information through all mediums (electronic, paper, and oral), as well as those who do not use electronic health records, and (iii) healthcare clearinghouses, i.e., entities that process information they receive from other entities into standard formats. The patients, under the Privacy Rule, can anytime request to examine and obtain copies of his or her medical files, as well as request corrections regarding their personal information, as well as perform any further editing process. The use and disclosure of PHI that is needed for patient care is one of the major issues that the Privacy Rule introduces, highlighting the need to train workforce and implement appropriate safeguards by the companies and organizations that are subject to the Privacy Rule. The final update of the Rule that came into force in 2002, aimed at clarifying several of the Rule's provisions to address unintended negative effects of the Rule on the healthcare quality and access to health-related data, under the Standards for privacy of identifiable health information.

The HIPAA Security Rule [33] sets national standards for the protection of PHI that is created, received, used, or maintained by a covered entity. The Security Rule poses safeguards that are addressed to covered entities to protect the PHI's: (i) confidentiality (PHI is not available or disclosed to unauthorized entities), (ii) integrity (PHI is not modified by unauthorized entities), and (iii) availability (PHI is accessible and useable only by an authorized covered entity). To evaluate whether the security rules of a covered entity are fulfilled or not, the HHS Office for Civil Rights and the Office of the National Coordinator for Health Information Technology have launched a Security Risk Assessment (SRA) tool [34]. The SRA tool is addressed to all the healthcare covered entities to assist them during the risk assessment procedure. The latter tracks down areas within a covered entity that are highly likely to present breach of PHI. The NIST HIPAA Security Toolkit Application is an additional tool that was launched under the HIPAA Security Rule to inform the covered entities to better understand the requirements posed by the Security Rule and effectively address those requirements according to the covered entity, as well as evaluate the requirements on the covered entity's environment. Several conferences have been held by HIPAA since 2010 toward the safeguarding of healthcare information to inform the audience about the risk analysis requirements.

The HIPAA Breach Notification Rule [24] is the corresponding FTC's Health Breach Notification Rule in the case of HIPAA covered entities, e.g., companies or organizations that are subject to HIPAA. The Breach Notification Rule requires that all covered entities that have been compromised by a security breach must assess the risk of the breach and provide notifications related to (i) the nature of the PHI, (ii) the person who impermissibly committed this action or the one whom the disclosure was made to, and (iii) the type of breach (i.e., whether the data were viewed or acquired). According to HIPAA, a breach is defined as any unauthorized attempt to compromise the security and privacy of an individual's PHI that is against the requirements that are set by the HIPAA Privacy Rule, with three exceptions in case of (i) unintentional acquisition, access, or use of the PHI by workforce members or authorized personnel, (ii) inadvertent disclosure of PHI, and (iii) good faith belief from the covered entity's side that the unauthorized person would not have been able to retain the PHI. The covered entities usually provide probabilities regarding the level of the data breach and must notify both the HSS Secretary in charge and the individuals whose PHI is affected. The media are also involved in certain circumstances, i.e., when the breach involves more than 500 individuals. A list with thousands of hacking incidents, thefts, and cases regarding unauthorized access that are currently under investigation by the Office of Civil Rights can be found in Ref. [31] where the reader can meet up-to-date breach cases that sometimes involve more than 1 million affected individuals.

The HIPAA Enforcement Rule [35] imposes provisions and audits regarding the penalties of companies or organizations that violate the HIPAA Security Rules. The HITECH Act sets four categories that quantify the level of violations against the HIPAA Rules (e.g., the Privacy, Security, and Breach Notification Rules) [35]: (i) Did not know, (ii) Reasonable Cause, (iii) Willful Neglect—Corrected, and (iv) Willful Neglect—Not Corrected, along with four types of financial penalties (that are defined for each category), with an annual maximum penalty of 1.5 million US dollars per category (penalties might get fined up to 50,000 US dollars per day). If a company or organization commits more than one violation, the individual penalties per violation are summed up until the violations are resolved by that company or organization.

4.6 Overlapping between EU and US protection laws

While both the US and the EU envisage to enhance the privacy protection of their citizens, the US data protection strategy is a sectoral-oriented mix of legislation, regulation, and self-regulation, which differs from the strategy that is followed by the EU Commission. Right after the EU's Data Protection Directive 95/46/EC took effect in 1998, the transfer of personal data from EU individuals to non-EU countries, which did not comply with the EU "adequate" standards of personal data protection, was prohibited. At that time, the restriction included the United States, a fact that posed serious threats in the transatlantic information flow which in turn created confusion in the EU–US parties as such a restriction could seriously harm the trade and investment relationships between the two parties. After continuous negotiations that took place in 2000 between the US DOC and the European Commission, the two parties agreed to bridge this gap by introducing a new Framework that would allow the US entities to meet the data protection requirements posed by the Data Protection Directive and thus gain access to EU data flows.

This Framework was the well-known Safe Harbor Framework that was introduced by the DOC in July 2000 [16]. Under the Safe Harbor Framework, the DOC published seven Safe Harbor Privacy Principles [16] to meet the EU's data protection requirements. At the same year, the Safe Harbor Privacy Principles were recognized by the European Commission, with some restrictions though. These restrictions limited the flow of personal data in cases where national security, public interest, or law enforcement issues were of great matter. The Safe Harbor Privacy Principles stated that each organization shall [16] (i) notify the individuals regarding the processing of their data, (ii) respect the individuals' decisions regarding the use and disclosure of their personal data, especially in cases where sensitive civilian data are involved, (iii) apply the two previous Principles to any third parties that are involved in any manipulation of personal data, (iv) take appropriate security measures to reduce the misuse and breach of personal data, (v) evaluate data quality measures, such as the relevance of personal information according to the purposes of processing, (vi) ensure that the individuals can anytime access their personal data and perform any changes they wish to, and (vii) include effective mechanisms for privacy law enforcement when any of the previous Principles is not followed. It is notable though that the participation in the Safe Harbor was open to those entities that were already compliant with the FTC Act, which enhances the protection law. Eventually, these Principles can be seen as a set of legislation parts that lie in the intersection between EU and US protection laws.

However, a critical report of the European Commission, in 2004, changed things [4]. The Commission noticed more than 400 US companies that were self-certified as Safe Harbor compliant without, however, establishing any transparency regarding the fulfillment of the Safe Harbor requirements. Similar reports in 2008 and 2013 evaluated several concerns regarding hundreds of companies that were falsely claiming or avoiding to be Safe Harbor compliant, including large companies, as well as companies ignoring requests from the citizens concerning the manipulation of their personal data. Two cases that came up in 2015 were enough for the Court of Justice of the European Union (CJEU) to immediately invalidate the Safe Harbor. The main reason was an investigation into the Facebook's data protection actions regarding a complaint that was brought to the Irish Data Protection Authority (DPA) by an Austrian citizen in 2015. The citizen claimed that his personal data were transferred from Facebook's EU-based servers to its servers located in the United States. CJEU, by taking into consideration the violation of the requirements posed by the Article 25 of the EU Data Protection Directive and the EU's Charter of Fundamental Rights, invalidated Safe Harbor. This happened in the light of the well-known "Snowden leaks" regarding global surveillance disclosures run by NSA in cooperation with governments and colossal companies based in the United States.

In the late 2013, US and EU discussions were initiated regarding the update of the Safe Harbor in the light of the criticisms and allegations of other secret intelligence surveillance programs of the United States in cooperation with EU involving large companies. In February 2016, the two parties launched the EU–US Privacy Shield [25], which completely replaced the Safe Harbor. Till then, the Privacy Shield continues to allow companies to transfer EU citizens' personal data to the United States with respect to additional requirements outlined by the European Court of Justice. The most important part of the Privacy Shield is the fact that it requires, in contrast to the Safety Harbor, the commitment of the US national security authorities and the governmental agencies to provide reports regarding the protection of the EU citizens. In fact, the Privacy Shield addresses the concerns of the CJEU, highlighting four fundamental pillars [25]: (i) the enhanced commitment of the US authorities concerning the purpose of processing data flows from the EU, (ii) the enhanced commitment of the DOC by continuously monitoring the compliance Privacy Shield through the FTC that has an active role on privacy law enforcement, (iii) the establishment of safeguards and transparency from the side of the national intelligence agencies (from both parties) that must provide adequate reports and reviews concerning the protection of personal data, and finally (iv) the enhancement of the citizens' involvement in any operation that involves the processing of his or her personal data.

The EU citizens can anytime refer to the European DPAs having any complaints regarding the processing of his or her personal data. The EU DPAs must cooperate with the FTC to resolve any complaints. The Privacy Shield states seven fundamental Principles, which state that the organizations must [25]: (i) inform the individuals about the purpose of processing their personal data and their rights for data access and further manipulation (Notice), (ii) let the individual decide whether he or she wishes to provide sensitive data (Choice), (iii) ensure and account for the transparency of any actions that are performed by any third parties (Accountability), (iv) develop security measures against data misuse and unauthorized access (Security), (v) ensure the quality of the data according to the purposes to which they have been collected (Data integrity), (vi) provide the individuals with the ability to access and perform any changes on their data (Access), and (vii) account for effective privacy law enforcement mechanisms (Recourse, Reinforcement, and Liability). These Principles lie in the intersection between the EU–US data protection laws. Furthermore, in 2012, the DOC and the European Commission and Swiss Administration launched the Swiss–US Privacy Shield Framework [26] considering for the development of a mechanism that would enable the transatlantic data flow of personal information from Swiss citizens with the purpose of enhancing the transatlantic commerce without of course affecting the corresponding laws of both parties. Although the Switzerland's Federal Act on Data Protection and the EU's data protection regulations are similar, the Swiss–US Privacy Shield Framework is based on the Principles of the Privacy Shield Framework to benefit from the *Switzerland's recognition of adequacy* [26].

The EU–US Privacy Rule is currently under the effect of more effective provisions in the light of the GDPR that came into force in May 2018. The GDPR sets higher standards for data protection than the former Directive. For example, the definition given for the term "processing data" under the GDPR is broad, including a wide range of activities, such as the collection, storage, and modification of personal data, and thus must be considered carefully in any provisions that will be made by the Privacy Shield. The GDPR, like the Data Protection Directive, allows information data flows to countries outside the EU, only if these countries are GDPR compliant, fulfills the "adequate" data protection requirements, and of course meets the seven Privacy Shield Principles. The adequate requirements consider for the decision of the European Commission the model clauses (i.e., contractual clauses that are liable, respect the subjects' rights, and cannot be modified) and consents of obligations [36]. The EU and US officials state that the Privacy Shield is now a stronger data protection initiative.

The Working Party (a team of EU DPAs under Article 29), however, still expresses concerns including (i) the insufficient protection during the onward transfer of personal data to third countries, (ii) the lack of mechanisms to track the removal of personal data, and (iii) the considerations regarding the adjustment of the Privacy Shield requirements according to the GDPR. To boost these concerns, the EU and US authorities are working toward the Data Privacy and Protection Agreement, the so-called Umbrella Agreement, for enhancing privacy law enforcement (in effect since 2016) [37] and the US Judicial Redress Act, which replaces the US Privacy Act of 1974 to EU citizens [38].

4.7 Global initiatives

Apart from the previous data protection initiatives, data protection remains an out-of-border concern that requires a broader, international cooperation of regional law enforcement agencies. This vision, however, is difficult due to the heterogeneity of the regional data protection legislations. Toward this, four major data protection international initiatives have been launched, namely (i) the United Nation's resolution 68/167, (ii) the International Data Protection Commissioner's (IDPC) Initiative, (iii) the Organisation for Economic Co-operation and Development (OECD) Guidelines on the Protection of Privacy and Transborder Flows of Personal Data, and (iv) the Council of Europe Convention (CoE Convention) 108. In this section, each one of these initiatives will be briefly discussed. The strengths and weaknesses of these initiatives along with future advances will be discussed in Section 4.8, under the context of a complete data protection framework.

Since 2015, the United Nations has an active role in data protection [39]. On the UN's General Assembly resolution 68/167 (70th plenary meeting in December 2013), the General Assembly expressed concerns regarding the negative impact of the rapid technological advancements in the privacy of personal data. In fact, they highlighted that the digital advancements in modern surveillance methods and new interception approaches pose serious threats in the human rights. A general call was sent to the Member States to take measures against such practices and actions, recalling Article 17 of the International Covenant on Civil and Political Rights, which states that [40] (i) no one has the right to interfere with the privacy, family, and home of any individual, as well as obscure the individual's honor and reputation and (ii) every individual has the right to be protected against such harmful attempts by the Union law. The human rights for data protection and privacy are also presented in Article 12 of the Universal Declaration of Human Rights [41]. This means that all 167 Member States of the United Nations must respect the privacy law and consider for the protection of public health data and reaffirm the right to privacy of their citizens. Apart from the call, the General Assembly requested additional monitoring for unauthorized surveillance and collection of personal data from the High Commissioner for Human Rights. In addition, the UN Human Rights Council appointed the Special Rapporteur [42], a natural person who reports on right to privacy, by (i) collecting information regarding the privacy of data from the States and any public or private organization located in the States, (ii) promoting data protection principles, and (iii) reporting any alleged violations and obstacles that he identifies in the States.

IDPC's Initiative is a promising framework that was developed by the DPAs of the UN States. The most important part of this initiative was the Montreux Declaration of 2006 [43], according to which the Data Protection Commissioners agree to (i) coordinate their law enforcement authorities for the development of common standards for ensuring secure exchange of personal information flows, (ii) assist countries that do not have any similar authorities, (iii) collaborate with the rest of the Commissioners and with DPOs of organizations, and (iv) promote the initiative's vision on nongovernmental organizations. Toward a global realization of these agreements, the Commissioners highlighted a set of 11 principles with respect to the former EU Data Protection Directive, APEC Privacy Framework and guidelines received from the UN. These principles are [43] (i) lawful processing, (ii) accuracy, (iii) transparency, (iv) purpose specification, (v) transparency, (vi) data security, (vii) responsibility, (viii) nondiscrimination, (ix) independent supervision, (x) adequate level of protection in case of processing from third parties, and (xi) individual participation (right for access).

In 2013, OECD issued the Guidelines on the Protection of Privacy and Transborder Flows of Personal Data [44]. These Guidelines were an updated version of the original Guidelines that were initially issued in 1980 taking into consideration the rapid technological advancements of our era. The Guidelines apply to the 34 OECD States and form the backbone of several regional data protection laws. These Guidelines can be summarized in eight principles that involve the [44] (i) limited collection of personal data (according to the purposes of processing), (ii) quality of personal data (e.g., relevant, accurate), (iii) purposes of the collection and processing of personal data, (iv) limited use of personal data (under the individual's or authorized approval), (v) establishment of security safeguards (e.g., risk assessment), (vi) openness of actions and policies with respect to personal data, and most importantly (vii) rights of individuals to obtain and further manipulate their personal data. Since 2016, 33 out of 34 OECD States have implemented these Guidelines, except of the United States, which follows a sectoral-oriented approach rather than a common protection law.

The CoE Convention 108 [11] is another promising international data protection initiative. Despite its "European nature," which came into force since 1981, it consists of 47 EU and non-EU Member States (with Uruguay being the first non-EU country to adopt the Convention in 2013 and four Asian countries being under evaluation since 2016). The CoE Convention Principles are applied on any kind of automated processing on personal data both in the public and private sectors. The Convention highlights once more (i) the need for qualified personal data in terms of lawful processing, adequacy, relevance, readable form, and accuracy, (ii) the establishment of safeguards for the subject's privacy, (iii) the crucial need for appropriate security measures in cross-border data flows and also offer its States the ability to provide extended protection without any limitation from the Convention Principles. The most interesting and unique aspect of CoE Convention is the fact that the agreement between its Member States and any countries wishing to join this global initiative is binding on signatories [39].

4.8 Toward a more complete data protection framework

The harmonization of national data protection laws is a great challenge. The four major international data protection initiatives are promising toward the establishment of a global data protection framework.

The UN initiative has a broad impact in more than 150 nations with a long history on preserving the human rights and freedom. This enhances the impact of any data protection legislation that is about to be issued by the UN General Assembly. However, a proposed legislation not only needs to be translated to different languages but also needs to be appropriately adjusted to be easily comprehended by the participants. This comes from the fact that the regulations that are often introduced by the UN are straightforward and could be characterized as "high level," for several UN Member States. In addition, the lack of democratic mechanisms from the major UN members, i.e., the United States, Russia, France, the United Kingdom, and China, hampers the adoption of a global data protection legislation due to the democratic heterogeneities that are present among these Members. The fact that these Members hold the "veto power" over the rest of the Member States sometimes poses significant obstacles toward the realization of this vision [39]. Furthermore, it is difficult for the UN to deal with issues regarding data protection threats that arise from non-UN Member States. This is of course a general problem that hampers the vision of any international data protection initiative.

IDPC's initiative consists of a group of International Data Protection and Privacy Commissioners with significant broad impact and strong universal profile. The Commissioners are experts on the development of data protection laws and are part of a convention that expresses a promising will to harmonize data protection laws. The open-access nature of the initiative promotes the impact of more effective data protection legislations of nations that wish to be part of this initiative. Besides, the initiative has been already requested from the Council of Europe to invite non-EU Member States with appropriate data protection regulations, as Members of the convention. This call involves and promotes the vision of the Council of Europe Convention 108. However, the initiative suffers from the nonbinding nature, i.e., the lack of binding on signatories. In addition, the lack of a formal and a structural format of the initiative's context introduces difficulties during the incorporation of future provisions.

The Privacy Guidelines issued by OECD are widely acceptable and provide integrated tools to support long-term peace and thus ensure the privacy and rights of the individuals. This is a key strength of the OECD Guidelines along with the support they offer to regional and local entities for conflict management and for enhancing their data protection strategy. This enables the strategic alignment of national policies that in turn promotes the development of integrated and joint approaches for a common data protection framework. Furthermore, the Guidelines include Principles that are widely acceptable and effectively target on bridging the heterogeneity of the data protection regulations. Another important key strength is the fact that the international factors have power over the regional and local structures that promotes peace and in fact creates "pressure for peace" [44]. This is a more generic scope but directly affects the factors of privacy and protection. On the other hand, this creates problems in the relationships among groups of nations that share different strategical plans with others nations.[4] In addition, the Guidelines suffer from the nonbinding nature, as well as the lack of any data minimization mechanisms for reducing the risk of personal data breach. Moreover, OECD is the initiative with the least number of participating nations, including mostly the developed ones.

The Council of Europe Data Protection Convention (CoE Convention) exhibits the least weaknesses among the four major international data protection initiatives. The CoE Convention has been recognized by the International Data Protection and Privacy Commissionaires as the best global data protection framework. This arises from the fact that the Convention is the only initiative that involves the binding agreement of its Members. This means that the Members are bound to adopt the data protection Principles and collaborate with the rest of the Members through the conventions open-process scope. Transparency is thus the key strength along with its wide binding acceptance from 47 countries. Apart from its European nature, the CoE Convention adopts key data protection principles that overcome several privacy challenges that are expressed by the rest of the initiatives. The main drawback of the Convention is the inability of its Principles to be adjusted into strictly

---

[4] An example of a conflict is the different attitude of nations toward Syria's government crackdown in 2011 [45].

Table 4.1 Quantification of the strengths and weaknesses of the four major international data protection initiatives across different data protection characteristics.

Characteristic	UN resolution 68/167	OECD Guidelines	CoE Convention 108	IDPC
Broad impact	Strong	Strong	Strong	Weak
Number of participants	Strong	Weak	Moderate	Weak
Binding on signatures	Weak	Weak	Strong	Weak
Managing out-of-border data flow	Moderate	Weak	Strong	Moderate
Accounting for data minimization	Weak	Weak	Strong	Weak
Principles context	Moderate	Strong	Strong	Strong
Ability to adjust in strict legislations	Weak	Moderate	Weak	Weak
Reducing the trade-off between data protection and surveillance	Strong	Weak	Strong	Weak
Scalability	Moderate	Moderate	Strong	Weak
Interoperability	Moderate	Moderate	Moderate	Moderate
Protection in case of data breach	Moderate	Moderate	Strong	Moderate

heterogeneous data protection environments, such as the United States's sectoral-oriented legislation that follows a much different strategical approach than the Convention's vision.

The strengths and weaknesses of the aforementioned data protection initiatives are summarized in Table 4.1, for several data protection characteristics that lie in the backbone of any data protection regulation.

4.9 Conclusions

Remarkable regional data protection initiatives have been launched up to now, including GDPR that currently superseded the Data Protection Directive 95/46/EC in Europe, FTC Act, and the HIPAA in the United States, as well as partially international initiatives, such as the EU–US Privacy Shield, APEC Privacy Framework, and international frameworks, including the United Nation's resolution 68/167, IDPC's Initiative, OECD Guidelines on the Protection of Privacy and Transborder Flows of Personal Data, and the CoE Convention 108. There is no doubt that the supervised authorities of the world's nations share common goals toward the protection of their citizens' personal data and privacy. Healthcare is just one aspect of these large-scale initiatives, most of which are currently under provision and additional adjustments.

GDPR is an ambitious and powerful data protection legislation that is based on the "data protection by design and by default" principle, which introduces the new principles of data pseudonymization and data minimization. In contrast with the former Data Protection Directive 95/46/EC, GDPR requires that all entities who wish to process personal data must use only a portion of them according to the purposes of processing, as well as take appropriate security measures to protect the subject's identity, such as data pseudonymization. The great thing with the GDPR is that it has been carefully built on EU legislations and follows the same path with other national data protection laws. It is notable that many international governments including those of Australia, South Africa, and Japan have several similarities with GDPR [46]. More specifically, GDPR exhibits similarities with (i) the Australian Privacy Act of 1988 [47] regarding the privacy by design principle and data breach notifications, (ii) the South African Personal Information Act of 2013 [48] under which the involved authorities have expressed their will to follow a common route with GDPR, and (iii) the updated Japanese Act on the Protection of Personal Information of 2017 [49] regarding data pseudonymization and the requirement to keep records of processing (to whom the data were transferred, from whom the data were obtained, how the data were transferred). So, the question still remains: Does the GDPR exhibit any similarities with the US laws?

The United States has adopted a sectoral-based strategy for privacy law development that involves a mix of legislation, regulation, and self-regulation mechanisms that differs from the strategy followed by the European Commission. The US privacy law enforcement is rigorous, and this is proven by the FTC that serves as the "primary statue" of law enforcement under the FTC Act. The FTC has international impact due to the fact that it is involved as a privacy law enforcement authority in APEC CPBR System for controlling data flows in the Asia–Pacific region. As far as the healthcare section is concerned, the US Department of HHS has launched HIPAA. Most importantly, the HHS has developed two fundamental HIPAA Rules: (i) the HIPAA Privacy Rule that sets national standards for the protection of healthcare electronic transactions and the (ii) HIPAA Security Rule that sets national standards for the protection of personal health information that is created, received, used, or maintained by a covered entity. These Rules still remain as the basis of health-related data protection in the United States along with less known HIPAA Rules, such as the HIPAA Breach Notification Rule that offers guidelines in case of data breach and the HIPAA Enforcement Rule that poses serious penalties in case of data protection violations under the HIPAA Rules.

An interesting challenge in privacy law making is the EU–US data protection law harmonization. In 2016, the two parties agreed to share the same data protection goals and adopted the EU–US Privacy Shield. The Privacy Shield completely replaced the ambiguous Safe Harbor and its Privacy Principles after significant weaknesses were revealed in the light of unauthorized surveillance programs and massive cases of privacy breach. The Privacy Shield has developed seven Privacy

Principles that comprise the backbone of the EU—US data protection laws and form a major step toward the harmonization of data protection regulations across EU and the United States. The rigorous nature of the US data protection law system enhances the regular adoption of provisions.

However, several provisions in the Privacy Shield are necessary, especially in the light of GDPR. GDPR permits information data flows to countries outside the EU, only if these countries fulfill the "adequate" data protection requirements. In addition, GDPR introduces updated definitions of the data protection actors and the purposes of processing that require the establishment of new safeguards in different areas of interest, including healthcare, trades, etc. The EU and US officials state that the Privacy Shield is a powerful data protection initiative and indeed several concerns of the European Court of Justice regarding the Safe Harbor have been already addressed. However, it is still not clear whether the Privacy Shield meets the requirements posed by the EU Working Parties regarding surveillance programs, as well as, whether the Privacy Shield will consider for any new provisions in the near future.

The international integrated approaches toward a global data protection framework envisage to overcome the heterogeneity of the data protection legislations across the nations. Toward this vision, remarkable initiatives have been launched, including the United Nation's resolution 68/167, IDPC's Initiative, OECD Guidelines on the Protection of Privacy and Transborder Flows of Personal Data, and the CoE Convention. Each one of these initiatives is promising, with broad impact in the majority of the nations and all of them share common principles (Fig. 4.2). On the other hand, these integrated approaches can have negative effects and reduce

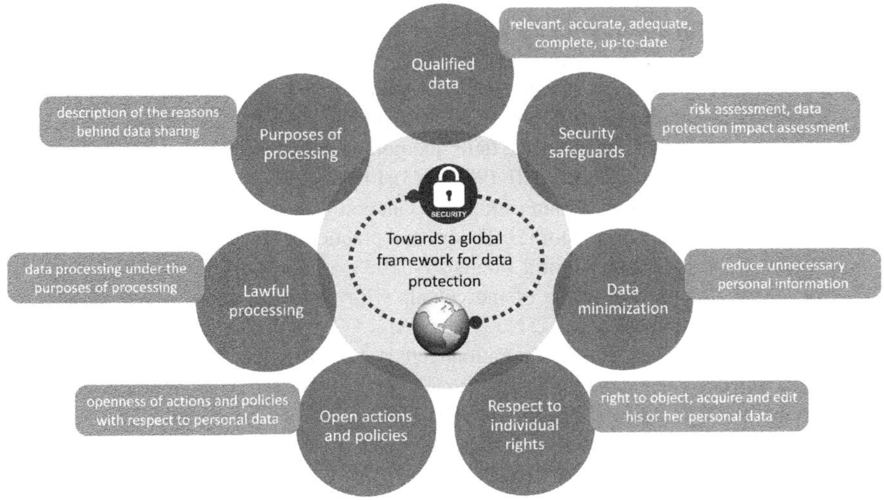

FIGURE 4.2

Seven data protection principles that lie in the intersection of the four major international data protection initiatives.

their impact to the citizens of different nations. This usually occurs when the political interests of different national governments diverge from a common political route. Such cases create conflicts between the nations that choose to follow different strategical approaches. Although the idea of a global data protection framework is still away, significant attempts have been made toward this. The key to success simply lies in the cooperation of the supervised authorities of each nation.

References

[1] Hustinx P. EU data protection law: the review of directive 95/46/EC and the proposed general data protection regulation. Collected Courses of the European University Institute's Academy of European Law, 24th Session on European Union Law; 2013. p. 1–12.

[2] Goldfarb A, Tucker CE. Privacy regulation and online advertising. Manag Sci 2011; 57(1):57–71.

[3] Gubbi J, Buyya R, Marusic S, Palaniswami M. Internet of Things (IoT): a vision, architectural elements, and future directions. Fut Gener Comput Syst 2013;29(7):1645–60.

[4] The implementation of commission decision 520/2000/EC on the adequate protection of personal data provided by the safe harbour privacy principles and related frequently asked questions issued by the US department of commerce. Commission Staff Working Document, European Commission; 2004. SEC(2004)1323/1.

[5] Lyon D. Surveillance, snowden, and big data: capacities, consequences, critique. Big Data Soc 2014;1(2). 2053951714541861.

[6] Directive 95/46/EC of the European Parliament and of the Council of 24 October 1995 on the protection of individuals with regard to the processing of personal data and on the free movement of such data. Off J Eur Union 1995;281:31–50.

[7] Regulation (EU) 2016/679 of the European Parliament and of the Council of 27 April 2016 on the protection of natural persons with regard to the processing of personal data and on the free movement of such data, and repealing Directive 95/46/EC (General Data Protection Regulation). Off J Eur Union 2016;119:1–88.

[8] Atchinson BK, Fox DM. From the field: the politics of the health insurance portability and accountability act. Health Aff 1997;16(3):146–50.

[9] Universal declaration of human rights. UN General Assembly; 1948.

[10] Van Dijk P, Hoof GJ, Van Hoof GJ. Theory and practice of the European Convention on Human Rights. Martinus Nijhoff Publishers; 1998.

[11] Convention for the protection of individuals with regard to automatic processing of personal data. Strasbourg: EST. No. 108; 1981.

[12] U. S. Food and Drug Administration/Center for Drug Evaluation and Research. Worsening depression and suicidality in patients being treated with antidepressant medications: FDA public health advisory. Washington, DC: Author; 2004.

[13] Posner RA. The federal trade commission. Univ Chic Law Rev 1969;37(1):47–89.

[14] Khatri V, Brown CV. Designing data governance. Commun ACM 2010 2010;53(1): 148–52.

[15] Takabi H, Joshi JB, Ahn GJ. Security and privacy challenges in cloud computing environments. IEEE Secur Priv 2010;8(6):24–31.

[16] Weiss MA, Archick K. US-EU data privacy: from safe harbor to privacy shield. CRS Report. 2016.

[17] De Hert P, Papakonstantinou V. The proposed data protection regulation replacing Directive 95/46/EC: a sound system for the protection of individuals. Comput Law Secur Rev 2012;28(2):130–42.

[18] Treaty of Lisbon amending the treaty on European Union and the treaty establishing the European community. Off J Eur Union 2007;306:1–271.

[19] The European citizens' initiative. Link: http://ec.europa.eu/citizens-initiative/public/welcome.

[20] Charter of fundamental rights of the European Union. Off J Eur Union 2012;326: 391–407.

[21] Consolidated version of the treaty on the functioning of the European Union. Off J Eur Union 2012;326:47–390.

[22] Consolidated version of the Treaty on European Union. Off J Eur Union 2012;326: 13–390.

[23] Neale AD. The antitrust laws of the United States of America. Cambridge University Press; 1960. p. 68–72.

[24] The HIPAA Breach Notification Rule, 45 CFR §§ 164.400-414. 2016. Link: https://www.hhs.gov/hipaa/for-professionals/breach-notification/index.html.

[25] Commission implementing decision (EU) 2016/1250 of 12 July 2016 pursuant to Directive 95/46/EC of the European Parliament and of the Council on the adequacy of the protection provided by the EU-U.S. Privacy Shield (notified under document C(2016) 4176) C/2016/4176. Off J Eur Union 2016;207:1–112.

[26] Swiss-U.S.. Privacy shield framework principles. U.S. Department of Commerce; 2017. Link: https://www.trade.gov/td/services/odsi/swiss-us-privacyshield-framework.pdf.

[27] Asia Pacific Economic Cooperation (APEC) privacy framework. Asia Pacific Economic Cooperation Secretariat. 2005. p. 81. Link: http://publications.apec.org/-/media/APEC/Publications/2005/12/APEC-Privacy-Framework/05_ecsg_privacyframewk.pdf.

[28] Privacy & data security update: an overview of the commission's enforcement, policy initiatives, and Consumer Outreach and Business Guidance in the Areas of Privacy and Data Security: January 2017 – December 2017. 2017. Link: https://www.ftc.gov/system/files/documents/reports/privacy-data-security-update-2017-overview-commissions-enforcement-policy-initiatives-consumer/privacy_and_data_security_update_2017.pdf.

[29] Breaches of unsecured protected health information affecting 500 or more individuals. October 2018. Link: https://ocrportal.hhs.gov/ocr/breach/breach_report.jsf.

[30] HITECH Act enforcement Interim Final Rule. U.S. Department of Health and Human Services; 2009. Link: https://www.hhs.gov/hipaa/for-professionals/special-topics/hitech-act-enforcement-interim-final-rule/index.html.

[31] Hudson KL, Holohan MK, Collins FS. Keeping pace with the times—the genetic information nondiscrimination act of 2008. N Eng J Med 2008;358(25):2661–3.

[32] Health Insurance Portability and Accountability Act Privacy Rule, 45 CFR §§ 160, Subparts C, D, and E. 2000. Link: https://www.hhs.gov/hipaa/for-professionals/privacy/laws-regulations/index.html.

[33] Health Insurance Portability and Accountability Act Security Rule, 45 CFR §§ 160. 2003. Link: https://www.hhs.gov/hipaa/for-professionals/security/laws-regulations/index.html.

[34] Security Risk Assessment Tool (SRA Tool). U.S. Department of Health and Human Services; 2018. Link: https://www.healthit.gov/topic/privacy-security-and-hipaa/security-risk-assessment.

[35] Health Insurance Portability and Accountability Act Enforcement Rule, 45 CFR §§ 160, Subparts C, D, and E. 2016. Link: https://www.hhs.gov/hipaa/for-professionals/special-topics/enforcement-rule/index.html.

[36] Morrison M, Bell J, George C, Harmon S, Munsie M, Kaye J. The European General Data Protection Regulation: challenges and considerations for iPSC researchers and biobanks. Regen Med 2017;12(6):693−703.

[37] Council Decision (EU) 2016/920 of 20 May 2016 on the signing, on behalf of the European Union, of the agreement between the United States of America and the European Union on the protection of personal information relating to the prevention, investigation, detection, and prosecution of criminal offences. Off J Eur Union 2016;154:1−2.

[38] Langheinrich M. Privacy by design—principles of privacy-aware ubiquitous systems. Springer, Berlin, Heidelberg. In: International conference on ubiquitous computing; 2001. p. 273−91.

[39] Data protection regulations and international data flows: implications for trade and development. In: United Nations conference on trade and development (UNCTAD); 2016.

[40] Joseph S, Castan M. The international covenant on civil and political rights: cases, materials, and commentary. Oxford University Press; 2013.

[41] Universal Declaration of Human Rights (UDHR). United Nations (UN) General Assembly, Paris. 1948. Link: http://www.un.org/en/udhrbook/pdf/udhr_booklet_en_web.pdf.

[42] Resolution adopted by the General Assembly on 18 December 2008: 63/155. Intensification of efforts to eliminate all forms of violence against women. United Nations (UN) General Assembly. 2009. Link: http://www.un.org/en/ga/search/view_doc.asp?symbol=A/RES/63/155.

[43] Montreux declaration. 27th international conference of Data Protection and Privacy Commissioners. Montreux; 2006. Link: https://edps.europa.eu/sites/edp/files/publication/05-09-16_montreux_declaration_en.pdf.

[44] Woodward R. The organisation for economic co-operation and development (OECD). Routledge; 2009.

[45] Holliday J. The struggle for Syria in 2011, vol. 16. Institute for the Study of War; 2011.

[46] The international impact and opportunity of the General Data Protection Regulation. NTT Security; 2017. Link: https://www.nttsecurity.com/docs/librariesprovider3/resources/global_thought_leadership_gdpr_uea_v4.

[47] Privacy Act. 1988. Link: https://www.legislation.gov.au/Details/C2014C00076.

[48] Protection of Personal Information Act 4. 2013. Link: http://www.justice.gov.za/inforeg/docs/InfoRegSA-POPIA-act2013-004.pdf.

[49] Amended Act on the Protection of Personal Information. Personal Information Protection Commission, Japan. 2016. Link: https://www.ppc.go.jp/files/pdf/Act_on_the_Protection_of_Personal_Information.pdf.

CHAPTER

Medical data harmonization

5

Chapter outline

5.1 Overview

The heterogeneity of data among biobanks, cohorts, and other sources of medical data is a critical scientific limitation which poses significant obstacles in the effective analysis of such data, yielding clinical studies with poor statistical power and, thus, inaccurate disease outcomes [1–3]. In computer science, data harmonization is an emerging technique which aims to overcome the structural heterogeneities that are present among the medical data derived from multiple sources by producing homogenized versions of the heterogeneous data that share a common medical domain (context). The overall idea of data harmonization is to transform the heterogeneous data into a common format with the exact same parameters and range values, using data-driven, and other computational approaches, such as lexical and semantic

Medical Data Sharing, Harmonization and Analytics. https://doi.org/10.1016/B978-0-12-816507-2.00005-0

matching, to enable the integrative analysis of the heterogeneous data and, therefore, enhance the statistical power of the clinical studies which make use of such data. Based on the above concept, data harmonization can enable the interlinking and subsequent integration of clinical data to deal with the unmet needs in various diseases.

According to the literature, a variety of computational methods for medical data harmonization has been proposed so far [4–10]. The majority of the current efforts toward data harmonization involve the definition of a global standard (common) procedure for data collection which is ideal in the case of prospective data (i.e., data that will be updated in the future). The difficult and most challenging part though is the need to harmonize retrospective data (i.e., data that have been already collected in the past with the absence of a predefined standard data collection protocol). In general, there are two conceptual approaches/strategies to accomplish data harmonization, namely the stringent and the flexible strategy [1–3]. The former limits the harmonization process only on data that will be (or have been) collected under common measurement procedures (standards), whereas the latter approach extends the harmonization process to include data that have been already collected under different measurement procedures or protocols [1–3]. In this chapter, emphasis is given on flexible methods that enable the harmonization of retrospective clinical data due to its underlying complexity and its clinical importance.

In flexible prospective data harmonization, the experts of a particular domain agree on a set of standard variables that will serve as the core variables during the prospective data collection process, whereas in flexible retrospective data harmonization, the same set of variables is used by a study to derive the percentage of variables that are matched (aligned) with the variables of the set of standard variables [1,3]. The reference model is usually expressed in a semantic form through an ontology which provides a hierarchical representation of a specific clinical domain based on a set of entities (i.e., classes and subclasses) and object properties (i.e., parameters) that explicitly describe the knowledge of a particular clinical domain [2,7–12]. The majority of the methods for retrospective data harmonization make use of a predefined, standardized model which describes the requirements of a particular clinical domain and serves as a common template (i.e., a gold standard) for harmonization [1,3]. This standard template includes a set of clinical variables (parameters) that are able to describe the domain knowledge of a disease of interest. Of course the existence of a standard template is not always necessary especially in the case when the scope of data harmonization is to seek for variable associations between two (or more) heterogeneous datasets which are clinically relevant. In that case, one dataset can be considered as the reference model.

A robust data harmonization method involves the application of lexical and semantic matching algorithms. A lexical matching algorithm uses string similarity techniques [13,14] to identify common terminologies (i.e., exact sequences or similar block sequences) that are present between the terms of the standard model and those of the original dataset. External vocabularies can also be used to enrich the clinical domain knowledge and thus enhance the accuracy of the overall lexical matching process through the identification of homonyms or synonyms. On the other hand, the semantic matching method [15–18] uses semantic relationships that exist between the terminologies, apart from the lexical matching process that is already included, to reduce the information loss and enhance the overall data harmonization process. This can be accomplished through the construction of ontologies which represent the clinical domain knowledge of interest in the form of entities (e.g., classes) and object properties (e.g., "includes," "has," "consists of") [15–18]. Semantic matching uses a standard (or reference) model which is usually expressed in the form of an ontology, where the classes are considered as categories, e.g., "Clinical tests," that might consist of further subclasses, e.g., "Blood Tests," etc. Each class can include a set of variables which are related to the class they belong to in terms of common meaning or concept [15–18]. For example, the class "Blood tests" includes the variables "age," "gender," "hemoglobin levels," etc. This can lead to the semantic interoperability of the variables which might not be lexically identical but share a common concept.

Another family of data harmonization methods is statistical harmonization which involves the application of linear and nonlinear statistical models to investigate the effect of different latent factors on a set of one or more items [19–21]. In statistical theory, the items can be considered as all types of informative variables (e.g., depression) that are observed and the latent factors as variables that are not directly observed but are rather inferred by the items [19–21]. The purpose of statistical data harmonization is to homogenize scales that measure the same item and transform them into a common metric of the same scale, where the types of the items might vary from discrete and ordinal to continuous [19–21]. For example, a clinical center might record the cholesterol levels using the scale low, medium, high, whereas another clinical center may record the same levels using the scale 0, 1, and 2. Thus, statistical harmonization tries to recode the variables that belong to the same construct so that they are commensurately scaled at the end [19–21]. Of course, the detection of variables that express different (or common) scales and belong to the same construct is challenging because there is no prior knowledge regarding the names of the items like in lexical or semantic matching [19–21]. Even if two variables (items) describe the same construct, it is not always proper to match these two variables because there might be differences in the population characteristics between the clinical studies, e.g., differences in the education level, ethnicity, gender, etc. Such differences need to be controlled during the statistical analysis process, and thus the complexity of the harmonization process is greatly enhanced. Moreover, the types of the items directly affect the type of the statistical model to be used. Toward this direction, a variety of statistical methods has been proposed so far, especially for the harmonization of psychometric and cognitive items with different measurement scales across clinical data, including [19–21] (i) simple linear factor analysis (LFA) for continuous items, (ii) 2-parameter and multiparameter logistic item response theory (IRT) analysis for binary items, (iii) generalized linear factor analysis (GLFA) for mixtures of continuous and discrete items, and (iv) moderated nonlinear factor analysis (MNFA) for mixtures of continuous and discrete items with nonlinear dependencies, among others.

The prospects of data harmonization in the medical domain are many. Harmonization can overcome the lack of common (standard) data collection protocols, a fact that introduces biases during the collection of medical data among multiple sources of medical data (e.g., clinical centers), especially during the recording of the measurement units of various laboratory-related attributes. This can be accomplished by either normalizing the measurement units according to a predefined range that is already defined in the standard model or based on statistical approaches, through a procedure which is known as data standardization (see Chapter 3, Section 3.3). In contrast to data standardization, data harmonization is a more generalized strategy which aims to first align the structure of the datasets and then apply data standardization as part of the normalization process. In fact, data harmonization involves (i) the identification of terminology matches among the heterogeneous data through a procedure which is known as terminology mapping and (ii) the application of data standardization to normalize the measurement scales across the matched terms through a procedure which is known as terminology alignment. The fact that the data harmonization process can be performed in a semiautomated manner reduces the time effort that is needed by the clinicians to manually homogenize their clinical data, which is rather impossible in the case of large datasets, where the number of parameters is vast, like in omics data [22–24]. The clinician's involvement can be further reduced by constructing machine learning algorithms which are able to learn from external sources, such as medical index vocabularies, through the development of semantic interlinking mechanisms which are able to speed up the data harmonization process. The fact that data harmonization allows the integration of heterogeneous data can provide great insight on the assessment of the unmet needs in various diseases [1] through (i) the development of more robust risk stratification models for the early identification of high-risk individuals, (ii) the identification of new prominent biomarkers for the prediction of disease outcomes, and (iii) the development of new, targeted, therapeutic treatments and health policies.

There is no doubt that the current advances in data harmonization are very promising, with numerous clinical and scientific prospects deriving from the subsequent integration and analysis of heterogeneous data from multiple medical data sources. The majority of the data harmonization methods is semiautomated as it involves a clinician who is able to verify the automatically derived associations by picking up the most appropriate matches for each terminology (based on a reference model), a process which subsequently increases the impact of data harmonization in healthcare. However, the scientific challenges toward the realization of such an attempt are many and will be the point of discussion in this chapter along with the presentation of technical approaches and existing tools and frameworks that envisage to enhance the interoperability among relevant, heterogeneous data sources.

5.2 The origins and prospects of harmonizing datasets

So, how can we enable the integration and subsequent analysis of heterogeneous clinical data that are obtained from multiple data sources? This is a crucial scientific question which has led to the definition of a research field that is responsible for the development of new practices toward the homogenization of data with structural heterogeneities (e.g., incompatibilities), namely data harmonization. The fundamental basis of data harmonization is data sharing (see Chapter 3), which has a leading role before the data harmonization process, because it is necessary to develop appropriate data sharing mechanisms that allow for the interconnection of clinical data across multiple data sources. This mechanism, however, must be able to fulfill all the legal and ethical requirements that ensure the protection and privacy of the patient data. On the other hand, although data sharing enables the interlinking of clinical data from various cohorts and clinical centers worldwide, the heterogeneities that exist between them hamper the execution of the data analytics workflows and pipelines. As a matter of fact, data sharing and data harmonization complete each other, acting as "communication vessels" within a federated healthcare platform.

Why is it necessary to harmonize the data? As it has been already mentioned, data harmonization can enable the integration of clinical data from cohorts, and other similar sources of medical data, that coexist under a specific clinical domain (e.g., that describe a common disease of interest) which can reveal valuable clinical information regarding a disease's onset and progress over time. Indeed, the integration of harmonized data from multiple data sources can significantly enhance the population subgroups that participate in the clinical studies and thus yield more powerful patient stratification models which are able to precisely identify groups of individuals (subjects) that are more prone to the development of a disease outcome [25,26]. These individuals are referred to as high-risk individuals who belong to specific subgroups and, thus, the early diagnosis of such subgroups is of great importance for enhancing the quality of the existing healthcare systems. Because data harmonization can transform the clinical data into standardized formats, it can enable the interlinking of electronic health records with electronic health record systems worldwide [27–29].

The integration of harmonized data can also lead to the identification of prominent clinical factors having significant contribution in the prediction of one or more disease outcomes, as well as the confirmation of the existing prominent clinical predictors, especially in the case where the type of the disease under investigation is rare and/or chronic [1]. These clinical predictors are referred to as biomarkers and are able to reveal the underlying mechanisms of various diseases. The clinical importance of biomarkers is high in the case of genetic data where phenotypes and genetic variants are present [30]. Moreover, the outcomes of clinical studies that make use of integrated, harmonized data enable the development of more targeted therapeutic treatments for the different population subgroups that can greatly enhance or replace the existing treatments and thus shed light into the progress of a disease over time, as well as promote the establishment of new healthcare policies by the healthcare stakeholders.

Apart from the integration of harmonized data into centralized databases, another important advantage lies on the interlinking of harmonized data that are stored in distributed clinical databases which can enable the development of data analytics tools for analyzing the data in distributed environments [31]. In general, a centralized data repository is more vulnerable to security threats and privacy breach, and thus it is not always feasible to maintain the data under a common database [32]. A solution to this is to store the clinical data in distributed databases and then interlink these databases with multiple authentication levels for increased security (see Chapter 6, Section 6.1). So, the only way to analyze the data that are stored in distributed sites is to develop new machine learning and statistical models or extend the existing ones so that they are able to incrementally process the individual data on each site. For example, assume that a researcher wishes to compute population characteristics (e.g., descriptive statistics) across the distributed data or, in a more demanding case, assume that the researcher wishes to run a risk stratification model for predicting disease outcomes across distributed data. These two scenarios can only be feasible in the case where the individual data, on each site, have a common format, i.e., a common set of parameters and standardized values. It is obvious now that this limitation can be effectively addressed by harmonizing the individual datasets on each site so that the data model can be able to adapt on the same set of parameters.

5.3 Cohort integration requirements

Let us assume that a clinical cohort study aims to examine the underlying mechanisms of a disease's onset and progress (e.g., develop risk stratification models) or examine the effect of a newly proposed drug on a target population through the integration of a set of *N* clinical cohorts which include clinical data from patients that have been diagnosed with a common disease or medical condition [33,34]. This is the most common scenario in the case of large-scale clinical studies. To realize such a scenario, it is first important to interlink the individual cohort data. Therefore, the first and foremost requirement is data sharing. Of course, the cohort data must fulfill all the requirements that are posed by the data protection regulations otherwise sharing would be pointless. Thus, a predefined requirement for data sharing is the establishment of a framework for data protection that would ensure the sustainability and the delivery of the cohort data (see Chapter 4). Once the data from the individual cohorts are shared, they need to be brought into a common structural format for coanalysis, and thus the second requirement is harmonization.

To enable the harmonization of individual cohort data, it is necessary first to develop a standard model which will describe the minimum requirements that are needed to exclusively describe the domain knowledge of the disease of interest. In the case where only associations need to be computed between the individual cohort data, one cohort dataset can be used as a standard model that can be aligned (harmonized) with each one of the remaining cohort datasets. These minimum requirements consist of a set of medical parameters, value ranges, and descriptions that can effectively describe the related domain. This set of parameters is used as a reference template according to which the data will be harmonized. Therefore, the third requirement is the existence of a reference model for the disease under investigation. The methods that can be used for data harmonization are presented in Section 5.6. Data analytics workflows can then be applied on the harmonized data to extract valuable information regarding the disease's onset and progress. Finally, the transparency of the workflows and any related operations that are performed during the data harmonization and the data analytics procedures needs to be reassured.

5.4 Barriers toward medical data harmonization

The lack of a standard reference model often obscures the harmonization process, making the adoption of the majority of the data harmonization methods difficult to be implemented. In addition, the medical terms and the acronyms that are often adopted by the majority of the clinical centers during the data collection process are difficult to be parsed and sometimes unable to be matched with standard medical terms and indices (e.g., the use of the acronym "HGB" instead of "hemoglobin" or "haemoglobin" or any other use of acronyms during the definition of the attributes), a fact that obscures the accuracy of the harmonization process due to the underlying information loss. A solution to this would involve the clinician's effort during the terminology mapping process so that he/she would be able to verify the validity of the terms that were marked as homogeneous. On the other hand, the absence of timestamps during the collection of prospective data and the existence of erroneously parsed fields during the data collection process are additional factors that obscure the structural alignment process. Therefore, the application of a data curation workflow (see Chapter 3, Section 3.3) is an important preharmonization requirement to fix problematic fields (e.g., outliers, incompatibilities, unknown symbols) that are present within the original data.

The lack of terminology descriptions along with the absence of information regarding the meaning of the range values on each attribute, especially in the case of attributes with categorical values, hampers the data standardization process. For example, a clinical center may record the state of a medical condition using the coding term "normal" or "abnormal." Another clinical center can record the same condition using the binary values 0 and 1, respectively. This knowledge should be clearly indicated before the harmonization process. A similar example occurs when a clinical center records the levels of a laboratory measure as "low," "normal," or "high," whereas another center may use the values 1, 2, and 3, respectively, to indicate these measurement levels and another one may use the terms "low" or "high," skipping the "normal" level. These again are important factors that should be taken into consideration before the harmonization process. As for the attributes with continuous values, the measurement units (or normalized units) should be clearly stated. For example, a clinical center may record a laboratory measure in "mg/mL," whereas another clinical center may record the same value in "mg/dL," "μmol/L," or "g/L."

So what if the parameters which are present in the standard template are not representative or limited to only a small portion of the domain's knowledge? This is a critical limitation that enhances the loss of information during the harmonization process and specifically during the terminology mapping process. For example, a retrospective dataset may include a set of 100 attributes, whereas the standard model may only include a set of 50 related parameters, where the relevance of an attribute is trivial because a medical condition (e.g., cryoglobulinemia) can be followed by a set of related symptomatology (e.g., fever, weight loss). One way to reduce this type of information loss is to define a semantic representation of the standard model, where each parameter is assigned to a category (or class). In the previous example, the parameters "fever" and "weight loss" can be assigned to the category "symptomatology." A similar example occurs in the case where a clinical dataset includes more than one demographic-related parameters (e.g., education level) and/or laboratory tests (e.g., blood tests) that could be parsed in the categories "Demographics" and "Laboratory tests," respectively. If any detected symptomatology is assigned in the homonymous category, instead of being ignored, then the overall information loss would be greatly reduced.

The majority of these barriers can be overcome in the case of the prospective data collection process where the prospective data can be recorded through appropriate digital data entry forms that already include these standard measurement units and range values, as well as the terminologies for each type of attribute. The data entry form can be used as a standard template like the one which is used in the case of the retrospective data harmonization process. In all cases, the scope of harmonization should be well defined. However, apart from the technical challenges that are met during data harmonization, the most prominent factor that facilitates data harmonization is the establishment of a legitimate environment that enables the sharing of data from multiple data sources. Data harmonization is in line with data sharing, and, thus, the lack of a legitimate data sharing mechanism would make data harmonization pointless.

5.5 The two fundamental types of harmonization

5.5.1 The stringent approach

The stringent approach is an ideal strategy which constrains the harmonization process to clinical data that have been collected under common collection criteria and operating procedures [2,3], where the common data collection criteria refer to the adoption of identical study specifications (uniform measures) between the clinical studies that participate in the data harmonization process. These specifications include [2] (i) common inclusion and exclusion criteria for the definition of the population subgroups, (ii) common follow-up time periods, and (iii) a common set of qualitative and quantitative measures (e.g., therapies), among others. These specifications together constitute a data collection protocol and are exclusively designed by domain experts who are able to identify (i) the domain of the field of interest (type of study), (ii) the set of measures that should be collected for the specified study, and (iii) the standardized measurement units for the recommended set of measures for the particular type of study. The range of diseases that can be covered by the stringent approach can be vast as long as the data follow the same standard operating procedures.

According to the stringent approach, the clinical studies that participate in the data harmonization process must be initially designed to meet these specifications to be harmonized and finally synthesized, otherwise the data harmonization process will fail. It is obvious that these requirements are strict and limited to only a small portion of clinical centers and other similar sources of clinical data that adopt common data collection criteria and standard procedures. Of course, the majority of the clinical centers do not follow identical procedures for the data generation process and thus stringent harmonization remains a conceptual and ideal strategy for the scientific community. This is highly present in the case of retrospective data harmonization, where the stringent approach is useless because the data have been collected in the past, where a standard data collection protocol is usually absent. The stringent approach would be meaningful in the case of a prospective study of perhaps in a cross-sectional study which focuses on data that have been obtained at a specific time point although it would require a substantial amount of time to be prosperous.

5.5.2 The flexible approach

The stringent method is a strict and a rather ideal approach that significantly limits the statistical power of the data harmonization process because it obscures the integrity of the produced harmonized data through the underlying information loss and limits the harmonization to a small portion of data that have been collected under the same standard operating procedure. An alternative approach that aims to deal with the limitations that are posed by the stringent approach is flexible harmonization [2]. As its name implies, the flexible approach allows a certain level of heterogeneity between the data which participate in the harmonization process instead of the stringent case where the complete absence of heterogeneity among the individual data is required. Therefore, the flexible approach can support the harmonization of both prospective and retrospective data as far as the level of compatibility between them is well defined. Through this manner, the flexible methodology envisages to enable the harmonization of data that do not necessarily need to be homogeneous or obtained under a common data collection protocol criterion with equal-sized populations.

In flexible harmonization, the level of heterogeneity of the data directly affects the percentage of harmonized variables across them. This implies that the amount of flexibility is constrained to a specific set of requirements that need to be defined. That is, the set of clinically relevant parameters (factors) that will be common among the heterogeneous data. Of course, the clinical domain where the data that participate in flexible harmonization belong to must be common. To facilitate flexible harmonization, the clinical experts must first define a set parameters (variables) that will serve as the core set for the domain of interest allowing for a specific level of flexibility regarding the data collection protocol and the standard operating procedures [2]. Therefore, flexible harmonization is constrained to specific outcomes that are defined by the clinical experts. In the prospective case, the core set of variables is defined and agreed to by the experts so as to allow a specific level of flexibility during the recording of the follow-up data. In the retrospective case, the core parameters are combined together with pairing rules to identify potential associations with those from the heterogeneous data and thus quantify the harmonization accuracy.

The flexible strategy is far more realistic and has a much higher clinical value and overall applicability than the stringent approach, although, in both cases, certain compatibility criteria must be carefully defined so that harmonization can be feasible. The compatibility criteria are expressed in the form of a set of standard variables, i.e., a core set of variables that describe the requirements of the clinical domain of interest. In both cases, however, the standard model is defined by the clinical experts in the field in such a way to (i) be in line with the majority of the parameters within the data that are collected by different clinical centers and (ii) explicitly describe the domain knowledge of the disease under investigation. This means that the clinical experts select the variables of the standard model by taking into consideration (i) the contribution of each variable toward the efficient description of the disease's domain knowledge and (ii) the extent to which these variables are present in the majority of the data that exist under each clinical center. Additional information regarding the format (and the type) of the medical data that will be involved in the harmonization process, along with the ethical and legal concerns, the quality of the data, as well as the precise definition of the objectives, i.e., the reason behind data harmonization, should also be taken into consideration for the realization of the flexible strategy [2,3].

The level of heterogeneity and diversity between the data is more or less trivial and can be reflected by the structure of the core set of variables. For example, the clinical experts of a specific domain might consider that a set of N variables is enough to describe the knowledge of the domain under investigation, whereas another group of experts, on the same field, might consider the need to add more (or less) variables in the core set. Therefore, the experts must agree on a core set of variables that overlap with the majority of the data to increase the harmonization accuracy. This can be extremely difficult especially in the case of retrospective flexible harmonization where the data might have been collected under diverse protocols and might exist under different identifiers, as well. The same stands for the objectives of the data harmonization process. The data providers must clarify the scope of data harmonization, as well as provide any kind of information regarding the study design and the specifications that were used for the data collection process which are valuable for the definition of a more accurate core set of variables.

5.6 Novel technical approaches and tools

5.6.1 Lexical matching

The lexical matching method is a string matching approach that is mainly used in linguistics and natural language processing which quantifies the similarity between a pair of strings based on their inverse string distance score which in turn is interpreted as a similarity score. The overall idea behind lexical matching is based on the fact the higher the inverse distance score (or the lower the distance score) between two strings, the higher the similarity score between them and, therefore, the higher their overlap in terms of syntactic relevance. The lexical matching method for lexical analysis has been initially presented in Section 3.4 as part of data standardization to identify lexically similar terminologies between the variables of a standard model, which was subsequently used as a template for standardization, with those from the input data. Because data standardization can be considered as a preharmonization step, the same methodology can be adopted to identify common terminologies among heterogeneous data. In fact, the core set of variables that was mentioned in the previous section is the same with the standard model that was mentioned in Section 3.4. Thus, string matching methods can be used to seek for common terms.

The most common method for lexical matching involves the computation of the edit distance between two strings, assume x and y. Recall from Section 3.4 that the edit distance is given by

$$d_{x,y}(i,j) = \begin{cases} i & , \quad i = 0 \\ j & , \quad j = 0 \\ d[i-1, i-1] & , \quad i,j > 0 \text{ and } x_i = y_j \\ \min\begin{Bmatrix} d[i-1, j-1] + 1 \\ d[i-1, j] + 1 \\ d[i, j-1+1] \end{Bmatrix} & , \quad \text{o.w.} \end{cases} \tag{5.1}$$

where $d_{x,y}(i,j)$ is the distance between the first i characters of x and the first j characters of y. In fact, the edit distance aims to transform x into y by performing three possible types of operations, namely (i) insertion, (ii) deletion, and (iii) substitution. Assume a string $x =$ "abc" with size 3. Insertion involves the addition of a new character, assume d, into x, so that $x =$ "abcd." Deletion involves the removal of an existing character from the string, assume c, so that $x =$ "abd." Substitution involves the replacement of an existing character, assume d, by c, so that $x =$ "abd" becomes $x =$ "abc." As it has been already mentioned in Section 3.4, the Jaro (and Jaro–Winkler) distance [13,14,35,36] and the Levenshtein distance scores [37] are the most common methods for calculating the edit distance, although the latter is much closer to the definition of the edit distance against the former method which takes into consideration the number of transpositions between two strings. The latter is defined as half the number of matching characters between x and y. In fact, the Jaro distance first computes the number of matching characters, between x and y, assume m, as well as the number of transpositions, assume t. Then, according to (5.1), the Jaro distance measures the edit distance between x and y by computing the average of the percentage of matched characters in each string with the percentage of the transpositions in the number of matching characters. Thus, the higher the Jaro distance (i.e., the closer to 1), the more similar the two strings are.

The pseudocode for computing the edit distance is given by Algorithm 5.1.

The algorithm generates a distance matrix where each cell corresponds to a distance score. An example of the edit distance matrix for the strings x = "lymphadenopathy" and y = "lymphoma" is presented in Table 5.1. The edit distance value is expected to be 8 and the number of operations is expected to be the following:

1. lymph**a**denopathy - > lymphdenopathy (delete "a")
2. lymph**d**enopathy - > lymphenopathy (delete "d")
3. lymph**e**nopathy - > lymphnopathy (delete "e")
4. lymph**n**opathy - > lymphopathy (delete "n")
5. lymphopathy - > lymphomathy (substitute "p" with "m")
6. lymphoma**t**hy - > lymphomahy (delete "t")
7. lymphoma**h**y - > lymphomay (delete "h")
8. lymphoma**y** - > lymphoma (delete "y")

Note that in this example, no insertions are needed. The resulting distance matrix after the execution of Algorithm 5.1 is presented in Table 5.1. The desired distance is the value in the last cell of the table, i.e., cell (15,8) which is 8 (as expected). This denotes that the total number of operations that is needed to transform x into y is 8. The values which are depicted in bold are equal to the consecutive costs of the operations. More specifically, the zeros in the cells (1,1), (2,2), (3,3), (4,4), and (5,5)

Algorithm 5.1 A PSEUDOCODE FOR COMPUTING THE EDIT DISTANCE.

```
def edit_distance(x,y):
  x_len = length(x); #the number of characters in x.
  for i = 1:x_len
     x[i,0] = 0;
  y_len = len(y); #the number of characters in y.
  for i = 1:y_len
     y[i,0] = 0;
  for i = 1:x_len
     for j = 1:y_len
        if(x[i] == y[j]):
           d[i,j] = d[i-1,j-1];
        else:
           d[i,j] = min(d[i-1,j]+1, d[i,j-1]+1, d[i-1, j-1]+1);
return d[x_len, y_len];
```

Table 5.1 An example of an edit distance table for the strings "lymphadenopathy" and "lymphoma."

	-	**l**	**y**	**m**	**p**	**h**	**o**	**m**	**a**
-	0	1	2	3	4	5	6	7	8
l	1	**0**	1	2	3	4	5	6	7
y	2	1	**0**	1	2	3	4	5	6
m	3	2	1	**0**	1	2	3	4	5
p	4	3	2	1	**0**	1	2	3	4
h	5	4	3	2	1	**0**	1	2	3
a	6	5	4	3	2	**1**	2	3	2
d	7	6	5	4	3	**2**	3	4	3
e	8	7	6	5	4	**3**	3	4	4
n	9	8	7	6	5	**4**	4	4	5
o	10	9	8	7	6	5	**4**	5	5
p	11	10	9	8	7	6	5	**5**	6
a	12	11	10	9	8	7	6	6	**5**
t	13	12	11	10	9	8	7	7	**6**
h	14	13	12	11	10	9	8	8	**7**
y	15	14	13	12	11	10	9	9	**8**

denote that the first five characters in both strings, i.e., the characters "l," "y," "m," "p," and "h" are equal, and thus the cost is 0 because neither of the three types of operations (insertion, deletion, and substitution) is applied. The values 1, 2, 3, and 4 in the cells (6,5), (7,5), (8,5), and (9,5), respectively, denote that the four characters "a," "d," "e," and "n" in x shall be deleted with a total cost of 4. The value 4 in cell (10,5) denotes that the character "o" is equal in both strings, and thus it is the same as before because the operation has a 0 cost ($4 + 0$). The value 5 in cell (11,6) denotes that "p" shall be replaced by "m," and thus a substitution operation is applied adding an extra 1 in the cost ($4 + 1$). The value 5 in cell (11,7) denotes that "a" is equal and thus the cost remains the same. The values 6, 7, and 8 in the cells (12,8), (13,8), (14,8), and (15,8) denote that the characters "t," "h," and "y" should be removed and thus 3 deletions are applied yielding a final cost 8 ($5 + 3$) (See Tables 5.2 and 5.3).

The Jaro distance [13,14] is useful for quantifying the similarity between two strings in the interval [0, 1] against the Levenshtein distance, which measures the total number of different characters and thus is preferred in the design of pairing rules for lexical matching. For example, in the case where $x =$ "lymphadenopathy" and $y =$ "lymphoma," the Jaro distance is equal to 0.7329, whereas the Levenshtein distance is equal to 8 and thus is less informative. The Jaro–Winkler distance [35,36] is a weighted version of the Jaro distance which uses a prefix scale p to give more weight to strings that match from the beginning for a length l. This property, however, is statistically weak because, in the majority of cases, it yields falsified pairs. For example, in the previous example, the Jaro–Winkler distance

Table 5.2 An example of an LCS table for the strings "Raynaud" and "Reynaud."

		0	1	2	3	4	5	6	7
		-	**R**	**a**	**y**	**n**	**a**	**u**	**d**
0	-	0	0	0	0	0	0	0	0
1	**R**	0	**1**	0	0	0	0	0	0
2	**e**	0	0	0	0	0	0	0	0
3	**y**	0	0	0	**1**	0	0	0	0
4	**n**	0	0	0	1	**2**	2	2	2
5	**a**	0	0	0	1	2	**3**	3	3
6	**u**	0	0	0	1	2	3	4	**4**
7	**d**	0	0	0	1	2	3	4	**5**

Table 5.3 A summary of the fundamental frameworks for data harmonization.

Tool	Data harmonization strategy	Application
DataSHaPER [4]	Uses a DataSchema as a reference model to harmonize heterogeneous data schemas according to the user-defined DataSchema through the development of pairing rules.	A 36% compatibility for creating a harmonized database across 53 of the world's largest longitudinal population-based epidemiological studies [6].
SORTA [7]	Uses lexical matching to align phenotype data from heterogeneous biobanks according to international coding systems.	Matched 5210 entries in the LifeLines biobank [38] (97% recall) and 315 entries in the DUMR [39] (58% recall).
BiobankConnect [5]	Uses lexical and semantic matching to align heterogeneous biobanks according to a desired set of predefined elements.	An average precision 0.745 toward the harmonization of data across six biobanks (7461 terms) with 32 desired elements [5].
S-Match [9]	Uses semantic matching to quantify the semantic relations that exist between the elements of two lightweight ontologies into four different categories.	A 46% precision on the correct identification of semantic matches in the TaxMe2 dataset [40] as part of the Ontology Alignment Evaluation Initiative [9].
FOAM [8]	Trains HMMs on sequence profiles that exist in international registries to align heterogeneous sequence profiles.	A functional ontology that includes a set of more than 70,000 trained HMMs targeting 2870 different KOs [8].

would be 0.8664 because the subsequence "lymph" is common from the beginning and thus is given more weight due to its large length. The Jaro–Winkler distance would be useful in the case where two strings differ in the spelling near the end, e.g., when x = "Raynaud" and y = "Raynaud's," where, in that case, the Jaro–Winkler distance is 0.977, whereas the Jaro distance is equal to 0.926, thus giving more emphasis to the length of the common prefix.

Apart from these methods, however, there are additional methods that share the same scope. The Hamming distance is a method that can be used to compute the edit distance between two equal-sized strings, assume x and y, using only substitutions. The Hamming distance is defined as the minimum number of substitutions that are required to transform x into y (or y into x). A useful property of the Hamming distance is the fact that it can provide information regarding the number of different characters between two equally sized strings. For example, in the case where x = "Raynaud" and y = "Reynaud," the Hamming distance is equal to 1 because the strings differ in one character. In contrast to the Levenshtein distance, the Hamming distance does not involve any deletions or insertions, and thus its application is limited.

The Jaccard distance score [41] is equal to

$$d_J = 1 - J = 1 - \frac{(x \cap y)}{(x \cup y)}, \tag{5.2}$$

where $x \cap y$ is the number of common n grams that are located in both x and y, whereas $x \cup y$ is the total number of n grams that can be made by x and y. The n gram [41] is defined as a sequence of n characters. For example, the 1-gram of the string "Raynaud" is {"R", "a", "y", "n", "a", "u", "d"}, whereas the 2 gram of the same string is {"Ra", "ay", "na", "au", "ud"}. Assume that x = "Raynaud" and y = "Reynaud." The Jaccard index, J, in the case of n = 1 (1-gram sequence) is equal to 6/7 = 0.8571 and the distance d_J is 0.1428, whereas in the case of n = 2 (2-gram sequence) the Jaccard index is 3/5 = 0.75 and d_J is 0.4 which is larger than in the previous case. Therefore, the n grams have a direct effect on the string similarity score.

The Dice distance score [41] is a more strict lexical matching measure which is defined by

$$d_D = 1 - D = 1 - \frac{2(x \cup y)}{|x| + |y|}, \tag{5.3}$$

where $x \cup y$ is the number of common words between x and y, and $|x|$ and $|y|$ are the number of words in x and y, respectively. The Dice coefficient is useful in the case where a string consists of more than one words. For example, in the previous case, the Dice coefficient, D, would be zero because the terms "Raynaud" and "Reynaud" are not common. However, in the case where x = "monoclonal gammopathy" and y = "monoclonal gamopathy," the Dice coefficient is 2/4 = 0.5 because the word "monoclonal" is the only common word between the two strings and hence the distance will be the same.

An alternative method that can be used to solve the edit distance problem is to find the longest common subsequence (LCS) [42,43] between x and y, that is the largest sequence of common characters between them. The solution to the LCS problem is a computationally complex approach that seeks for the common subsequence between x and y with the highest length, where the common subsequence does not need to be consecutive. In this case, the distance can be defined as

$$d_{\text{LCS}} = M + N - 2|\text{LCS}(x, y)|\,, \tag{5.4}$$

where M, N are the lengths of the strings x and y, respectively, and $|\text{LCS}(x, y)|$ is the length of the LCS between x and y. A pseudocode that computes the length of the LCS between two strings x and y is given in Algorithm 5.1. The algorithm iterates across each character of x and seeks for common characters in y until every character is parsed and thus builds an M x N table, i.e., an LCS table. If the strings are empty, the length of the LCS is zero. If the characters are equal, the length of the LCS is equal to the length of the (i-1, j-1) suffixes of the LCS table incremented by one, otherwise it is equal to the maximum length of the (i-1, j) and (i, j-1) suffixes of the LCS table. This table can be used to solve (5.1).

An example of the LCS table for the strings x = "Raynaud" and y = "Reynaud" is presented in Table 5.1. The length of the LCS is the last element of the table, i.e., the cell (7,7) which is equal to 5. This is valid because the LCS between x and y is "ynaud" which has a length 5. A procedure known as backtracking can then be used to derive the characters of the LCS from the LCS table. The path that is used to derive the LCS is depicted in gray highlight. The bold values in the table denote the length of the subsequence that is common in both strings. Thus, the cell (3,3) is 1 because the "y" is the first common subsequence. The cell (4,4) is 2 because the "yn" is the second common subsequence, the cell (5,5) is 3 because the "yna" is the third common subsequence, the cell (6,6) is 4 because the "ynau" is the fourth common subsequence, and the cell (7,7) is 5 because the "ynaud" is the second common subsequence.

Will the algorithm know how to derive these characters as only the lengths are recorded? This can be formulated by a procedure which is known as backtracking [42,43]. The backtracking procedure starts from the final cell of the table and follows a bottom-up approach by searching (every time) for the nonzero neighboring value that is smaller by the current value. In the case where more than one values are smaller, the path that leads to the diagonal value is followed until the diagonal cell where the characters are equal is met. In our example, the trackback path is depicted in dark gray background with purple arcs. The bottom-up approach starts from the cell (7,7) and finds the neighboring value which is located right above in the cell (6,7). Then, the algorithm finds that the next value is 3 and thus follows the path $\{(5,7) \rightarrow (4,7) \rightarrow (3,7)\}$ that leads to the diagonal element. In a similar manner, the algorithm proceeds to the diagonal cells (4,4) and (3,3) until no larger value is met. The whole path can be summarized as follows: $\{(7,7) \rightarrow (6,6) \rightarrow (5,7) \rightarrow (5,6) \rightarrow (5,5) \rightarrow (4,4) \rightarrow (3,3)\}$, which yields the LCS: "ynaud."

Algorithm 5.2 A PSEUDOCODE FOR COMPUTING THE LENGTH OF THE LCS BETWEEN TWO STRINGS.

```
def LCS(x,y):
  x_len = len(x);
  for i = 1:x_len
    for j = 1:y_len
      if(i == 0 | j == 0):
        L[i,j] = 0;
      elif(x[i-1] == y[j-1]):
        L[i,j] = L[i-1,j-1]+1;
      else:
        L[i,j] = max(L[i-1,j], L[i,j-1]);
return L;
```

Algorithm 5.3 A PSEUDOCODE FOR RECURSIVELY COMPUTING THE LENGTH OF THE LCS BETWEEN TWO STRINGS.

```
x_len = len(x);
def LCS(x, y, x_len, y_len):
  if x[x_len - 1] == y[y_len - 1]:
    return LCS(xx, yy, x_len-1, y_len-1)+1;
  else:
    return max(LCS(xx,y, x_len, y_len-1), LCS(x,yy, x_len-1, y_len));
```

However, the solution of the LCS problem involves the computation of every possible subsequence between x and y, yielding high computational complexity, especially in the case where the size of the string is large. The computational complexity of Algorithm 5.2 can be greatly reduced by solving the LCS problem in a recursive manner through dynamic programming. By solving the LCS problem in a smart, recursive way (as in Algorithm 5.3), the computational complexity can be reduced from $O(2^n)$, where n is the total number of possible subsequences, to just $O(MN)$, where M and N are the lengths of x and y, respectively.

5.6.2 Semantic matching

The lexical matching method can sometimes ignore terminologies that share a common meaning or relationship. For instance, the terms "Blood tests" and "Hematological tests" are lexically heterogeneous but share a common basis because they describe the exact same type of laboratory test. The accuracy of the lexical matching process, however, can be reduced by using external vocabularies which include medical dictionaries that can be used to identify and match synonymous or homonymous terminologies. However, in the case where a dataset, assume A, includes the blood test–related variables hemoglobin, white blood cell count, and number of platelets, and another dataset, assume B, includes the blood test–related variables erythrocyte sedimentation rate and cholesterol levels, there is no lexical matching algorithm that is capable of capturing the lexical similarity between these terms although they might express the same concept (i.e., they are both related to blood tests). As a result, the absence of knowledge regarding the relationships between the variables reduces the percentage of the matched terms and thus the harmonization performance.

What if we could somehow use this knowledge to distil the relationships between the variables that share same context? This can be accomplished by constructing a hierarchical data presentation model or a semantic presentation of the data, where the relationships (i.e., the object properties) between the variables will be well defined and then use these relations to match the semantic presentations of the data instead of matching the variables of the data themselves. One way to construct a semantic presentation of the data is to construct an ontology [1,15–18]. In an ontology [1,15–18], the data are described in the form of entities and object properties, where the entities are classes and subclasses and the object properties are the relationships between them. An example of the format of an ontology is presented in Fig. 5.1. The main class in the ontology is the "Patient." The "Patient" is connected to the subclass "Laboratory tests" through the object property "has," i.e., the patient has laboratory tests. A subclass can also consist of further subclasses. For example, the subclass "Laboratory tests" consists of the subclasses "Blood tests," "Oral tests," "Urine tests," and "Ocular tests," where the object property "consist of" is used to denote this relationship.

A subclass can also include variables where the relationship between the subclasses and the variables are denoted by the object property "include." In this example, the subclass "Blood tests" includes the variables "hemoglobin" and "white blood cell." In a similar manner, the subclass "Urine tests" includes the variables "urine pH" and "urine gravity flow," and the subclass "Ocular tests" includes the variables "Rose Bengal score" and the "ocular staining score (OSS)." To demonstrate the structural complexity that an ontology might have, the subclass "Schirmer's test" has been added under the subclass "Oral tests" which is part of the subclass "Laboratory tests." The subclass "Schirmer's test" includes two variables, namely the "date" when the test was conducted along with the test's "score." In general, the levels of an ontology can be larger, especially in disease-oriented ontologies where the domain knowledge is vast. Semantic web technologies provide a rigorous solution to automatically integrate disparate information sources and database schemas [44]. The World Wide Web Consortium (W3C) [45] provides solutions to the expression of both data and rules for reasoning. The semantic models can be expressed in different formats including (i) the Resource Description Framework (RDF) [46], (ii) the Web Ontology Language (OWL) [47], and (iii) the Extensible Markup Language (XML) [48], among others.

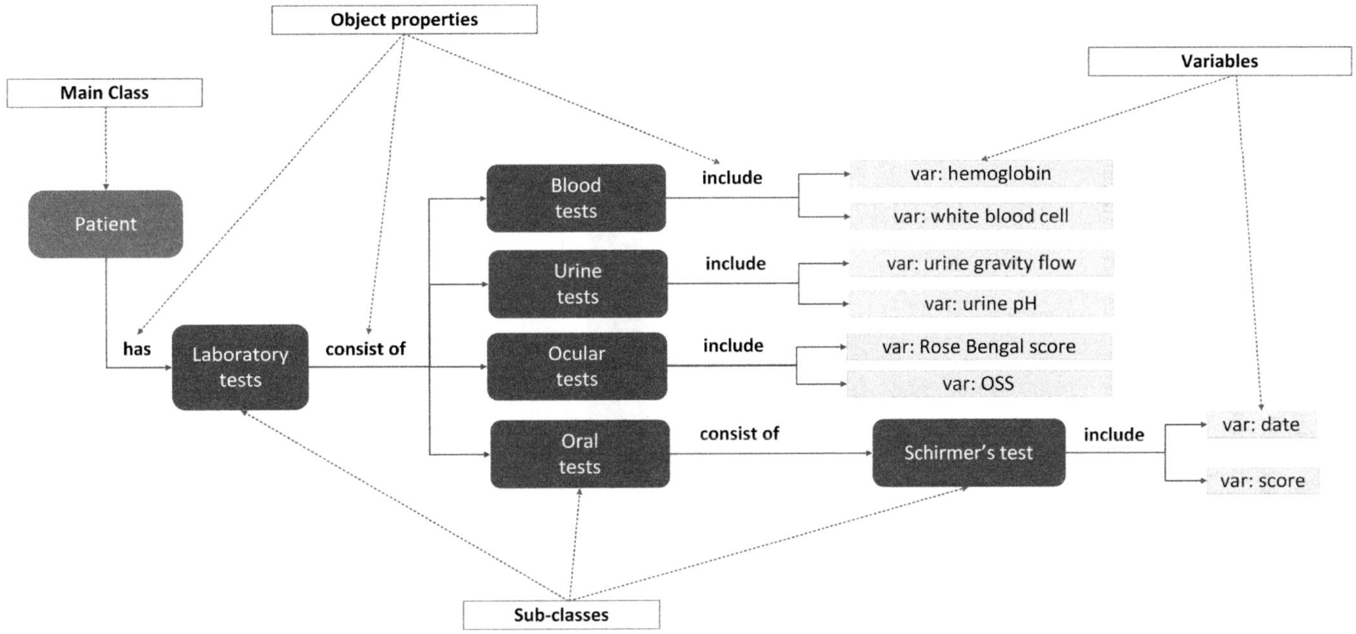

FIGURE 5.1

The fundamental components of an ontology.

Now that we have seen the fundamental components of an ontology, we can proceed with an example of the semantic matching process. Fig. 5.2 presents an extended version of the ontology which is depicted in Fig. 5.1 being a much closer instance of a disease-oriented ontology. For simplicity, the labels of the object properties have been ignored as they have been already presented in Fig. 5.1. In this ontology, the main class "Patient" includes four additional subclasses, namely the (i) "Demographics," (ii) "Treatments," (iii) "Symptoms," and (iv) "Biopsies." The subclass "Demographics" consists of three variables, namely the "gender," "education level," and "age." The subclass "Laboratory tests" has been already presented in Fig. 5.1.

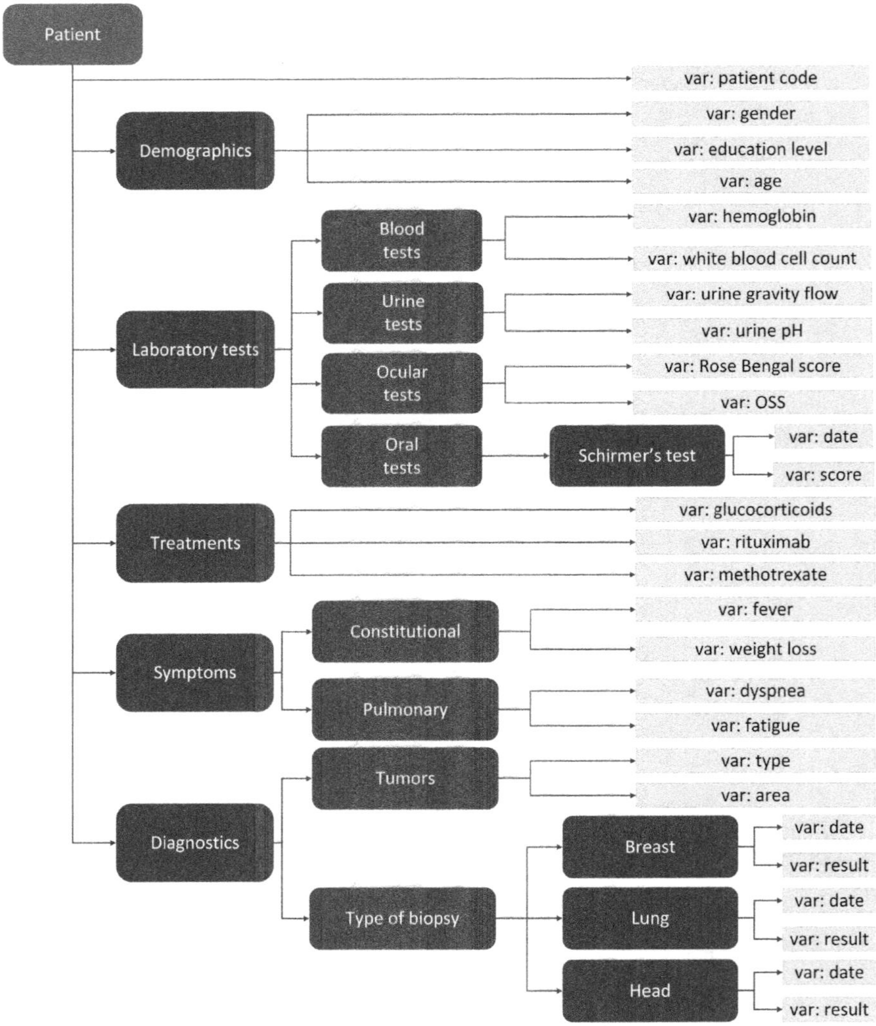

FIGURE 5.2

A larger instance of a disease-oriented ontology.

The subclass "Treatments" includes examples of prescribed drugs, such as the "glucocorticoids," the "rituximab," and the "methotrexate." The subclass "Symptoms" includes symptomatology that is related to the "constitutional" and the "pulmonary" domains, such as the "fever," the "weight loss," the "dyspnea," and the "fatigue." Finally, the subclass "Diagnostics" includes two subclasses, namely the "Tumors" and the "Type of biopsy." The former includes information regarding the "type" and the "area" of possibly identified tumors, whereas the latter includes information regarding the "date" and the "result" of three different types of common biopsies (subclasses), namely the "Breast," "Lung," and "Head." The main class can also include variables, apart from classes, such as the "patient code" in our case.

At this point, we will define a second ontology to demonstrate the semantic matching procedure, where the new ontology will be matched with the ontology of Fig. 5.2. The new ontology (Fig. 5.3) includes five subclasses, namely the (i) "Demographics," (ii) "Clinical tests," (iii) "Symptomatology," (iv) "Diagnostics," and (v) "Other." The subclasses "Demographics," "Symptomatology," and "Other" include only variables, whereas the rest of them include additional subclasses and variables. The subclass "Demographics" includes four variables, namely the "sex," "age," "ethnicity," and "date of first visit." The subclass "Symptomatology" includes four variables which are related to symptoms, including the fever, weight loss, dyspnea, and fatigue. The subclass "Other" includes five variables, namely the "bDMARDs," "rituximab," "infliximab," "cDMARDs," and any "other" drug name that has been prescribed to the patient.

Let us assume that these ontologies represent the database schemas for patient data management of two different clinical centers, assume Clinical center *A* and Clinical center *B*, respectively. An illustration of the semantic matching procedure is depicted in Fig. 5.4. The ontology on the right side is the ontology that was presented in Fig. 5.2. The ontology on the right side of Fig. 5.4 is the ontology of Fig. 5.3. The semantic matching process combines the lexical matching procedure with the semantic relations that exist between the lexically matched variables. The matched pairs of variables are depicted in green font with green links. Regarding the subclass "Demographics," the semantic matching process matches the terms "gender" and "sex" as homonymous, and the term "age" which is present in both subclasses. Because these variables are semantically related with the subclass "Demographics" through the object property "include," in both ontologies, then these two subclasses are also matched and are depicted in green font.

Now comes the interesting part: because the variables "gender" and "age" in the first ontology are matched with the variables "sex" and "age" from the second ontology, the subclasses "Demographics," in both ontologies, can also be matched due to the "include" object property that connects the matched variables with these subclasses. The fact that these two subclasses are semantically matched can reduce the information loss because the variable "education level" of the first ontology and the variables "ethnicity" and "date of first visit" in the second ontology can be brought together under the same class. This example illustrates exactly how semantic matching can significantly reduce the information loss during the data harmonization

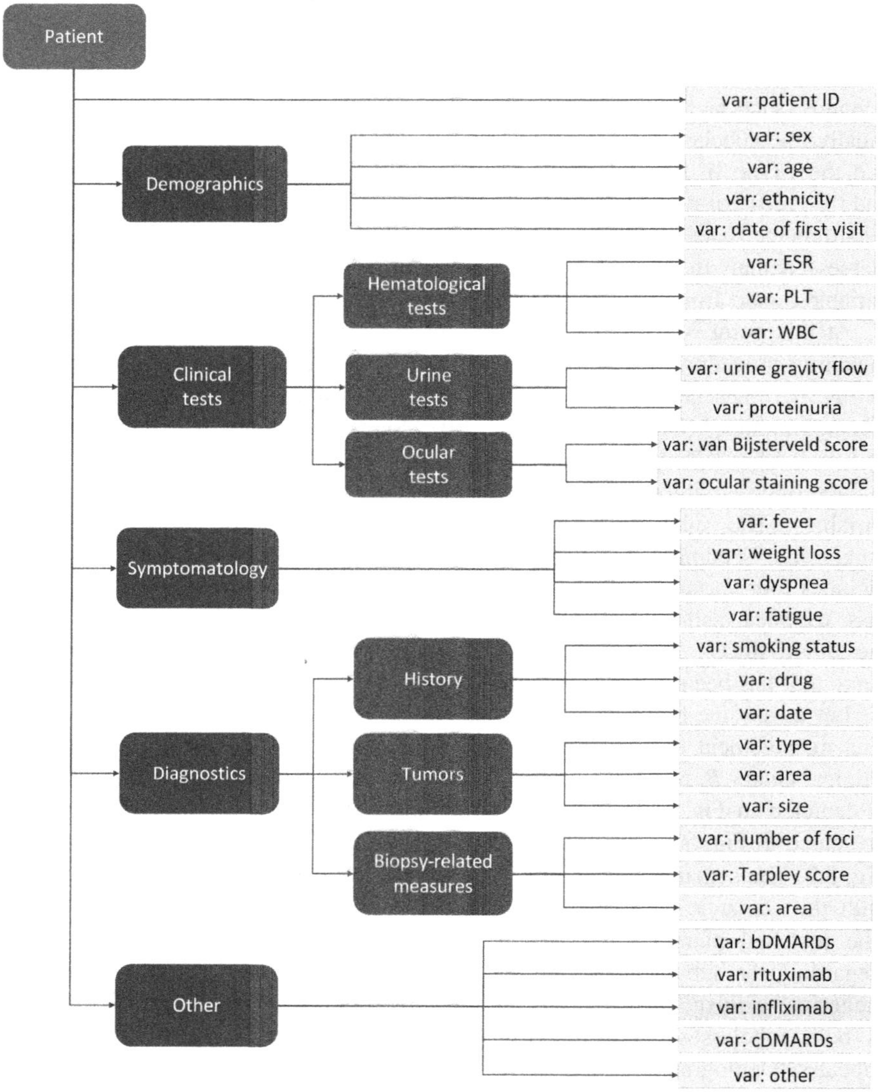

FIGURE 5.3

A second ontology.

process by taking advantage of the object properties between the lexically matched variables. Another interesting part with semantic matching is its ability to bring variables that share a common meaning but are not part of any subclass under a common subclass. This occurs in the case where the terms "patient code" and "patient ID" are matched. In that case, the semantic algorithm can generate a new subclass, say "Personal information," where the two terms can be brought together.

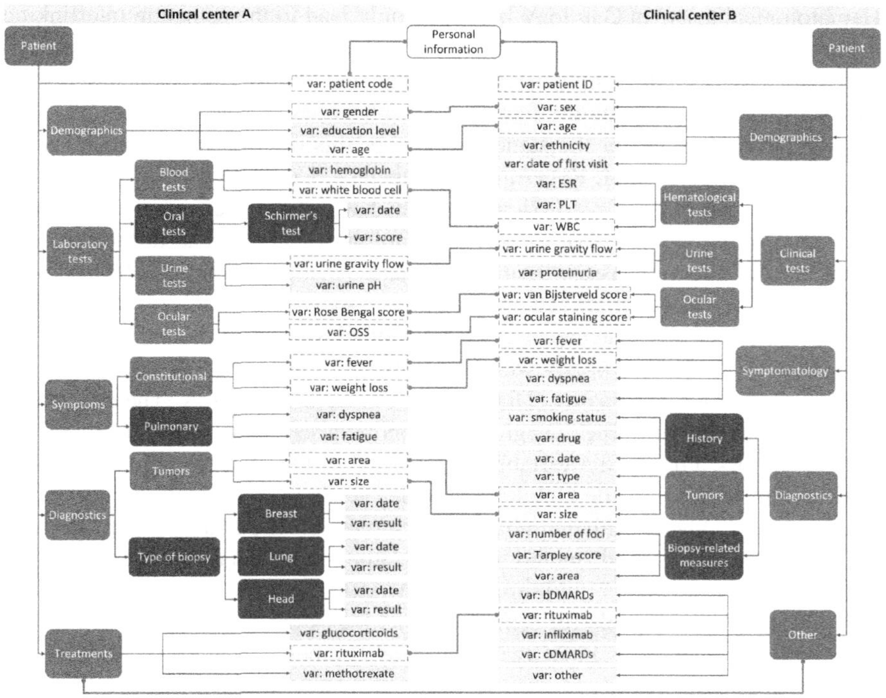

FIGURE 5.4

An illustration of the semantic matching procedure between the two ontologies.

As we already mentioned in Section 5.6.1, the lexical matching algorithm can be greatly enhanced with external medical vocabularies which can serve as medical index repositories. Indeed, the existence of an external vocabulary can provide great insight toward the identification of terms that are not lexically identical but share the same context (e.g., when they are synonymous or homonymous). For example, the term "white blood cell" of Ontology A can be lexically matched with the term "WBC" of the Ontology B, through a medical index repository that includes all possible synonyms and acronyms for the term "white blood cell." Thus, the link between these two variables can in turn lead to the semantic interlinking of the subclasses "Blood tests" and "Hematological tests." A similar example in the case of the term "OSS" which is matched with the term "ocular staining score" as the former is the latter's acronym. An instance of a lexically synonymous case occurs in the case of the term "Rose Bengal score" of Ontology A which can be matched with the term "van Bijsterveld score" in Ontology B because these two medical terms are synonymous. In fact, the Rose Bengal score in the literature is also referred to as the "van Bijsterveld score," and, thus, a medical index repository can assist the lexical matching process.

The identified match between the variable "white blood cell" of the subclass "Blood tests" in Ontology A with the variable "WBC" of the subclass

"Hematological tests" in Ontology B can not only lead to the semantic interlinking of these twosubclasses but can also lead to the interlinking of the subclasses "Laboratory tests" and "Clinical tests" through the "consist of" object property. This is also strengthened by the fact that the subclasses "Urine tests" and "Ocular tests" are also matched for the same reason. For this reason, these subclasses are also depicted in green font. Regarding the subclass "Constitutional," the variables "fever" and "weight loss" are lexically matched with the exact same terms in the subclass "Symptomatology" of the second ontology. This can lead to the semantic interlinking of the subclasses "Symptoms" and "Symptomatology." In a similar manner, the lexical link between the variables "area" and "size" of the subclass "Tumors" are lexically matched and the inner subclasses are semantically matched through the "consist of" property which in turns yields an additional semantic interlinking between the subclasses "Diagnostics," in both ontologies. The rest of the subclasses that belong to the "Diagnostics" can be also brought together to reduce information loss. Finally, the variable "rituximab" of Ontology A is matched with its homonymous term in the Ontology B which leads to the semantic interlinking of the subclass "Others" with the subclass "Treatments." As a matter of fact, the variables "glucocorticoids" and "methotrexate" in Ontology A can be concatenated with the variables "bDMARDs," "infiximab," "cDMARDs," and "other" in Ontology A.

A simplified pseudocode for semantic matching is presented in Algorithm 5.4. The algorithm receives as input two ontologies, namely ontology_A and ontology_B, and tries to solve the semantic matching problem. More specifically, the algorithm first discriminates the "include" and "consist of" object properties to isolate the variables (i.e., var_A, and var_B) from the (sub-) classes (i.e., class_A and class_B). Then, the algorithm iterates on the variables of the first ontology and tries to identify lexical matches with those from the second ontology. Once a lexical match is found, the algorithm stores the matched variables in a list along with their hierarchical level (class, subclass, etc.) in each ontology and then matches the classes (or subclasses) where these variables belong to.

The pseudocode for lexical matching is presented in Algorithm 5.5.

5.6.3 Statistical harmonization

5.6.3.1 Item response theory

As we have seen so far, the lexical and semantic matching methods work well with gold standard models and pairing rules to quantify the associations between the variables. Apart from the use of a gold standard models, like ontologies, there are other mathematical approaches that make use of statistical methods to quantify the associations between the variables in heterogeneous data to harmonize measures that share a common basis across them and thus commensurate these measures. An example of such an approach is the IRT, which is also known in the literature as latent trait theory [19–21,49]. IRT is widely used in hypothesis testing to develop scales (or latent traits) that measure the same constructs to quantify the information value of one (or more) item(s), where the constructs are the measurement levels (e.g., low, medium, high), and the "item" includes all types of informative items (e.g., depression).

Algorithm 5.4 A PSEUDOCODE FOR SEMANTIC MATCHING.

```
def semantic_matching(ontology_A, ontology_B):
    obj_prop_A = ontology_A.object_properties; #retrieve the object properties of
    ontology A
    obj_prop_B = ontology_B.object_properties; #retrieve the object properties of
    ontology B
    var_A = [obj_prop_A[i] for i, x in obj_prop_A if x == "include"]; #variables in
    ontology A.
    var_B = [obj_prop_B[i] for i, x in obj_prop_B if x == "include"]; #variables in
    ontology B
    class_A = [obj_prop_A[i] for i, x in obj_prop_A if x == "consist of"]; #classes in
    ontology A
    class_B = [obj_prop_B[i] for i, x in obj_prop_B if x == "consist of"]; #classes in
    ontology B
    c1 = 0; #initialize a simple counter
    for i in len(var_A):
      for j in len(var_B):
        if (match(var_A[i], var_B[j]) == 1) | (match(var_A[i], var_B[j]) == 2):
          matched_vars[k] = list([var_A[i], var_B[j]]); #keep the variables
          matched_inds[k] = list([i, j]); #keep the indices of the variables
          matched_class[k] = class_A[i] Uclass_B[j]; #unify the contents of the matched
classes
          c1 += 1; #update the counter
end
```

Algorithm 5.5 A PSEUDOCODE FOR LEXICAL MATCHING.

```
def match(var_A, var_B):
    dist = distance(var_A, var_B); #compute a lexical distance score
    synsets = nltk.synsets(var_B); #compute the synsets
    if (var_A in synsets):
       return 1;
    elif (dist >= th): #a typical threshold for ignoring lexically non-similar pairs
       return 2;
    else:
       return 0;
```

It is obvious that the items that are available for each construct may vary in the scale level. The purpose of IRT in the field of statistical data harmonization is to homogenize scales (traits) that measure the same item (e.g., depression) in heterogeneous data to transform it into a common metric at the same scale [21]. The overall scope of IRT is to enable the integrative data analysis (IDA) of heterogeneous data from multiple studies which is feasible only when the measures that are available in each study must [21] (i) reflect the same construct (or latent trait), (ii) have the same meaning, and (iii) have the same scale (i.e., measurement equivalence). In general, the type of an item can either be a continuous or a discrete variable. An example of a continuous variable is age (which is usually measured in years), whereas an example of a discrete variable is the cholesterol level (e.g., low, medium, high). The variables can be count numbers, ordinal numbers, binary values, etc. A significant challenge in IRT is the underlying need to account for different effects during the hypothesis testing procedure, including regional, age, and gender effects, among others.

IRT uses a logistic function to model the relationship between a binary item, assume $\boldsymbol{x_i}$, drawn by a set of k-binary items, assume $\{\boldsymbol{x}_1, \boldsymbol{x}_2, \ldots, \boldsymbol{x}_k\}$, and a single latent factor, assume $\boldsymbol{f_n}$, with a score $\boldsymbol{\theta_n}$, as follows:

$$P_i(\boldsymbol{\theta_n}) = P(\boldsymbol{x_i} = 1 | \boldsymbol{\theta_n}, a_i, b_i) = \frac{1}{1 + exp(-a_i(\boldsymbol{\theta_n} - b_i))}, \tag{5.5}$$

where a_i and b_i are the discrimination and difficulty parameters of $\boldsymbol{x_i}$. The model in (5.5) is known as the 2-parameter logistic IRT or 2-PL IRT model. In fact, the 2-PL IRT model uses a Bernoulli distribution as a binary distribution to estimate the conditional probability that the item score is equal to 1, given the latent factor $\boldsymbol{\theta_n}$. A fundamental assumption here is that the latent factor follows a standard normal distribution with zero mean and variance equal to one, so that $\boldsymbol{\theta_n} \sim N(\boldsymbol{0}, \boldsymbol{1})$. The difficulty parameter indicates the extent to which the latent factor yields a 50% probability estimation in (5.4), whereas the discrimination parameter indicates the positive correlation between the item and the latent factor. The number of parameters in the IRT model directly affects the model's complexity.

For example, the 3-PL IRT model is an extension of (5.5):

$$P_i(\boldsymbol{\theta_n}) = P(\boldsymbol{x_i} = 1 | \boldsymbol{\theta_n}, a_i, b_i, c_i) = c_i + \frac{1 - c_i}{1 + exp(a_i(\boldsymbol{\theta_n} - b_i))}, \tag{5.6}$$

where c_i is the lower asymptote or pseudo-guessing parameter of $\boldsymbol{x_i}$ with values in the range [0,1]. In a similar manner, the 4-PL IRT model is an extension of (5.6):

$$P_i(\boldsymbol{\theta_n}) = P(\boldsymbol{x_i} = 1 | \boldsymbol{\theta_n}, a_i, b_i, c_i, d_i) = c_i + \frac{d_i - c_i}{1 + exp(a_i(\boldsymbol{\theta_n} - b_i))}, \tag{5.7}$$

where d_i is the upper asymptote or inattention level of the item $\boldsymbol{x_i}$ with values in the range [0, 1]. The effect of these parameters in the distribution's shape is depicted in Fig. 5.5. To better understand the effect of each parameter on the distribution's shape, we have simulated the distributions in (5.5)–(5.7) for different values of a,

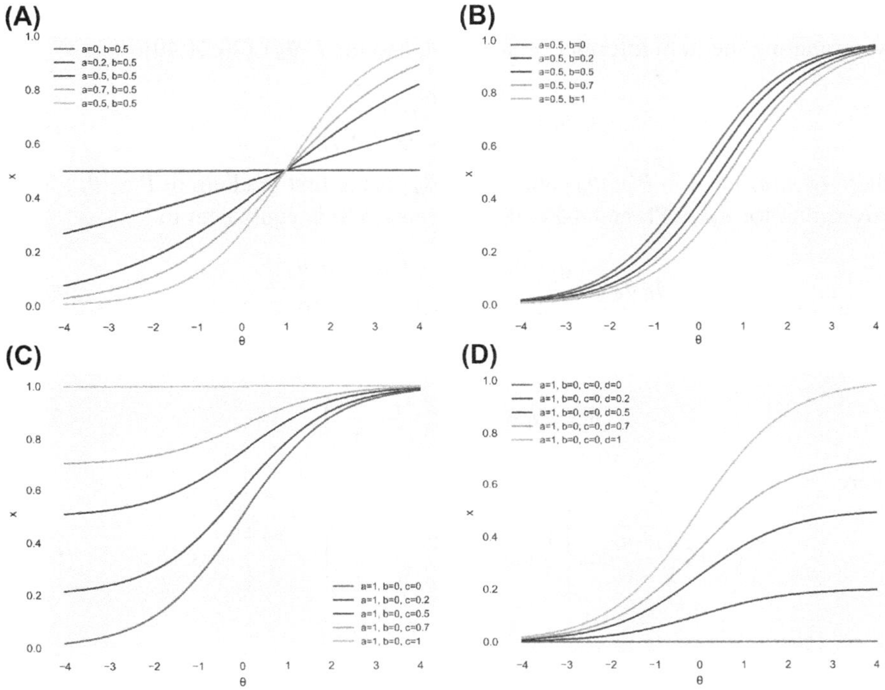

FIGURE 5.5

The effects of the (A) discrimination (a), (B) difficulty (c), (C) pseudo-guessing (c), and (D) upper asymptote (d) parameters in the shape of a standard logistic distribution for 1000 where $-4 \leq \theta \leq 4$ (horizontal axis). The vertical axis corresponds to the values of a 2-PL IRT model for the two parameters in (A) and (B), a 3-PL IRT model for the three parameters in (C), and a 4-PL IRT model for the four parameters in (D).

b, c, and d. In all cases, we assume a single item, so that $\boldsymbol{x_i}$ drops into x. The discrimination parameter affects the slope of the distribution and causes the distribution to stretch on the horizontal axis (Fig. 5.5A). The smaller the discrimination value, the larger the stretching, and for values equal to zero, the distribution is completely stretched and in fact overlaps with the horizontal axis. The difficulty parameter affects the slope of the distribution and causes the distribution to scale on the horizontal axis (Fig. 5.5B). Large upper-asymptote values cause the distribution to compress on the vertical scale (Fig. 5.5C), whereas high c values cause the distribute to decompress on the vertical scale (Fig. 5.5D). The probability of getting b is equal to $(1+c)/2$ which is equal to the point of the distribution where the slope is maximized.

So how can we use the IRT method to measure the differential item functioning (DIF) between two common items in different scaled data? The goal of IRT is to find the factor $\boldsymbol{\theta_n}$ that maximizes the information that is given by $\boldsymbol{x_i}$. One way to do this is to quantify how informative $\boldsymbol{x}_i$ is for different levels of $\boldsymbol{\theta}_n$. This can be accomplished by computing the item information function, assume $I_i(\boldsymbol{\theta}_n)$ [20,21,49]:

$$I_i(\boldsymbol{\theta}_n) = \frac{\nabla P(\boldsymbol{x}_i|\boldsymbol{\theta}_n)^2}{P(\boldsymbol{x}_i|\boldsymbol{\theta}_n)Q(\boldsymbol{x}_i|\boldsymbol{\theta}_n)}, \tag{5.8}$$

where $Q(\boldsymbol{x}_i|\boldsymbol{\theta}_n) = 1 - P(\boldsymbol{x}_i|\boldsymbol{\theta}_n)$ and $\nabla P(\boldsymbol{x}_i|\boldsymbol{\theta}_n)$ is the first gradient of $P(\boldsymbol{x}_i|\boldsymbol{\theta}_n)$. It is proven that for the 3-PL and 4-PL IRT models, (5.8) is equivalent to

$$I_i(\boldsymbol{\theta}_n) = \frac{a_i 2[P(\boldsymbol{\theta}_n) - c_i]^2[d_i - P(\boldsymbol{\theta}_n)]^2}{(d_i - c_i)^2 P(\boldsymbol{\theta}_n)[1 - P(\boldsymbol{\theta}_n)]}. \tag{5.9}$$

Maximizing (5.8) yields the following estimation:

$$\boldsymbol{\theta}_n^* = b_i + \frac{1}{a_i}\log\left(\frac{\boldsymbol{x}_i^* - c_i}{d_i - \boldsymbol{x}_i^*}\right), \tag{5.10}$$

where

$$\boldsymbol{x}^* = 2\sqrt{\frac{-u_i}{3}}\cos\left\{\frac{1}{3}\operatorname{acos}\left(-\frac{v_i}{2}\sqrt{\frac{27}{-u_i^3}}\right) + \frac{4\pi}{3}\right\} + 0.5, \tag{5.11}$$

$$u_i = -\frac{3}{4} + \frac{c_i - d_i - 2c_i d_i}{2}, \tag{5.12}$$

$$v_i = -\frac{c_i + d_i - 1}{4}. \tag{5.13}$$

In an attempt to demonstrate how the maximum information value can be used to measure the DIF between two common items in different datasets, we assume a given set of parameters, say $\{a, b, c, d\} = \{1.3, -1, 0.3, 0.9\}$, that corresponds to the distribution of the common item. We use the 4-PL IRT model to compute the maximum information value, θ, for the item, where the maximum information function is evaluated on 1000 evenly spaced samples that lie on the interval [-4, 4]. First, we calculate the item information function from (5.8), for the given set of parameters. The 4-PL distribution is depicted in Fig. 5.6A as a sigmoid function like those in Fig. 5.5. Then, we substitute the conditional probability $P(\theta)$ in (5.9) to calculate the distribution $I(\theta)$ which is depicted in Fig. 5.6B. The estimated value that maximizes the information function is given by (5.10) and is equal to -0.879. The red vertical line in Fig. 5.6B corresponds to the estimate value θ. This distribution can be used as a calibrated distribution, so that a new distribution that achieves a maximum information value equal to the calibrated can be considered as equivalent in terms of measuring the item in the same way.

A typical method for measuring the DIF is the likelihood ratio (LR) test [50]. According to the LR test, the likelihood that is obtained from the item in the first dataset, assume $P(\theta_0)$, is used as the null model and the model from the second dataset is used as the alternative one, assume $P(\theta_1)$. The LR method compares the maximum likelihood (ML) estimates that are obtained from the two models, assume θ_0 and θ_1,

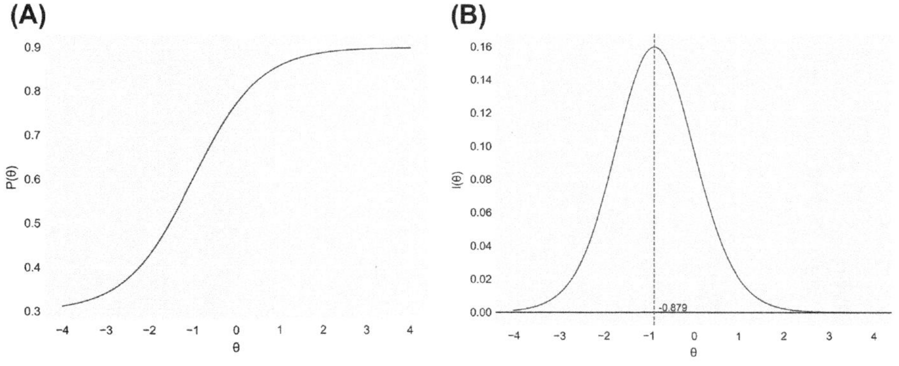

FIGURE 5.6

(A) The distribution of 4-PL IRT model, $P(\theta)$, with parameters $\{a, b, c, d\} = \{1.3, -1, 0.3, 0.9\}$ along with (B) the distribution of the latent factor for $-4 \leq \theta \leq 4$. The maximum information value is depicted in a *red [black in print version] vertical line.*

respectively, using the ML estimates as the unknown parameters in the distributions. Then, the LR method computes the LR, assume $\lambda = \theta_0/\theta_1$, which is related to the conventional chi-square statistical test [50], χ^2, through the following equation:

$$\chi^2 = -2\ln(\lambda). \tag{5.14}$$

Then, the null hypothesis is rejected when χ^2 is larger than the confidence level that corresponds to the chi-square distribution with k-degrees of freedom (k corresponds to the number of parameters that differ between the null and the alternative models). In that case, the ML estimates between these values differ significantly, and thus the information between them is less informative.

5.6.3.2 Linear factor and multiple factor analysis

In the case where the items are not discrete but continuous, LFA can be applied. Factor analysis is a general field in statistics which involves the calculation of joint variations between the variables (items) of interest and a predefined set of factors toward the identification of independent latent variables [19]. LFA is limited only on studying the effect of one latent factor on a set of items. For a given set of k-observations, assume $\{\boldsymbol{x}_1, \boldsymbol{x}_2, \ldots, \boldsymbol{x}_k\}$, and a single latent variable (factor), assume $\boldsymbol{f}_n$, the simple LFA method tries to model the observed item $\boldsymbol{x}_i$ as a linear combination of the single factor $\boldsymbol{f}_n$ as follows:

$$\boldsymbol{x}_i = w_i \boldsymbol{f}_n + \boldsymbol{e}_i, \tag{5.15}$$

where w_i is the interception parameter (weight) of the factor $\boldsymbol{f}_n$ on the observation $\boldsymbol{x}_i$, and e_i is the error residual. In fact, (5.15) is a simple linear regression model which models the relationship between the dependent variable (the item) and a single independent (exploratory) variable (the factor). An example of the LFA is depicted in Fig. 5.7 which demonstrates the relationship between a random item that corresponds to the IVSd (interventricular septal end diastole) and the age.

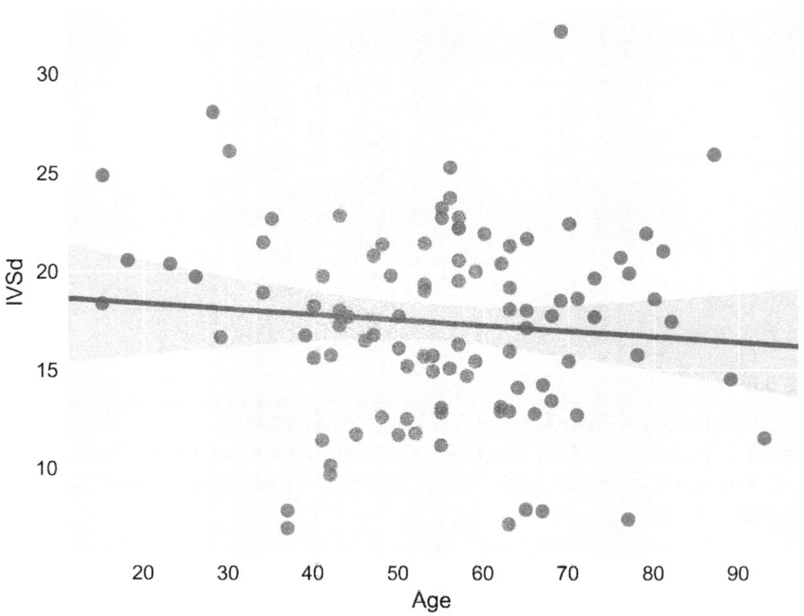

FIGURE 5.7

An example of a linear factor analysis where IVSd (interventricular septal end diastole) is the item (y-axis) and age is the factor (x-axis). The shaded region corresponds to the variance of the model.

In the majority of the cases, however, we are interested in studying the effect of one or more factors on one item. Thus, the simple LFA model can be extended to support more than one factor. For a given set of n-latent variables (factors), assume $\{\boldsymbol{f}_1, \boldsymbol{f}_2, \ldots, \boldsymbol{f}_n\}$, where $n < k$, the multiple factor analysis method tries to model the observed item $\boldsymbol{x}_i$ as a linear combination of the factors $\{\boldsymbol{f}_1, \boldsymbol{f}_2, \ldots, \boldsymbol{f}_n\}$ as follows:

$$\boldsymbol{x}_i = w_{i,1}\boldsymbol{f}_1 + w_{i,2}\boldsymbol{f}_2 + \ldots + w_{i,n}\boldsymbol{f}_n + e_i = \sum_{j=1}^{n} w_{ij}\boldsymbol{f}_j + e_i \ (i = 1, 2, \ldots, k), \quad (5.16)$$

where $w_{i,j}$ is the interception parameter (weight) of the factor $\boldsymbol{f}_j$ on the observation $\boldsymbol{x}_i$, and e_i is the error residual. In fact, (5.16) tries to model the observed item $\boldsymbol{x}_i$ on a set of factors $\boldsymbol{f}_j, \boldsymbol{j} \in \{1, 2, \ldots, \boldsymbol{n}\}$ through a linear combination that reflects the variability (or correlation) that exists between them [19]. This type of analysis is also known as multivariate linear regression [51]. An example of a multivariate linear regression analysis is depicted in Fig. 5.8 between the previous item (IVSd) and two additional random factors, namely the BMI (body mass index) and the LAVs (left atrial volume size).

A more generalized version of (5.16) can be written as follows:

$$\boldsymbol{x}_i = \boldsymbol{f}_i^T \boldsymbol{w} + e_i, \quad (5.17)$$

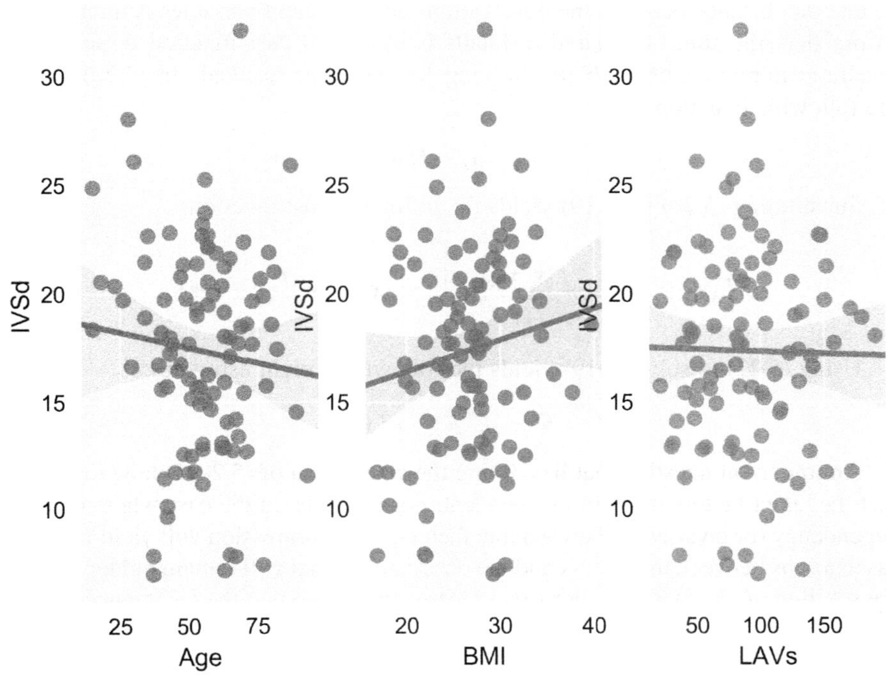

FIGURE 5.8

An example of a multiple-factor linear regression model where the item is again the IVSd (y-axis) and the factors (x-axis) are the (i) age, (ii) BMI (body mass index), and (iii) LAVs (left atrial volume size). The shaded regions on each subplot correspond to the variance of the corresponding model.

where $w = (w_1, w_2, \ldots, w_n)$ is the weight vector. The matrix presentation of (5.6) can be written as follows:

$$x = \sum_{i=1}^{k} \sum_{j=1}^{n} F_{ij} w_j + e_i = Fw + e, \tag{5.18}$$

where $x = (x_1, x_2, \ldots, x_k)$ is the vector with the observations (k denotes the number of items-regression equations and n the number of observations-factors),$F \in R^{kxn}$ is the design matrix, where the column f_j is the jth factor, $w \in R^{nx1}$ is the weight vector, and $e \in R^{nx1}$ is a vector with the error residuals for each x_i. The weight coefficients (interception parameters), say $\widehat{w}$, can be estimated by

$$\widehat{w} = argmin_w\, L(w), \tag{5.19}$$

where $L(w)$ is a cost function that minimizes the error residuals:

$$L(w) = X - Fw^2 = e^2, \tag{5.20}$$

so that the distance between the observations and the latent variables is minimized. In practice, the sum of squared residuals is used as a cost function to solve the quadratic minimization problem. Solving for the error residuals in (5.20) yields the following equation:

$$e_i = x_i - f_i^T w. \tag{5.21}$$

Substituting (5.20) in (5.19) yields the following cost function:

$$L(w) = \sum_{i=1}^{k} \left| x_i - \sum_{j=1}^{n} F_{ij} w_j \right| 2 = (x - Fw)^T (x - Fw). \tag{5.22}$$

Using (5.22) to solve (5.19) yields the following weight estimates:

$$\hat{w} = \left(F^T F\right)^{-1} F^T x. \tag{5.23}$$

An important question that lies during the calculation of (5.20) is how to ensure that the latent factors in F are independent. For example, in the case where there is dependency (or invariance) between the factors, linear regression will yield falsified associations between the factors and the observations that will remain hidden during the solution of (5.19). One way to avoid this is to assume that the covariance matrix of F, assume $\Sigma_F \in R^{nxn}$, is equal to the identity matrix I_n so that the interdependencies between the factors are zero. Moreover, in statistical studies, where the factorial invariance is examined in two or more groups which are defined based on age, gender, etc., the researchers need to ensure that the factor means and variances are equal among the groups. If this stands, then the factorial invariance is said to be strong otherwise the factorial invariance is considered to be partial [19,20]. In the case where the latent variables are of different scales, the z-score value can be computed to transform the variables so that they have zero mean and variance equal to one. The z-score of the item x_i, assume z_i, is defined as follows:

$$z_i = \frac{x_i - \mu_i}{\sigma_i}, \tag{5.24}$$

where μ_i is the mean vector of the ith observation and σ_i is the standard deviation. Thus, (5.17) becomes

$$Z = Fw + e, \tag{5.25}$$

where $Z \in R^{kxm}$ is the normalized version of X. Although linear factor and multiple factor analyses provide an easy tool for modeling linear relationships between the items and the latent factors, the fact that it is limited to continuous variables hampers its adoption in cases where the variables are discrete (e.g., binary).

5.6.3.3 Generalized linear factor analysis

The GLFA is an extension of the conventional LFA in the case where the set of the items includes variables that are not only continuous but also discrete, including binary, ordinal, and count variables [19,51]. The GLFA model is based on the definition of the generalized linear model (GLM), where each item x_i is assumed to be derived from a distribution that belongs to the exponential family. The GLM approach is based on the fact that the mean of the exponential distribution is associated with the latent factors that are stored in matrix F. So, one way to identify this relationship is to estimate the mean and the variance of the distribution that is related to the independent factors. For a given set of k observations, assume $\{x_1, x_2, \ldots, x_k\}$, and n-latent variables (factors), assume $\{f_1, f_2, \ldots, f_n\}$, where $n < k$, the mean vector, assume $\boldsymbol{\mu}$, can be estimated as follows:

$$\boldsymbol{\mu} = \boldsymbol{E}(\boldsymbol{x}|\boldsymbol{F}) = \boldsymbol{g}^{-1}(\boldsymbol{Fw} + \boldsymbol{e}), \tag{5.26}$$

where $\boldsymbol{g}^{-1}()$ is the inverse link function, and $\boldsymbol{Fw} + \boldsymbol{e}$ is the linear regression equation in (5.8). The variance vector, assume $\boldsymbol{v}$, can be estimated as a function of $\boldsymbol{\mu}$:

$$\boldsymbol{v}(\boldsymbol{\mu}) = \boldsymbol{v}\left(\boldsymbol{g}^{-1}(\boldsymbol{Fw} + \boldsymbol{e})\right). \tag{5.27}$$

As we already mentioned, the GLM uses the (inverse) link function to model the relationship between the items and the factors, where the link functions belong to the exponential family. The latter includes a variety of distributions, varying from the normal and exponential distributions to the Poisson and Bernoulli distributions [19,51]. The exponential family is convenient because it includes different types of distributions that support (i) items with real values, including the normal and gamma distributions, (ii) items with count of occurrences, such as the Poisson distribution, and (iii) binary items, such as the Bernoulli and the binomial distribution. In the case where the link function is equal to the Bernoulli function, the GLM model is equal to the 2-PL IRT model, because the Bernoulli distribution uses a logit function, and thus the GLM can be used to reproduce the 2-PL IRT model. More specifically, in the case of the Bernoulli distribution, the link function, $\boldsymbol{g}(\boldsymbol{\mu})$, is given by the logit (log-odds or logarithm of the odds) function:

$$\boldsymbol{g}(\boldsymbol{\mu}) = ln\left(\frac{\boldsymbol{\mu}}{1 - \boldsymbol{\mu}}\right). \tag{5.28}$$

Substituting (5.28) in (5.27) yields the following mean estimation:

$$\boldsymbol{\mu} = \frac{1}{1 + \exp[-(\boldsymbol{Fw} + \boldsymbol{e})]}, \tag{5.29}$$

where the exponential in the denominator can be rewritten as follows:

$$\exp[-(\boldsymbol{Fw} + \boldsymbol{e})] = exp\left[-\boldsymbol{w}\left(\boldsymbol{F} + \boldsymbol{w}^{-1}\boldsymbol{e}\right)\right], \tag{5.30}$$

which is equal to the 2-PL IRT model in (5.5). The most common way to estimate the weight vector $\boldsymbol{w}$ is to solve the ML problem [19,51].

5.6.3.4 Moderated nonlinear factor analysis

Although the single and GLFAs provide a common approach for modeling linear associations between the observations and the independent factors that can be easily understood, they are only constrained to continuous items (and discrete) with linear associations between them and to item responses with distributions that are limited to the exponential family [19,52]. In addition, MNFA considers the presence of nonlinearities between the items and the parameters. The distribution of the latent factors, however, is once more assumed to be normal. The interesting part with MNFA though is the fact that it accounts for "external" (or "exogenous") variables, which are referred to as moderator variables, which affect the parameters of the item's response and thus introduce biases during the factor analysis [19]. For example, in GLFA, the parameters that formulate the item's distribution (e.g., the interception and loading factors) are assumed to be equal among different population subgroups (e.g., between males and females). This means that GLFA assumes that the factors are invariant among the population subgroups which are ideal because the items and the factors, as well, are affected by one or more covariates (moderator variables) [19]. Examples of moderator variables include the age, gender, nationality, and any other discrete or continuous variable that affects the model's distribution parameters. The difference in the model's distribution is reflected by the model's parameters which include the mean, the variance, and the weights of the factors.

Let us assume a set of q moderators, say $\{m_1, m_2, \ldots, m_q\}$. Let us also assume that the effect of the moderator m_j on the ith factor is defined as m_{ij}. Then, the mean of the ith factor, say μ_i, can be modeled as follows:

$$\mu_i = a_0 + \sum_{j=1}^{q} a_j m_{ij}, \tag{5.31}$$

where a_j is the effect of the moderator m_{ij} on the mean of the ith factor's distribution and a_0 is the effect of the moderator on the mean when the moderator is equal to zero.

In a similar manner, the variance of the ith factor, say v_i, can be modeled as follows:

$$v_i = z_0 \exp\left(\sum_{j=1}^{q} z_j m_{ij}\right), \tag{5.32}$$

where z_j is the effect of the moderator m_{ij} on the variance of the ith factor's distribution, and z_0 is the effect of the moderator on the variance when the moderator is equal to zero. Typical values for a_0 and z_0 are 0 and 1, respectively, to normalize the scale of the factor under examination. Eq. (5.31) models the mean as a linear function, whereas (5.32) models the variance through a logistic function. The logistic model is a nonlinear model similar to the one that is used by IRT model to study the relationship of one or more factors on the item's response. Until now, we have seen how the moderators can affect the mean and the variance of the factors through Eqs. (5.31) and (5.32). To understand the effect of the moderators on the items this time, the GLFA model of (5.26) will be rewritten in the following analytic form:

$$\mu_{ij} = \boldsymbol{g}^{-1}(e_{ij} + w_{ij}\boldsymbol{f}_j), \tag{5.33}$$

where μ_{ij} is the mean of the jth factor given the ith item, w_{ij} is the weight (loading) parameter of factor f_j on the ith item, and e_{ij} is the interception parameter of factor f_j on the ith item. In fact, (5.33) introduces the effect of the moderator as an additional dimension on the item's response through the j subscript. The interception factor can then be rewritten as follows:

$$e_{ij} = e_{0i} + \sum_{l=1}^{q} e_{li} m_{lj}, \tag{5.34}$$

where e_{li} is the interception parameter for the ith item as a function of the moderator m_l on the jth factor, and e_{0i} is the intercept value of the ith item when all moderators are zero. In a similar manner, the weight (loading) parameter can be expressed as follows:

$$w_{ij} = w_{0i} + \sum_{l=1}^{q} w_{li} m_{lj}, \tag{5.35}$$

where w_{li} is the weight (loading) parameter for the ith item as a function of the moderator m_l on the jth factor, and w_{0i} is the weight value of the ith item when all moderators are zero. It is proven that Eqs. (5.31), (5.33), (5.34), and (5.35) can be properly used to construct LFA and IRT models that were previously described in Sections 5.6.1 and 5.6.2 [19–21,51]. Indeed, in the case of the GLFA, the link function in (5.33) is assumed to be the identity function and thus

$$\mu_{ij} = e_{ij} + w_{ij} \boldsymbol{f}_j. \tag{5.36}$$

In the case where the link function is a logistic function, i.e., in the case where the item's response is binary, (5.36) becomes

$$\mu_{ij} = \frac{1}{1 + exp(-(e_i + w_i f_j))}, \tag{5.37}$$

which yields the 2-PL IRT model that we have already met in Section 5.6.1.

To better understand the difference between GLFA and MNFA, we consider the problem of studying depression using a set of three factors, namely the stress level, the alcohol consumption, the drug involvement. The item here is depression, and the latent factors are the stress level, the alcohol consumption, and the drug involvement. The GLFA approach models the item using the general schema that is depicted in Fig. 5.9A which uses a linear function between the item and the two factors. On the other hand, the MNFA is able to model the item using one or more moderators (covariates) (Fig. 5.9B) that affect the mean and variance of the item and the three latent factors. In our example, GLFA models the depression as a linear function of the stress level, the alcohol consumption, and the drug involvement (Fig. 5.9C), whereas the MNFA models the depression as a linear function of the stress level, the alcohol consumption, and the drug involvement by accounting for the moderators age and gender (Fig. 5.9D).

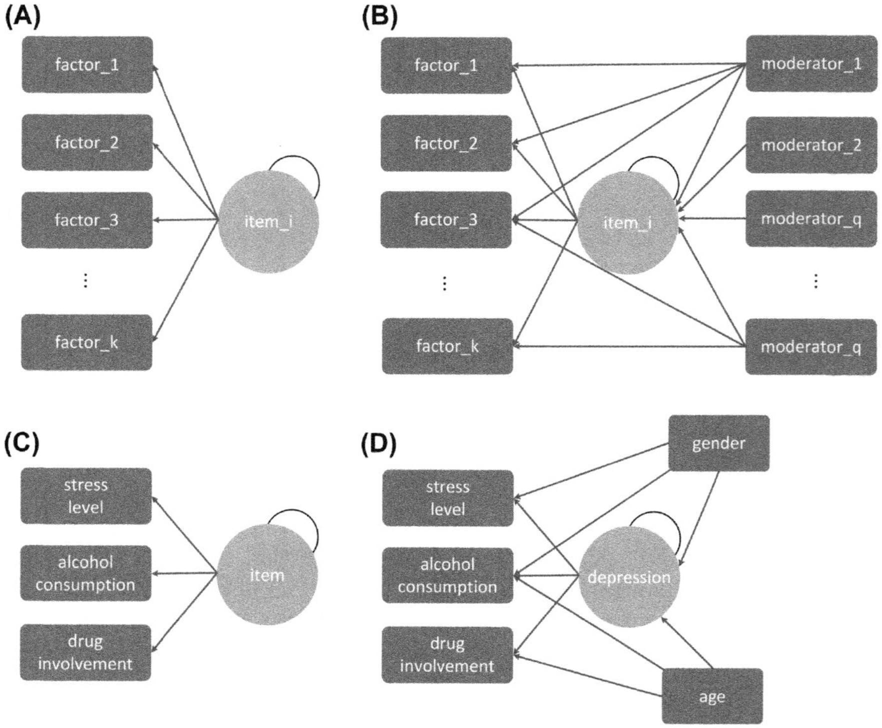

FIGURE 5.9

Differences between the GLFA and the MNFA methods for a single item. (A) The general schema of GLFA with k factors, (B) the general schema of MNFA with k factors and q moderators, (C) an example with three factors for modeling depression through GLFA, and (D) an example of three factors and two moderators for modeling depression through MNFA.

Thus, the MNFA can model the DIF between two item responses by taking into consideration the effect that a set of one or more moderator variables would have in a set of one or more factors, given the item's response. In the case where a common item is considered across two studies, the DIF can be measured between the common item on the two studies in different population subgroups. In our previous example, MNFA can be used to measure the DIF between depression in two different studies using the three latent factors as in the previous sections but this time across the males and females and different age groups. MNFA has been employed in Ref. [19] to normalize measurement scales in the alcohol involvement between young children and adolescents using data that were obtained from two longitudinal studies involving interviews and annual assessments. The results indicated a positive and statistically significant relationship between alcohol involvement and five indicators (disorder, consequences, heavy use, use frequency, and positive alcohol expectancies) [19].

5.7 Existing frameworks

Several data harmonization frameworks have been launched to enable the IDA of heterogeneous medical data, such as clinical and genomic data, the majority of which is presented below, including the DataSHaPER (DataSchema and Harmonization Platform for Epidemiological Research) [4] framework, the SORTA (System for Ontology-based Re-coding and Technical Annotation of biomedical phenotype data) tool [7], the BiobankConnect software tool [5], the S-match semantic matching framework [9], and finally the FOAM (framework for ontology alignment and matching) [8].

The DataSHaPER (DataSchema and Harmonization Platform for Epidemiological Research) framework [4] was developed under the BioSHaRE project [53] to enable the harmonization of heterogeneous biobanks. The DataSHaPER uses a DataSchema (which is also referred to as "Generic DataSchema") as a reference model which consists of a core set of more than 180 variables that belong to 45 domains and 13 themes [53]. To understand the difference in the definition of a domain and a theme, an example of a theme is the class "Laboratory tests" that we have already seen in Section 5.8. An example of a domain is the subclass "Blood tests" and an example of a variable is "hemoglobin" that belongs to the "Blood tests." The DataSchema also includes 3 modules which are more abstract entities that include the domains, the themes, and the variables. The DataSHaPER framework supports the harmonization of both prospective and retrospective studies only when the reference model is exclusively defined in a DataSchema format. The harmonization strategy involves the execution of a three-step procedure which involves [53]

- the development of pairing rules that quantify the ability of each variable in the study to generate the variables of the "DataSchema,"
- the application of the pairing rules to classify each variable in the study into "complete," "partial," or "impossible" according to its ability to generate the variables of the "Generic DataSchema," and
- the development of a processing algorithm that can automatically generate the variables of the "DataSchema" from the variables that have been marked as potentially matched.

The processing algorithm that enables data harmonization is executed through the Opal software [54] as soon as the harmonized DataSchema is constructed and distributed to the biobanks. The harmonized DataSchema includes a core set of variables which is related with the domain of interest and are selected according to seven criteria [4]. The biobank data are stored in dedicated Opal servers [54] which process the data to compute the harmonized DataSchema specifications. Then, the harmonized DataSchema specifications from each individual study are converted to the variables in the DataSchema format [4]. The individual harmonized datasets are stored in the Opal servers of each biobank and delivered in remote clients through the Mica web portal [55]. The overall procedure takes place under a secure cloud infrastructure known as DataSHIELD (Data Aggregation Through Anonymous Summary-statistics from Harmonized Individual-levEL Databases) [56] which will be extensively described later on, in Section 6.1 (Chapter 6). The DataSHaPER approach has been applied in Ref. [4] to harmonize eight retrospective cohort studies with more than 200,000 individuals under the Healthy Obese Project (HOP), yielding a 70% harmonization accuracy toward the identification of variables with shared terminology.

The BiobankConnect software [5] was launched under the BioSHaRE EU-funded project [53] to enable the integration of heterogeneous biobanks. The BiobankConnect software uses lexical matching combined with semantic matching to enable the semiautomated harmonization of terms in heterogeneous biobanks. The software quantifies the potential of harmonizing the terms among the heterogeneous data with the desired elements that are defined by the user by searching for lexical matches between the terms of the raw data with existing terms along with semantic information which is derived by the subclasses and the object properties of the ontologies. The user annotates the desired elements, i.e., the core set of elements, through the Bio-portal [57] which serves as a widely used repository for biomedical ontologies. The software then seeks for lexical matches between the terms of the raw data with those from the existing ontologies and provides a matching score that reflects the percentage of relevant matches. The BiobankConnect software tool has been used to harmonize data across six biobanks including 7461 terms based on a schema of 32 desired elements yielding 0.74 and 0.75 precision rates in two ranks that were defined by the experts [5].

The SORTA tool [7] is an example of another tool that was developed under the BioSHaRE EU-funded project [4], toward the integration of phenotypes across heterogeneous biobanks. While the BiobankConnect software involves the mapping of the heterogeneous data structures (elements) into a target schema, the SORTA tool deals with the heterogeneity of the data in a more high-level manner through the coding (or recoding) of the data values into international coding systems, such as the SNOMED-CT [58], the ICD-10 [59], and the Human Phenotype Ontology (HPO) [60]. More specifically, the SORTA tool seeks for matches between the data values and a target, user-specified, coding system. The coding system can be defined in terms of an ontology or a .csv document. The SORTA tool then uses lexical matching algorithms, such as the *n*-gram, to quantify the lexical similarity between the data values to provide a sorted list of candidate elements per data value. SORTA has been used to match 5210 unique entries in the LifeLines biobank [7,38] and 315 unique entries in the Dutch Uniform Multicenter Registration system for genetic disorders and malformation syndromes [7,39] in line with the HPO, yielding 97% and 58% recall ratios, respectively.

The S-Match tool [9] is an open source semantic matching framework which provides a rigorous solution for the semantic interoperability problem using lightweight ontologies. A lightweight ontology is defined as a taxonomy, where the natural elements are described in a formal way using formal expressions in the form of treelike structures, where each node in the tree has a specific meaning or concept [9]. The S-Match uses semantic matching algorithms that quantify the semantic relations between the elements (nodes) of two lightweight ontologies into four categories, namely (i) equivalent, (ii) less general, (iii) more general, and (iv) disjointless. First, the algorithm computes the concepts and the meaning of each label, and then it computes the relations between the concepts of the labels and the nodes. The nodes are constructed in the form of trees, and the meaning of each "child" node is related with the meaning of its "parent" node. Once a new ontology is given as input, the algorithm seeks for an existing conceptual relation between the concepts of the labels. If no semantic relations exist, the algorithm uses syntactic level matchers, like the Edit distance, to seek for lexical matches. The algorithm outputs a file with the identified semantic and syntactic relations between the concepts of the labels of the two lightweight ontologies. The performance of the tool was tested on the TaxMe2 dataset [40] which is a benchmark under the Ontology Alignment Evaluation Initiative, yielding 46% precision toward the correct identification of the semantic relations.

The Functional Ontology Assignments for Metagenomes (FOAM) [8] is an example of a computational framework that focuses on the classification of gene functions which are related with microorganisms using Hidden Markov Models (HMMs). The HMMs are trained on the sequence profiles that exist on the Kyoto Encyclopedia of Genes and Genomes (KOGG) orthologs (KOs), which are part of the Gene Ontology [61], to enhance the accuracy of the classification process toward the successful alignment of the sequence profiles with the targeted KOs. In fact, the KOGG is used as an external vocabulary to enrich the existing information from the sequence profiles. The whole process is semiautomated because the user needs to first define the target space for the alignment. FOAM also provides a functional ontology that describes the hierarchy of the groups that are involved during the training of the HMMs. Currently, the framework includes more than 70,000 HMMs targeting 2870 different KOs [8].

5.8 Toward a generalized harmonization strategy

It is obvious that the existence of a core set of terms which will serve as a standard template is the most prominent method toward the development of a robust data harmonization process against the conventional statistical harmonization methods which is limited to the standardization of measurement scales between a predefined set of equivalent items and a set of latent factors among the heterogeneous data. Meanwhile, the overall idea of a generalized data harmonization strategy is a great scientific challenge. The success of the data harmonization process is not only based on the conformity of the lexical and semantic matching algorithms that is used to detect matched terms but also on the structure of the standard reference model. In fact, the reference model varies according to the domain of interest which hampers the development of a generalized approach for data harmonization. For example, the pairing rules that need to be developed to identify relevant terms between two heterogeneous datasets that belong to the endocrine domain is different from the pairing rules that need to be developed to detect matched variables among two heterogeneous datasets in the respiratory domain.

A solution to this is to define the core set of terms according to international medical standards that promote global health interoperability, such as the Health Level Seven (HL7) international standard [62]. HL7 incorporates the Clinical Document Architecture (CDA) component which serves as an XML-based standard that includes clinical information regarding the structure and the encoding of medical records to enhance the sharing of patient medical records. The CDA is in turn based on the referenced information model which serves as a semantic model consisting of a set of structural components and semantic relations that is used to represent clinical notes in the form of an XML document (see Section 2.4.3 for more information regarding the HL7). The International Classification of Diseases (ICD) that was developed under the aegis of the World Health Organization is an international standard that is used to code and classify disease manifestations, treatments, diagnostic criteria, etc., mainly in epidemiology. Apart from the HL7 standard which is mostly used to enable the interoperability of the clinical information that is exchanged between clinical centers, an exemplar of a clinical code set repository is the Systematized Nomenclature of Medicine—Clinical Terms (SNOMED-CT). SNOMED-CT [58] serves as a widely used code set vocabulary in the clinical domain, supporting the HL7 interoperability standard and ICD terms and expanding them to include anatomical information, as well as information regarding clinical outcomes and diagnoses.

As a matter of fact, a clinical expert can construct a reference model for a domain of interest using a set of international medical terms that enhance the interoperability of the harmonization process. Through the development of lexical matching algorithms that are able to detect lexically similar terms using external knowledge, i.e., information from external vocabularies, such as the SNOMED-CT, the statistical power of the data harmonization process is increased in terms of efficiency and robustness. However, such an attempt would increase the computational cost of the harmonization process because searching a code set of thousands of medical terms is not always feasible and might lead to inaccuracies due to the presence of duplicated or highly similar medical terms. This can be reduced by constraining the search into a subset of relevant terms which is more or less similar to the definition of the core set of terms. The clinical value of the standard model lies in the case where the disease is rare. In that case, the medical terms are often ignored by the international medical standards making the definition of the core set of terms a great challenge.

The fact that the core set of terms can be easily transformed into a semantic data representation model, i.e., an ontology, increases its impact because it can be linked with existing biomedical ontologies, such as those that are part of the National Center for Biomedical Ontology (NCBO) Bio-portal repository [57]. The Bio-portal currently includes 773 ontologies with more than 9 million classes and billions of direct annotations. The portal offers also the ability to examine mappings between classes in different ontologies. Given a set of multiple biomedical ontologies for a specific domain of interest, one can create a knowledge repository that could be used to automate the harmonization process using machine learning models that will be able to learn from the mapping information that is derived by the ontology matching process.

5.9 Conclusions

Data harmonization is a promising field with numerous applications in the medical domain. The success of data harmonization is a great scientific challenge which involves the collaboration of both the clinical and the technical experts. Data harmonization is perhaps the only strategy that can enable the coanalysis of heterogeneous data that belong to a common domain yielding scientific outcomes with increased statistical power. The prospects of data harmonization are yet many with the most important one being the integration of heterogeneous clinical databases that can be used to address the unmet needs in various diseases, especially in the rare diseases where the absence of data collection protocols and the lack of international standard medical terms and integration mechanisms are absent. The clinical need for harmonizing retrospective data through the flexible strategy is more crucial than ever to maximize the available information from heterogeneous data across clinical databases against the stringent approach which is limited on data that were gathered under common collection protocols and criteria.

Data harmonization differs a lot from simply putting the data together. During manual data integration, several variables must be removed due to the differences and incompatibilities in the measurement units, the different type of data collection protocols, etc. As a result, there is a limited exploitation due to the small subset of original data which limits the potential of new scientific discoveries. In addition, the fact that the data integration process is manual along with the nomenclature impose a high risk for mistakes and obscure the definition of the data sharing principles because data integration requires direct data access. On the other hand, data harmonization involves processes that transform the variables into compatible ones to make them compatible. The legal and ethical issues are well defined in advance through data governance mechanisms. Data harmonization limits the direct access to the data because only the data schema is required to transform the variables. Moreover, the fact that the data harmonization procedure is semiautomated in the majority of the existing tools and frameworks reduces the risk for manual mistakes. Besides, the fact that it uses interoperable data schemas can overcome the nomenclature factor during the analysis.

The majority of the existing approaches for achieving medical data harmonization use a standard set of core terms that describe the domain knowledge of the disease under investigation. Then, lexical matching methods are used to detect lexical similarities between the terms of the reference model and the terms of the heterogeneous dataset. Of course, during the detection of matched terms between two datasets, one dataset can be used as the reference dataset, i.e., the target one, which is much similar to the definition of the reference model. By transforming the reference model into a semantic model using structured schemas, such as classes and subclasses along with object properties, one can take advantage of the object properties that connect the lexically matched terms to match the parent classes yielding harmonized datasets with reduced information loss. On the other hand, statistical harmonization tries to standardize heterogeneous measurement scales between terms, i.e., items which are predefined as common and latent factors. This approach, however, is very constrained because it requires prior knowledge regarding the terms that are common between the data instead of seeking for common terms and focuses only on the standardization of the measurement scales among the common terms. In addition, the statistical approaches require direct access to data, whereas the lexical and semantic matching approaches require only the data schema.

Until now, well-established frameworks have been launched that envisage to make data harmonization a true story toward the interlinking of heterogeneous data that exist in biobanks and epidemiology studies. All of them highlight the existence of a knowledge repository where interoperable sets of data elements (e.g., international medical terms) are used as a reference model to enable the matching of heterogeneous data elements. Tools like the BiobankConnect [5], the S-Match [9], and the SORTA [7] combine semantic and lexical matching methods to align heterogeneous data structures, as well as quantify the relevance of the matched terms according to a core set of terms. The importance of the reference model toward rigorous retrospective data harmonization has been also highlighted by the DataSHaPER [4] framework which involves the definition of a DataSchema that serves as a set of core elements. The DataSchema enables the development of pairing rules that quantify the similarity between heterogeneous data elements. The FOAM tool [8] is an example of a promising computational-oriented approach which focuses on the training of HMMs to identify target data elements among heterogeneous sequence profiles.

As we have already mentioned, data harmonization can address the unmet needs in chronic and rare diseases by enabling the interlinking and subsequent coanalysis of heterogeneous cohort data. An example of a promising initiative that deals with the harmonization of longitudinal cohorts of patients with chronic and rare diseases is the HarmonicSS project [63]. HarmonicSS envisages to harmonize and coanalyze more than 20 longitudinal cohorts of patients who have been diagnosed with a rare autoimmune disease known as primary Sjögren's syndrome (pSS). In short, pSS is a chronic autoimmune disease causing severe salivary gland dysfunction yielding clinical manifestations which vary from dry eyes and dry mouth to severe rheumatoid disorders and lymphoma development [64]. HarmonicSS adopts a semiautomated semantic matching approach to align heterogeneous pSS-related terms with the terms of a pSS reference model [65] which was developed in cooperation with the clinical experts of the project. The pSS reference model consists of a set of parameters that efficiently describe the pSS-domain knowledge [65] including six classes (i.e., laboratory tests, medical conditions, demographics, lifestyle, SS disease activity indices, and interventions), where each class includes additional subclasses, e.g., the class laboratory tests includes blood tests, oral tests, ocular tests, urine tests, biopsies, and even further subclasses, e.g., the subclass blood tests consists of the lipid tests, hematological tests, serum protein tests, complement tests, etc. The semantic matching process is applied through a user-friendly interface, where the clinical and the technical experts can align terms that share similar concepts by defining mapping scenarios (e.g., a Lab-test outcome yes/no scenario) along with the related value mappings (e.g., set "0" to "no" and "1" to "yes"), as well as evaluate suggested terminology mappings and finally extract the mapping rules into various formats (e.g., JSON).

With the majority of the existing data harmonization tools and frameworks being semiautomated, emphasis must be given on the development of new strategies to eliminate the "semi-" term. A promising solution would be to create a repository with a collection of biomedical ontologies that lie under a specific medical domain. This would greatly increase the interoperability of the harmonization process because the available core set of terms would cover a much larger portion of the domain. Another idea that is more straightforward would be to enrich this repository with information regarding the mapping of heterogeneous data schemas with the ontologies to introduce the "smart" repositories. A fundamental objective of the "smart" repository is the fact that it could be used to train a proper machine learning algorithm that could learn from the existing knowledge to automatically align the data schema of an upcoming heterogeneous dataset. This would greatly reduce the time effort needed for semiautomated harmonization and enhance the applicability of such an approach across different domains by including the mapping information of the related ontologies.

There is no doubt that data harmonization has a leading role in the coanalysis of heterogeneous medical data. Harmonization is the "key" factor that can enable the integrative analysis of data from heterogeneous data sources and thus envisages to make the sharing of data meaningful by distilling the power of data sharing into the construction of a set of interoperable and homogeneous data schemas that can be used to deal with the unmet needs in rare and chronic diseases, as well as in biobanks, omics registries, and cohorts. The success of data harmonization toward this direction has been proven by the existence of several initiatives which have demonstrated promising results toward the harmonization of biobanks, electronic health records, and cohorts, in various medical domains worldwide. Practices and actions need to be taken from healthcare stakeholders to invoke the inclusion of biomedical ontologies from rare diseases, as well as the update of the existing ones in the international repositories to enable the development of data harmonization tools that will be able to make the interlinking of clinical centers worthy for the public.

References

[1] Kourou KD, Pezoulas VC, Georga EI, Exarchos TP, Tsanakas P, Tsiknakis M, et al. Cohort harmonization and integrative analysis from a biomedical engineering perspective. IEEE Rev Biomed Eng 2019;12:303–18.

[2] Fortier I, Doiron D, Burton P, Raina P. Invited commentary: consolidating data harmonization-how to obtain quality and applicability? Am J Epidemiol 2011;174(3): 261–4.

[3] Fortier I, Raina P, van den Heuvel ER, Griffith LE, Craig C, Saliba M. Maelstrom research guidelines for rigorous retrospective data harmonization. Int J Epidemiol 2017;46:103–5.
[4] Fortier I, Burton PR, Robson PJ, Ferretti V, Little J, L'Heureux, et al. Quality, quantity and harmony: the DataSHaPER approach to integrating data across bioclinical studies. Int J Epidemiol 2010;39:1383–893.
[5] Pang C, Hendriksen D, Dijkstra M, van der Velde KJ, Kuiper J, Hillege HL, et al. BiobankConnect: software to rapidly connect data elements for pooled analysis across biobanks using ontological and lexical indexing. J Am Med Inform Assoc 2014;22:65–75.
[6] Fortier I, Doiron D, Little J, Ferreti V, Little J, L'Heureux, et al. Is rigorous retrospective harmonization possible? Application of the DataSHaPER approach across 53 large studies. Int J Epidemiol 2011;40:1314–28.
[7] Pang C, Sollie A, Sijtsma A, Hendriksen D, Charbon B, de Haan M, et al. SORTA: a system for ontology-based re-coding and technical annotation of biomedical phenotype data. Database 2015;2015:1–13.
[8] Ehrig M, Sure Y. Foam–framework for ontology alignment and mapping results of the ontology alignment evaluation initiative. In Proc Integr Ontol Workshop 2005:72–6.
[9] Giunchiglia F, Autayeu A, Pane J. S-Match: an open source framework formatching lightweight ontologies. Semantic Web 2012;3:307–17.
[10] Pang C, van Enckevort D, de Haan M, Kelpin F, Jetten J, Hendriksen D, et al. MOLGENIS/connect: a system for semiautomatic integration of heterogeneous phenotype data with applications in biobanks. Bioinformatics 2016;32(14):2176–83.
[11] Bandrowski A, Brinkman R, Brochhausen M, Brush MH, Bug B, Chibucos MC. The ontology for biomedical investigations. PLoS One 2016;11(4):e0154556.
[12] Shen F, Lee Y. Knowledge discovery from biomedical ontologies in cross domains. PLoS One 2016;11(8):e0160005.
[13] Pradhan N, Gyanchandani M, Wadhvani R. A review on text similarity technique used in IR and its application. Int J Comput Appl 2015;120(9).
[14] Gandhi SJ, Thakor MM, Sheth J, Pandit HI, Patel HS. Comparison of string similarity algorithms to measure lexical similarity. J Syst Inf Technol 2017;10(2):139–54.
[15] Groß A, Pruski C, Rahm E. Evolution of biomedical ontologies and mappings: overview of recent approaches. Comput Struct Biotechnol J 2016;14:333–40.
[16] Faria D, Pesquita C, Mott I, Martins C, Couto FM, Cruz IF. Tackling the challenges of matching biomedical ontologies. J Biomed Semant 2018;9(1):4.
[17] Xue X, Hang Z, Tang Z. Interactive biomedical ontology matching. PLoS One 2019; 14(4):e0215147.
[18] Zhao M, Zhang S, Li W, Chen G. Matching biomedical ontologies based on formal concept analysis. J Biomed Semant 2018;9(1):11.
[19] Bauer DJ, Hussong AM. Psychometric approaches for developing commensurate measures across independent studies: traditional and new models. Psychol Methods 2009; 14(2):101–25.
[20] Griffith LE, Van Den Heuvel E, Fortier I, Sohel N, Hofer SM, Payette H, et al. Statistical approaches to harmonize data on cognitive measures in systematic reviews are rarely reported. J Clin Epidemiol 2015;68(2):154–62.
[21] Gibbons RD, Perraillon MC, Kim JB. Item response theory approaches to harmonization and research synthesis. Health Serv Outcomes Res Methodol 2014;14(4):213–31.

[22] Mehta N, Pandit A. Concurrence of big data analytics and healthcare: a systematic review. Int J Med Inform 2018;114:57–65.

[23] Senthilkumar SA, Rai BK, Meshram AA, Gunasekaran A, Chandrakumarmangalam S. Big data in healthcare management: a review of literature. Am J Theor Appl Bus 2018;4:57–69.

[24] Dihazi H, Asif AR, Beißbarth T, Bohrer R, Feussner K, Feussner I, et al. Integrative omics-from data to biology. Expert Rev Proteomics 2018;15(6):463–6.

[25] Woodford R, Spittal MJ, Milner A, McGill K, Kapur N, Pirkis J, et al. Accuracy of clinician predictions of future self-harm: a systematic review and meta-analysis of predictive studies. Suicide Life-Threat Behav 2019;49(1):23–40.

[26] Gligorijević V, Malod-Dognin N, Pržulj N. Integrative methods for analyzing big data in precision medicine. Proteomics 2016;16(5):741–58.

[27] Kahn MG, Callahan TJ, Barnard J, Bauck AE, Brown J, Davidson BN, et al. A harmonized data quality assessment terminology and framework for the secondary use of electronic health record data. Egems 2016;4(1).

[28] Katz A, Enns J, Wong ST, Williamson T, Singer A, McGrail K, Peterson S. Challenges associated with cross-jurisdictional analyses using administrative health data and primary care electronic medical records in Canada. Int J Popul Data Sci 2018;3(3).

[29] Bandyopadhyay A, Tingay K, Akbari A, Griffiths L, Cortina-Borja M, Bedford H, et al. Integrating electronic health records from different sources across the UK: lessons from a record linkage study. Int J Popul Data Sci 2018;3(2).

[30] Kalia M. Biomarkers for personalized oncology: recent advances and future challenges. Metabolism 2015;64(3):S16–21.

[31] Raghupathi W, Raghupathi V. Big data analytics in healthcare: promise and potential. Health Inf Sci Syst 2014;2:3.

[32] Brown JS, Holmes JH, Shah K, Hall K, Lazarus R, Platt R. Distributed health data networks: a practical and preferred approach to multi-institutional evaluations of comparative effectiveness, safety, and quality of care. Med Care 2010;48:S45–51.

[33] Song JW, Chung KC. Observational studies: cohort and case-control studies. Plast Reconstr Surg 2010;126(6):2234.

[34] Gamble JM. An introduction to the fundamentals of cohort and case–control studies. Can J Hosp Pharm 2014;67(5):366.

[35] Wang Y, Qin J, Wang W. Efficient approximate entity matching using jaro-winkler distance. In: International conference on web information systems engineering. Cham: Springer; 2017. p. 231–9.

[36] Dreßler K, Ngonga Ngomo AC. On the efficient execution of bounded jaro-winkler distances. Semantic Web 2017;8(2):185–96.

[37] Rao GA, Srinivas G, Rao KV, Reddy PP. Characteristic mining of mathematical formulas from document-A comparative study on sequence matcher and Levenshtein distance procedure. J Comp Sci Eng 2018;6(4):400–4.

[38] Scholtens S, Smidt N, Swertz MA, Bakker SJ, Dotinga A, Vonk JM, et al. Cohort profile: LifeLines, a three-generation cohort study and biobank. Int J Epidemiol 2014;44(4):1172–80.

[39] Zwamborn-Hanssen AMN, Bijlsma JB, Hennekam EF, Lindhout D, Beemer FA, Bakker E, et al. The Dutch uniform multicenter registration system for genetic disorders and malformation syndromes. Am J Med Genet 1997;70(4):444–7.

[40] Giunchiglia F, Yatskevich M, Avesani P, Shivaiko P. A large dataset for the evaluation of ontology matching. Knowl Eng Rev 2009;24(2):137–57.
[41] Nederstigt L, Vandi D, Frasincar F. A lexical approach for taxonomy mapping. J Web Eng 2016;15(1–2):84–109.
[42] Chen Y, Lu H, Li L. Automatic ICD-10 coding algorithm using an improved longest common subsequence based on semantic similarity. PLoS One 2017;12(3):e0173410.
[43] Pakray P, Bandyopadhyay S, Gelbukh A. Textual entailment using lexical and syntactic similarity. Int J Artif Intell Appl 2011;2(1):43–58.
[44] Berners-Lee T, Hendler J, Lassila O. The semantic web. Sci Am 2001;284:28–37.
[45] Bratt S. Semantic web and other W3C technologies to watch. Talks at W3C. 2007.
[46] Lassila O, Swick RR. Resource description framework (RDF) model and syntax specification. 1999.
[47] McGuinness DL, Van Harmelen F. OWL web ontology language overview. W3C recommendation. 2004. p. 10.
[48] Bray T, Paoli J, Sperberg-McQueen CM, Maler E, Yergeau F. Extensible markup language (XML) 1.0. 2000.
[49] Magis D. A note on the item information function of the four-parameter logistic model. Appl Psychol Meas 2013;37(4):304–15.
[50] Sur P, Chen Y, Candès EJ. The likelihood ratio test in high-dimensional logistic regression is asymptotically a rescaled chi-square. Probab Theory Relat Fields 2017:1–72.
[51] Chatfield C. Introduction to multivariate analysis. Chapman and Hall/CRC; 2018.
[52] Schumacker RE. Interaction and nonlinear effects in structural equation modeling. Routledge; 2017. ISBN: 9781315092614, https://doi.org/10.4324/9781315092614.
[53] Doiron D, Burton P, Marcon Y, Gaye A, Wolffenbuttel BH, Perola M, et al. Data harmonization and federated analysis of population-based studies: the BioSHaRE project. Emerg Themes Epidemiol 2013;10(1):12.
[54] Doiron D, Marcon Y, Fortier I, Burton P, Ferretti V. Software application profile: Opal and Mica: open-source software solutions for epidemiological data management, harmonization and dissemination. Int J Epidemiol 2017;46(5):1372–8.
[55] Doiron D, Raina P, L'Heureux F, Fortier I. Facilitating collaborative research: implementing a platform supporting data harmonization and pooling. Norsk Epidemiol 2012;21(2).
[56] Wilson RC, Butters OW, Avraam D, Baker J, Tedds JA, Turner A, et al. DataSHIELD-new directions and dimensions. Data Sci J 2017;16(21):1–21.
[57] Fergerson RW, Alexander PR, Dorf M, Gonçalves RS, Salvadores M, Skrenchuk A, et al. NCBO bioPortal version 4. In: Proceedings of the international conference of building officials (ICBO); 2015.
[58] Benson T, Grieve G. Principles of health interoperability: SNOMED CT, HL7 and FHIR. London: Springer-Verlag; 2016.
[59] Quan H, Sundararajan V, Halfon P, Fong A, Burnand B, Luthi JC, et al. Coding algorithms for defining comorbidities in ICD-9-CM and ICD-10 administrative data. Med Care 2005:1130–9.
[60] Köhler S, Doelken SC, Mungall CJ, Bauer S, Firth HV, Bailleul-Forestier I. The human phenotype ontology project: linking molecular biology and disease through phenotype data. Nucleic Acids Res 2013;42(D1):D966–74.
[61] Gene Ontology Consortium. The gene ontology resource: 20 years and still GOing strong. Nucleic Acids Res 2018;47(D1):D330–8.

[62] Saripalle R, Runyan C, Russell M. Using Hl7 FHIR to achieve interoperability in patient health record. J Biomed Inform 2019:103188.

[63] HarmonicSS: HARMONIzation and integrative analysis of regional, national and international Cohorts on primary Sjögren's Syndrome (pSS) towards improved stratification, treatment and health policy making. Link: www.harmonicss.eu/.

[64] Mavragani CP, Moutsopoulos HM. Sjögren syndrome. Can Med Assoc J 2014;186(15): E579–86.

[65] Pezoulas VC, Exarchos TP, Andronikou V, Varvarigou T, Tzioufas A, De Vita S, Fotiadis DI. Towards the establishment of a biomedical ontology for the primary Sjögren's Syndrome. In: Proceedings of the IEEE engineering in medicine and biology conference; 2018. p. 4089–92.

CHAPTER

Cloud infrastructures for data sharing

6

Chapter outline

6.1 Overview

Under the guidance of the National Institute of Standards and Technology (NIST), cloud computing has been the point of interest as an on-demand computing model that is able to provide computational resources to fulfill the computational needs of demanding services, i.e., services with high computational power [1]. To realize a computing model, it is necessary first to establish a cloud model. In general, there are four standard cloud models [1–8]: (i) the public cloud which makes its computational resources available to the general public, under low payment constraints, (ii) the private cloud which constrains the access to its computational resources (which are often high) only to a group of registered customers, e.g., an organization or a company, (iii) the community cloud which makes its computational resources available only to a particular group of organizations that share common policies and

Medical Data Sharing, Harmonization and Analytics. https://doi.org/10.1016/B978-0-12-816507-2.00006-2

requirements, and (iv) the hybrid cloud which is a combination of two or more of the previous cloud models. A cloud model is managed by a cloud provider who manages the cloud resources following provisioning rules.

The purpose of a cloud provider is to provide a powerful and reliable cloud platform but at the same time cost-efficient, a fact that introduces a cost-over-power trade-off. According to the services that a cloud platform can provide, there are three fundamental types of models [1–10]: (i) the Infrastructure as a Service (IaaS), (ii) the Platform as a Service (PaaS), and (iii) the Software as a Service (SaaS). The IaaS model provides all the necessary hardware resources, i.e., servers, networks, storage, processing units, which are often implemented in the form of virtual machines (VMs) and are hosted by the cloud provider. The hardware resources are provided to the users, on-demand (i.e., anytime), through the establishment of a secure Internet connection. An example of a popular IaaS provider is the Amazon EC2 [11]. The PaaS model which is being built on the infrastructure layer provides the operating system and any software components that are necessary for the development of the software services. A popular PaaS provider is the Microsoft Windows Azure [12]. Through the PaaS model, the end users can build and test web applications using software frameworks which are already installed in the cloud and thus nullify the need to install software on their computers. The SaaS model provides, on-demand, the web applications (e.g., Gmail) to the end users through the Internet [13].

The growing number of everyday remote physical objects with ubiquitous intelligence (i.e., the "smart" objects) that can be interconnected through the Internet and exchange information gives rise to a new era, namely the Internet of Things (IoT) era [14–17]. The IoT (also referred to as Internet of Objects) envisages to bridge the gap between the physical world and the information world [14–17]. Every "smart" object (i.e., an embedded system), such as a wireless sensor, a smartphone, or any electronic device that is able to interact with the Internet, can be interconnected to a global IoT network which opens a new world for the development of web services which envisage to improve the quality of our everyday lives in various domains [14,17], including healthcare, urban infrastructure, manufacturing, etc. In fact, it is estimated that by 2020, 83% of the enterprise workload will be executed in cloud computing platforms [18]. In addition, the number of IoT connections is estimated to reach the 100 billion by 2025 with the number of estimated "smart" devices being approximately 50 billion in 2020 [19]. The IoT is already part of our lives with ourselves being witnesses of the great number of innovative IoT applications varying from the "smart" homes and cities for energy savings to the healthcare improvement through the remote monitoring of health indicators (eHealth and mHealth) and the control of drug delivery [20–22].

The vast amount of IoT data that is collected every day at an unprecedented rate imposes the development of new strategies for big data management. These strategies involve the establishment of cloud environments which are able to provide computing services able to effectively deal with such big data using high computational resources under certain security protocols and data protection measures.

A cloud computing environment enables the on-demand access to a variety of computing resources (e.g., networks, servers, storage, applications) that can be rapidly provisioned and delivered to the end users [3,4,7–9]. This means that the end users can have access to a variety of web services without the need to install software on their computers. A cloud computing platform can be easily expanded in terms of hardware and software resources to fulfill the need of applications that require high computational power. In addition, a cloud computing platform can offer high-performance tools for big data analysis, as well as user interfaces for easier data management.

In healthcare, cloud computing has been proven to deal with the computational complexity of clinical research and data management problems which are crucial scientific barriers in the biomedical arena [7–9]. The existing healthcare cloud providers offer a vast amount of services which are addressed to healthcare providers, biotechnology and pharmaceutical companies, patients, public health policy makers, and other healthcare practitioners. In general, the healthcare applications involve the delivery of cloud computing services for [7–9] (i) medical data management varying from diagnostic imaging viewers to genomic data viewers, (ii) high-performance data analytics for the analysis of genomic and epidemiological data, as well as any other source of medical big data, (iii) data storage either in centralized, distributed, or decentralized medical databases, (iv) the deployment of highly secure cloud repositories that can be used for direct access to the stored medical data, (v) medical data management through user-friendly interfaces, and (vi) the delivery of health data to the patient through "smart" devices for personalized medicine, among others.

Cloud computing systems have been already adopted toward medical data sharing and analytics including the delivery of mobile health services [22], the analysis of genetic profiles [23] and medical images [24], the development of emergency medical systems [25], the management of different sources of big medical data [26], and the emergence of hesitant attempts toward blockchain-based decentralized cloud computing which makes use of decentralized networks (i.e., with the absence of a central authority) to transfer and manage medical records [27]. As it has already been stated [8,28], the worldwide medical consortiums which are actively involved in biomedical informatics with the purpose of sharing medical data and applications can take advantage of the concept of cloud computing as an exemplar for promoting biomedical research. There is also evidence [8,29] that cloud computing can reduce EHR (electronic health record) costs and promote its adoption by large-scale clinical research studies. Moreover, the governments and governmental agencies have invested in cloud computing [8,30–32] to update their healthcare system and enable the sharing of patient medical records across interconnected hospitals and clinical centers.

Undoubtedly, cloud computing can enhance the scalability, availability, flexibility, interoperability, and cost-effectiveness of any infrastructure. However, cloud computing is obscured by several technical and regulatory challenges, with data security being the dominant barrier toward its adoption. The lack of data security protocols (e.g., poor encryption and decryption keys), as well as the incompliance to standard security guidelines (e.g., for network attacks), can result to data leakage, data abuse, loss of data integrity, and control over the hosted data and the cloud applications. Moreover, a cloud computing system must be compliant with any data protection requirements and respect the intellectual property rights to ensure the privacy of the stored data and the regulatory compliance of the delivered services. Efficient provisioning of the cloud resources is necessary to provide on-demand services and enable the standardization with cloud resources in heterogeneous cloud computing systems to enable federated cloud computing.

6.2 The concept of cloud computing

In our rapidly advancing technological era, the IoT has undoubtedly been the subject of interest with more and more information being collected from embedded devices and exchanged through the Internet everyday [14–17]. The information exchanged within the IoT can be of any source, varying from sensors to smartphones and personal computers, as long as this source is able to connect to the Internet. As a matter of fact, the IoT can be seen as a vast network of interconnected devices that interact with each other through the Internet [14]. An emerging technology that considers for the computational resources which are required to realize this concept and thus host and deliver such services is the cloud computing [1,2]. According to the NIST, cloud computing is defined as [2] *"a model for enabling the on-demand network access to a shared pool of adjustable computing resources (e.g., networks, servers, storage, applications) that can be provisioned and released with minimal management effort or service provider interaction."* Cloud computing is a general-purpose computing model which offers on-demand hardware resources, such as central processing units (CPUs), memory units (RAM), and large storage capabilities, with the purpose of enabling a user or a company to use this model to fulfill the demands of their services. Examples of popular cloud providers include Microsoft, Google, and Amazon, among many others.

A cloud computing architecture can be realized in the form of a multilayer framework (Fig. 6.1). According to Fig. 6.1, a cloud computing conceptual architecture consists of three layers where each layer is related with a particular cloud computing service. At the bottom of the framework lies the hardware layer which consists of the hardware components, including the networks, servers, storage units, and VMs which are duplicates of physical computers. The hardware layer is part of the IaaS model [1–10] which provides on-demand access to the server and network infrastructures of the platform. Right above the hardware layer lies the back-end layer where the operating system (e.g., Linux, Windows) is established along with the software framework for the development of software services. The back-end layer is part of the PaaS model [1–10] and is used for building and delivering the applications. Furthermore, the back-end layer involves the development of security mechanisms (e.g., for user authentication). The third and final layer is the front-end

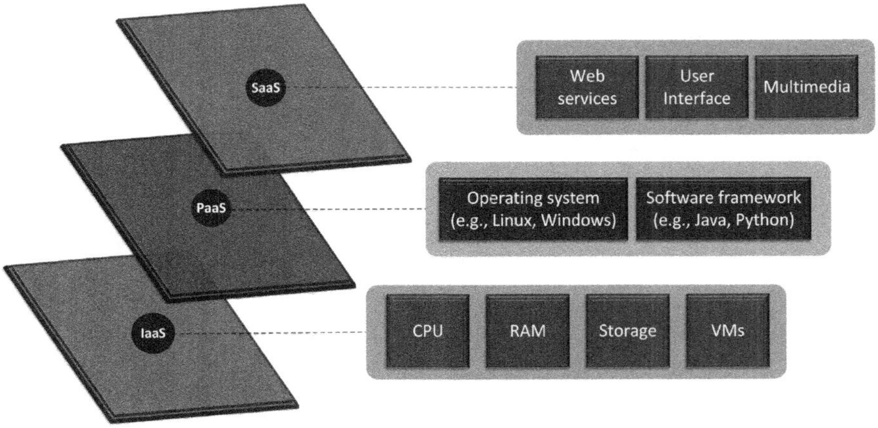

FIGURE 6.1

A conceptual cloud computing architecture depicting the three cloud computing services.

layer which is part of the SaaS model [1–10] and includes the user interface and the web and multimedia applications that are provided to the end user, on-demand. The end user is able to use any of the three cloud services according to his or her technical expertise.

Every cloud computing architecture is orchestrated by a cloud provider, also known as the infrastructure provider [1]. The cloud provider is responsible for a variety of functionalities concerning the overall cloud's infrastructure management. In computer science, a common way to describe the responsibilities of the cloud provider is performed through a use-case diagram [33] which depicts the interaction of the "actor" (in our case the cloud provider) with the "system" (in our case the cloud infrastructure) according to the different use cases that the "actor" is involved with. More specifically, the cloud provider (Fig. 6.2) is able to (i) manage the processing cores, the memory units, and the storage resources, (ii) configure the firewall system (and reconfigure it when necessary), (iii) monitor the information flow (e.g., the flow of sensitive data) within the cloud, (iv) add provisioning rules, i.e., rules for the effective execution of the operations within the cloud (performance improvement), (v) configure VMs for the implementation of software services, (vi) manage the cost of the cloud resources or any additional cloud resources, (vii) manage the pending approvals from the software providers regarding the demand of the services, (viii) maintain ordinary backups from the VMs, (ix) schedule the cloud operations according to their needs (this is related with the provision rules), (x) refresh the services catalog, i.e., the number and types of services of the cloud platform, (xi) register or block the service providers, and (xiii) configure the network interfaces.

A service provider serves the end user according to the computational resources that are provided by the cloud provider. Note, however, that a software provider and the cloud provider can be the same actor. A typical use-case diagram for the software provider is shown in Fig. 6.3. A service provider interacts with the back-end layer

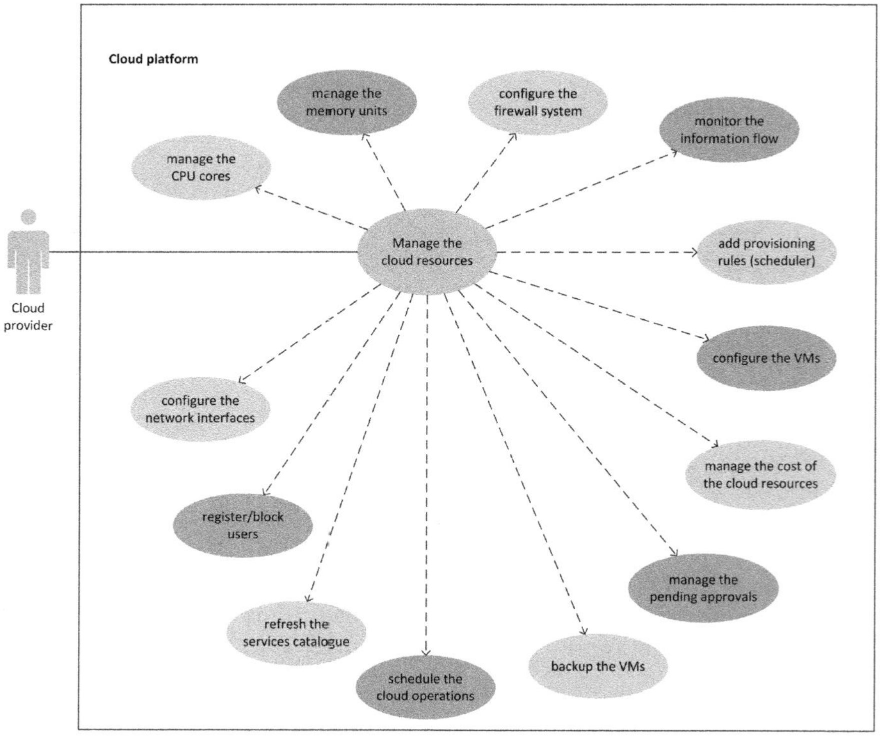

FIGURE 6.2

A typical use-case diagram for the cloud provider.

(Fig. 6.1) and is able to (i) define an operating system (e.g., Linux, Windows), (ii) develop web applications using Representational State Transfer (REST) services, (iii) develop software services for different tasks (e.g., compilers, schedulers), (iv) test the software services (debugging), (v) develop visualization components for the user interfaces, (vi) integrate software components, (vii) develop software architecture, (viii) conduct requirements analysis, and (ix) manage his or her user account settings. The software provider can be an external actor who is able to connect to the cloud platform and provide software services.

The end user is the costumer who is authorized to gain access into the platform and run the services. A typical use-case diagram for the end-user provider is depicted in Fig. 6.4. The end user interacts with the higher layer of Fig. 6.1, i.e., the front-end layer. More specifically, the end user is located on the front side of the latter layer and is able to use a variety of cloud services, such as (i) upload and download data on private cloud spaces, (ii) run web applications according to his or her interests, (iii) visualize summary data as a result of the services execution, (iv) request access to personal data or shared data, (v) edit and further manipulate his or her data (e.g., share with other users), (vi) request access to additional computational

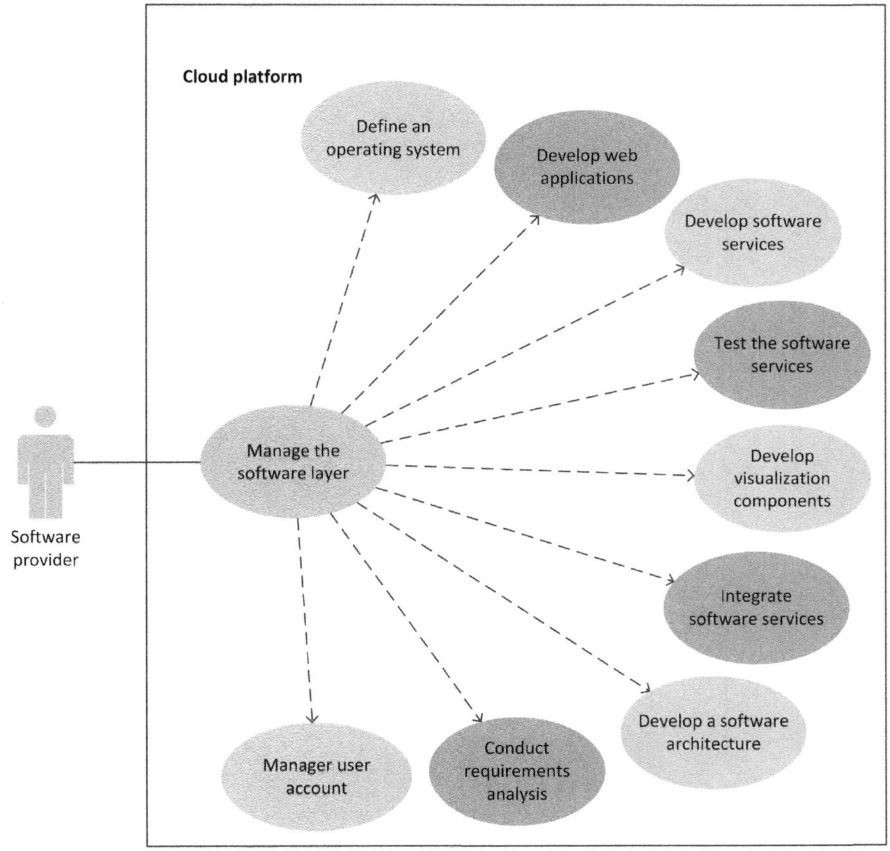

FIGURE 6.3

A typical use-case diagram for the software provider.

resources from the infrastructure and software providers, (vii) configure additional settings concerning the access rights, and (viii) manage his or her user account settings (e.g., credentials). The type and the quality of the cloud service and the cloud computing requirements are, in most cases, agreed on in a service level agreement (SLA) between the cloud provider and the end user [3,4,6].

So far, we have seen that the services (computational resources) that a cloud can provide can be divided into three categories which vary from hardware resources to software and network components. Apart from the three types of cloud services though, there are also different types of clouds that provide these services, in business terms. In general, there are four types of cloud computing deployment models, namely [1–8] (i) the public cloud, (ii) the private cloud, (iii) the community cloud, and (iv) the hybrid cloud. Each one has its own scope and benefits depending on whether the cloud provider wishes to invest on security and reliability instead of high computational power or whether the cloud provider wishes to make the services

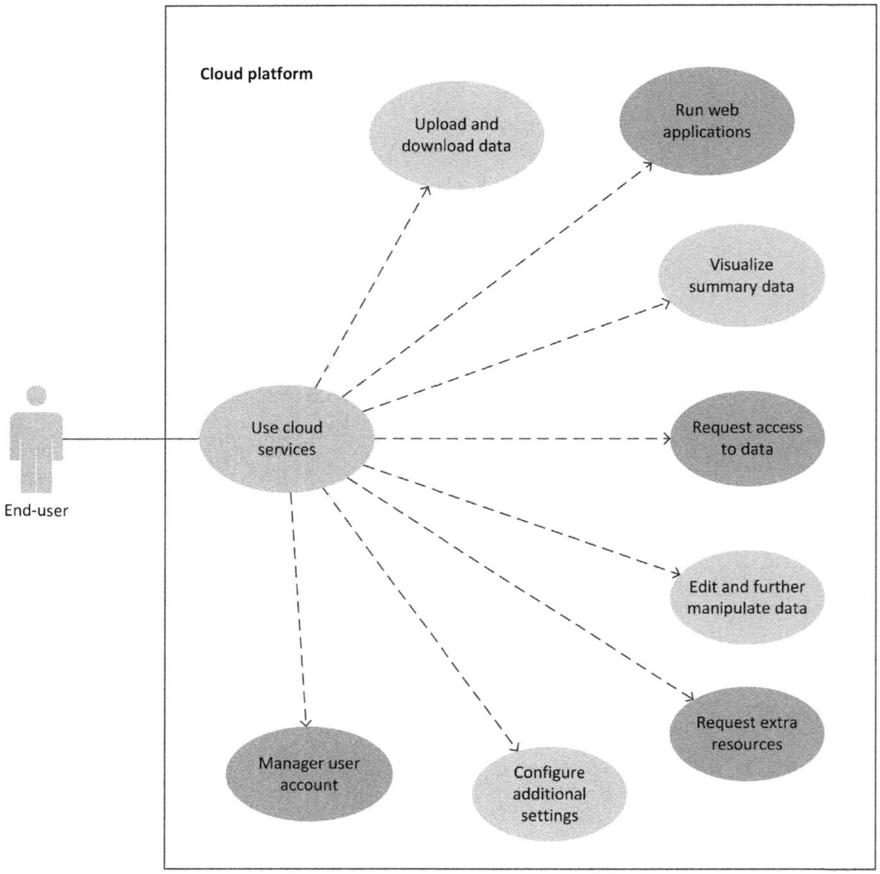

FIGURE 6.4

A typical use-case diagram for the end user of a cloud platform.

available to the public or to specific groups of consumers who might share the same policies or not.

A public cloud [1−8] makes the cloud services to the greater public without the need to invest large amounts of money on infrastructures or disclaim the responsibilities of an infrastructure provider. A public cloud, however, is often characterized by inadequate security measures and thus is more prone to data breach (Table 6.1). On the other hand, a private cloud [1−8], which is also referred to as an internal cloud, employs strict security measures and focuses more on the platform's reliability, as well as on offering an increased computational power over cost for the platform's services. The majority of the international firms deploy private clouds within their infrastructures due to the quality they offer in terms of data privacy and optimized support. Private clouds, however, require large capital investments

Table 6.1 A summary of the benefits and drawbacks among the four different types of cloud models.

	Public	Private	Community	Hybrid
Easily accessible	High	Moderate	Moderate	Moderate
Reliability	Low	High	Moderate	Moderate
Interoperability	Moderate	Moderate	High	Low
Computational power	Low	High	High	High
Scalability	Moderate	High	High	Moderate
Impact in industry	Moderate	Moderate	Low	High
Cost over power	Low	Low	High	High

and thus are addressed to only a specific group of customers, i.e., a single company or an organization or a third party.

A community cloud [1–8] is similar to a private cloud with the difference that the former aims to overcome the large capital investments that the latter requires. In a community cloud, the cloud services are shared among a specific group of customers that share common goals or interests which might vary from common policies and scope to common computational requirements. A community cloud can either be hosted by a member of the community or by a third-party vendor. In fact, the community cloud can be seen as a balanced solution between groups of people that require both high degree of control over performance and capital investments, in a collaborative manner. A straightforward solution that provides a formal way to deal with the trade-off between cost and computational power is the hybrid cloud (Table 6.1). The hybrid cloud [1–8] is a combination of the three previous types of clouds (public, private, or community clouds) which aims to overcome the limitations offered by each individual type of cloud. A hybrid cloud manages to provide a cloud environment with better security layers than those of a public cloud and at the same time offer on-demand services with high degree of control over performance (Table 6.1). The optimal split, however, between public and private clouds is a technical challenge. According to the latest 2017 Forbes statistics report [34], 50% of industrial healthcare companies use hybrid clouds, 26% user private clouds, and 24% public clouds, with the community clouds being excluded from the list.

In healthcare, the end user can be a healthcare provider, a physician, a health policy maker, a clinical researcher, a pharmaceutical company, or even the patient. The clinician can upload patient data (e.g., clinical images, genomic data) in the cloud and manage the data in the cloud repository through appropriate interfaces. The clinical researcher can execute web applications to process and analyze the clinical data, e.g., using machine learning models to develop prediction models for the disease onset and progress, conduct statistical tests among different factors (e.g., age, gender) within the populations, etc. The health policy maker can use the summary population statistics (metadata) to develop appropriate rules for the identification of high-risk individuals (e.g., individuals that are more prone to the development of cancer) and thus propose new health policies. The patients can manage (edit, delete) their data and get informed about their health status through appropriate health indicators (measures) that travel from the cloud to their "smart" devices (e.g., smartphones, smartwatches). The pharmaceutical companies can select specific groups of patients to propose or test a new drug (patient selection for multinational clinical trials).

A summary of the cloud computing providers that are actively involved in healthcare along with the cloud computing services that they offer is presented in Table 6.2. The majority of the cloud providers offer software solutions for the management of medical data for almost all types of sources (e.g., radiology and pathology images, genomic data, diabetes data, EHRs) to promote medical data sharing. In addition, these providers are compliant with data protection regulations (e.g., HIPAA compliant) and quality policies to ensure data governance. The most popular of them (i.e., Microsoft Azure, Amazon AWS, Google Cloud, Microsoft HealthVault) serve as IaaS, PaaS, and SaaS providers because they offer both hardware (IaaS) and software (PaaS) components, as well as high-performance software solutions (SaaS) for a variety of healthcare analytics applications that vary from medical imaging analysis to whole genome analysis, e.g., for the development of prediction models (for the onset and/or progress of a specific disease), the analysis of genetic profiles and genetic associations using high-performance computing power under distributed environments for faster execution. Other SaaS applications involve the development of mobile health applications for eHealth, like the GlucoMe for diabetic patients or the Amazon AWS and Microsoft Azure for Health for informing the end users about their lifestyle and fitness status, as well as the design and evaluation of clinical trials (e.g., IBM Clinical Development, Medidata Cloud) to evaluate the efficacy of a proposed drug in specific population subgroups and develop public health policies. All these cloud computing services might be used by healthcare stakeholders and practitioners.

6.3 Web services

As far as the implementation of the cloud computing services is concerned, the cloud providers usually create application programming interfaces (APIs) which provide access to the software components that lie in the PaaS layer. The APIs are mainly addressed to the software providers (Fig. 6.3) with the purpose of developing software applications. An API can be seen as a secure tunnel through which an application can access the software components, with the term "interface" denoting a virtual interface among the software components [47–51]. APIs can support the information exchange between two or more different software components, as well as the integration of new software components into existing (or new) software applications. In the latter case, the API is referred to as a "plug-in API" [49]. From a

Table 6.2 A summary of the major cloud computing providers that are currently used in healthcare along with the ensuing cloud computing services.

Cloud provider	Scope	Types of healthcare cloud services
Microsoft Azure Services for Health [12]	Provide an HIPAA-compliant cloud computing services to different stakeholders in healthcare for the collection, storage, retrieval, and analysis of sensitive healthcare data.	• Data sharing (any type) • Data governance • Data storage • Data management • Big data analytics • High-performance computing • Distributed computing • eHealth monitoring • Software development interfaces
Amazon Web Services [11]	Provide an HIPAA-compliant cloud computing services in healthcare and life sciences for healthcare providers, pharma, and biotech, to store, process, transmit, and analyze clinical information.	• Data sharing (any type) • Data governance • Data storage • Data management • High-performance computing • Distributed computing • eHealth monitoring • Software development interfaces
Google Cloud [35]	Provide an HIPAA-compliant cloud-based service for healthcare data sharing and big data analytics.	• Data sharing (any type) • Data governance • Data storage • Data management • Big data analytics • eHealth monitoring • Cloud Machine Learning • Software development interfaces
Microsoft HealthVault [36]	Provide a cloud-based platform for medical data sharing that is addressed to hospitals, pharmacies, lab testing companies, and patients.	• Data sharing (EHR) • Data governance • Data storage • Data management • eHealth monitoring
GlucoMe [37]	Provide a comprehensive, connected diabetes care platform that simplifies the way patients, caregivers, and medical professionals manage diabetes.	• Data sharing (diabetes data) • Data management • Big data analytics • Data monitoring • Smart glucose monitor • eHealth monitoring
MedCloud [38]	Provide a cloud computing platform which offers an integrated healthcare data management system that works in a secure and a reliable way.	• Data sharing (medical images) • Data governance • Data management • Data storage • eHealth monitoring

Continued

Table 6.2 A summary of the major cloud computing providers that are currently used in healthcare along with the ensuing cloud computing services.—*cont'd*

Cloud provider	Scope	Types of healthcare cloud services
Practise Fusion [39]	Provide a cloud-based solution for Electronic Health Record (EHR) management.	• Data sharing (EHR) • Data governance • Data storage • Data management
IBM Merge iConnect Cloud Archive [40]	Offer secure HIPAA-compliant, comprehensive cloud-based storage and access in medical images.	• Data sharing (medical images) • Data governance • Data storage • Data management
Merge eMix [41]	Offer cloud-computing services for the exchange of patient images and information with outside healthcare institutions, providers, and physicians.	• Data sharing (medical images) • Data governance • Data storage • Data management
RxEye Cloud [42]	Provide cloud-based services for efficient and secure remote viewing of medical images.	• Data sharing (radiology and pathology images) • Data governance • Data storage • Data management
IBM Explorys [43]	Provide an HIPAA-enabled cloud-based analytics platform that harnesses big data for clinical integration, predictive analytics.	• Data sharing (EHR) • Data governance • Data storage • Data management • Data analytics
eClinicalWorks [44]	Provide a cloud-based EHR management system which brings together healthcare practitioners.	• Data sharing (EHR) • Data governance • Data storage • Data management
IBM Clinical Development [45]	Provide an FDA-compliant cloud-based data management platform that allows the design and management of clinical trials.	• Data sharing (any type) • Data governance • Clinical trials design • Clinical trials management
Medidata Cloud [46]	Offer cloud-based solutions for advancing clinical testing and transform the global clinical development.	• Data sharing (genomic, images) • Data governance • Clinical trials planning • Clinical trials management

developer's point of view, the API can be seen as a collection of software libraries, i.e., sets of functions, data structures, classes, protocols, and additional object properties, which are properly brought together to develop one or more software applications through a web client. These operations are usually implemented in the form of a workflow, starting from the input and the execution of the core functionalities until the output is reached.

The API users can be divided into two main actors, i.e., the API provider who develops the methods to call the APIs and the API developer who programs the APIs in terms of coding for the SaaS layer [50,51]. The former actor is usually the infrastructure provider, whereas the latter is the software provider. The majority of the cloud-based services that lie in the SaaS layer are implemented in the form of remote calls to a web client, i.e., in the form of a REST API or a Simple Object Access Protocol (SOAP) API [52–54]. In general, REST APIs are preferred due to their easiness and flexibility against the SOAP APIs which work well in enterprise environments but are more complex [53]. Widely used protocols that are used for transferring information between the APIs include the JavaScript Object Notation (JSON) [55] which provides a human-readable textual representation of the web resources when a REST service is called and the remote procedure call [56] which supports the interprocess communication of software processes with different address spaces in distributed computing environments, among others.

What a typical REST API does is that it receives and handles remote user requests regarding the cloud services, normally the software applications which are located in the SaaS layer. A REST API, however, can also interact with the IaaS layer. For example, a secure REST API can be used to transfer a virtual image of the cloud infrastructure from the IaaS to another virtual space or to create new VMs. When a user requests access to a cloud service, normally by the web through a hypertext transfer protocol (HTTP), the REST API is executed in the background and delivers the results of the cloud service to the end user in a textual representation format, i.e., in a JSON format. A simple example of a REST API functionality is depicted in Fig. 6.5 where a REST API is used to deliver cloud healthcare services to the end users (clinical researchers, health policy makers, patients) through the web. According to Fig. 6.5, a clinical researcher can execute a data analytics service through the SaaS layer which in turn invokes the REST API to execute a service on the data that lie in the cloud's data warehouse.

The results of a REST API are usually returned to the end user through a typical JSON format, i.e., a human-readable textual representation of the output in the form of a data structure which includes a complete set of parameters that describe the outcomes of the service. An example of a simple JSON file is a data structure that includes the parameters "username," "password," and "access flag" of a user who wishes to gain access into a web platform. This JSON file could be the outcome of a user access management service that either approves ("access flag" = 1) or disapproves ("access flag" = 0) the users' access into the platform. The JSON format, however, is a low-level approach that is used to transfer any service-related information between the web browser that calls a cloud service and the cloud server. In fact, the JSON protocol is mainly addressed to the developers and technical experts and not an elegant to display the outcomes to the end user, especially if the latter is a clinician or a healthcare provider. An elegant way to display the results is through a graphical user interface (GUI). A GUI is usually provided by a proper visual analytics service that is located in the SaaS layer. The visual analytics service receives the JSON file and displays the results in a user-friendly way. Popular tools for the

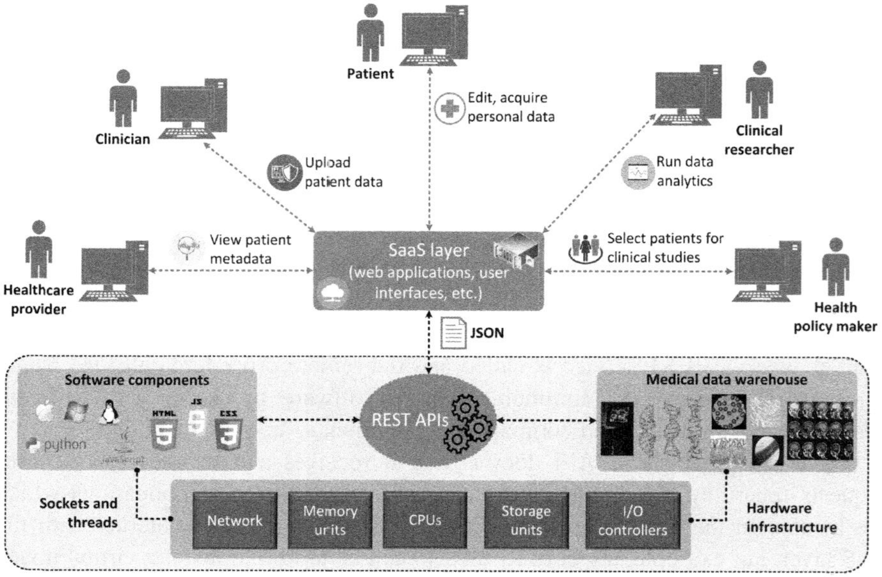

FIGURE 6.5

Delivery of healthcare cloud services to the end users through Representational State Transfer (REST) application programming interfaces (APIs). All the user requests that are made to the SaaS layer are secure (depicted with *green [gray in print version] dashed lines*). The REST APIs interact with the software components and the hardware infrastructure of the cloud to deliver the results of the services.

development of web GUIs include the triad [57] HTML (HyperText Markup Language), Javascript, and CSS (Cascading Style Sheets), as well as related software packages, such as Angular [58], Spring (Java) [59], Django (Python) [60], etc.

According to Fig. 6.5, the end users can send requests to the SaaS layer through secure Internet connections. This can be accomplished under the Secure Sockets Layer (SSL) protocol [61], a standard protocol which makes use of encryption algorithms to encrypt sensitive information (e.g., login credentials, credit card numbers, etc.) and prevent hackers from intercepting it. An encryption algorithm can be symmetric, asymmetric, or both [61,62]. Symmetric encryption uses a single public key with a typical size of 128 or 256 bits to both encrypt and decrypt the sensitive data, whereas asymmetric encryption uses separate encryption (public) and decryption (private) keys with a typical size of 1024 or 2048 bits. This means that in the symmetric encryption, both sides need to know the public key (and thus must be transferred), whereas in asymmetric encryption this does not occur because the public key can be distributed as long as the private key is secret. Public and private keys can be generated using cryptographic algorithms, such as the Rivest–Shamir–Adleman algorithm [62–64] which is based on the factorization problem of large integers and the Elliptic Curve Cryptography algorithm [62,63,65] which is based on the problem of seeking the discrete logarithm of a random elliptic curve over finite elements. Additional algorithms include the Digital Signature Algorithm [62,65] and the Elliptic Curve Digital Signature Algorithm [65], which, however, are mainly used for digital verification instead of encryption [62].

Asymmetric encryption is much safer because it uses different keys for encryption and decryption, a fact that makes difficult and even impossible to intercept the private key using only the public key [65]. However, the computational cost of the asymmetric encryption is much higher than the one for the symmetric encryption because the number of generated bits is almost 10 times larger than those in symmetric encryption, supporting 2^{1024} to 2^{2048} unique numbers. The SSL protocol uses both symmetric and asymmetric encryption [61,62]. Once a typical user sends a typical connection request through a browser, the following steps are executed for the establishment of a secure web server–client connection:

1. The cloud server sends a copy of its asymmetric public key to the browser.
2. The browser sends a symmetric session key based on the server's asymmetric public key for encryption.
3. The cloud server uses its asymmetric private key to get the browser's symmetric session key.
4. The cloud server and the browser can exchange any kind of sensitive information using the symmetric session key for both encryption and decryption. As a result, a secure tunnel is established.

A Transport Layer Security (TLS) protocol is simply an updated and more secure, broader, version of the SSL protocol [62,66,67]. The TLS protocol uses a Transmission Control Protocol [62] to ensure the security of the connection in IP communication networks. The connection and communication between a server and a web browser are performed on the top of TLS, through the well-known HTTP [62,68] which is an international standard protocol that sends and receives requests (e.g., the user's credentials) between two hosts (e.g., between the server host and the user's browser) through a secure tunnel. A HTTP that is established through an SSL/TLS protocol is referred to as an HTTP-over-SSL (or its predecessor TLS) or simply HTTP [62,68]. The HTTP indicator is usually displayed by a green bar on the server's host which denotes that the host holds a secure domain.

6.4 Architectures

In this section, the major types of the cloud computing services that were initially presented in Section 6.2 will be further discussed. Additional emphasis will be given on the Data as a Service (DaaS) model as part of the SaaS model for providing data services, on-demand, instead of software services only. The cloud computing services are presented along with their interactions in Fig. 6.6.

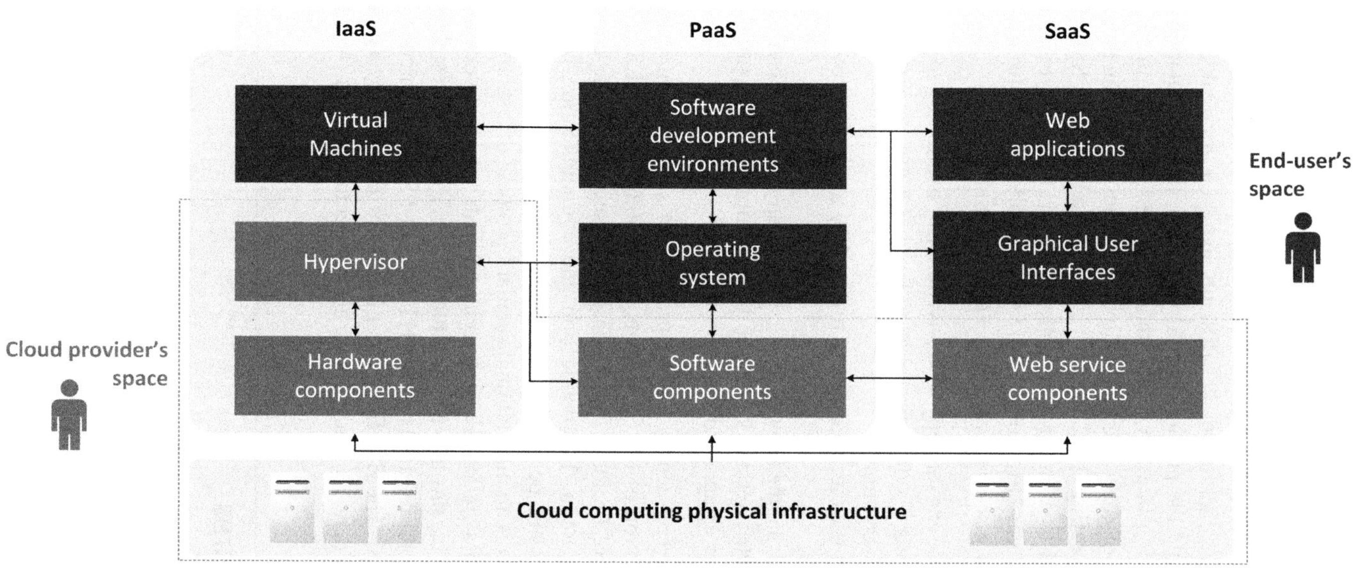

FIGURE 6.6

An illustration of the information flow across the three major cloud computing services.

6.4.1 Infrastructure as a service

An IaaS model provides all the necessary IT infrastructures for processing, such as storage units, networks, servers, etc. [1–10]. The infrastructure provider offers these services on-demand, according to the end-user's needs. For example, an end user with advanced technical skills can directly use an IaaS model to configure his or her own server and network applications for the development of software services. The end user can gain access to these services through the Internet and use the infrastructures in a decentralized manner. This means that the user makes uses of infrastructures that are not located in his or her premises but miles away. To make these physical resources available to remote locations through the Internet, the cloud provider uses a procedure known as virtualization. What virtualization does is that it defines a group of VMs which act as substitutes of physical computers and provide the operating system and related functionalities just like a physical computer system.

The VMs are monitored by a virtual machine monitor (VMM) which partitions the hardware layer and distributes its components into one or more VMs (Fig. 6.7) [69–71]. The VMM is also referred in the literature as a hypervisor [69–71], because it is the supervisor of the cloud resources. Each VM acts an individual component with its own software interface and a partitioned hardware layer. As soon as a VM receives a task, it uses the CPU cores of the hardware layer (direct execution) in an unprivileged mode, while the VMM maintains the privileged mode and manages the control of the physical resources [69–71]. The same stands for memory virtualization through the use of the "shadow page table" [69], which lets the VMM to manage the memory available to the VMs. The storage virtualization is accomplished by partitioning the disk into a specific number of components (Fig. 6.7). Each VM has its own operating system where the end user or the software provider can install additional software components. A popular virtualization infrastructure is the Kernel-based Virtual Machine (KVM), a Linux-based VM with hardware extensions that consists of a loadable kernel module that provides the core visualization infrastructure and a processor-specific module for handling the processes [72]. The main advantage of the KVM is that it can be turned into a hypervisor, i.e., a VMM. The storage requirements, the number of processors, the size of the memory units, and the network specifications are usually requested by the end user and are taken into account from the cloud provider during the construction of the VMs. The number of VMs varies according to the end-user's needs. In any case, the end user is not able to take control of the hardware layer but has the ability to install an operating system and/or any software components.

The benefits from using an IaaS cloud computing model in healthcare are many. The IaaS providers offer technical safeguards and data protection compliant (e.g., HIPAA compliant) infrastructures with high-performance hardware components that are not easy to afford and can be used for the secure storage of large amount of medical data from different types of sources (e.g., from imaging data to genomic data). In addition, the majority of the IaaS providers offer disaster recovery and backup plans to prevent data loss, as well as secure network facilities to prevent

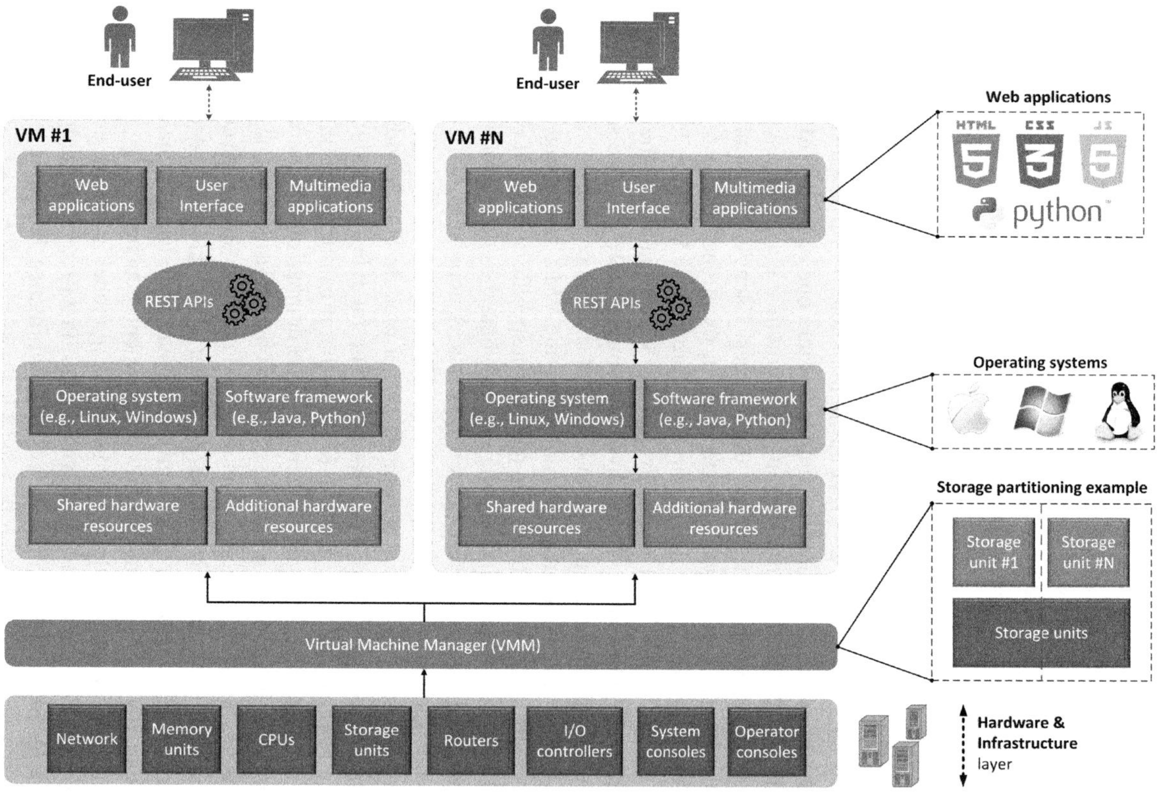

FIGURE 6.7

A high-level illustration of the virtualization process. The hypervisor partitions the hardware components for the construction of the virtual machines (VMs). The software components (e.g., operating system, software frameworks) are added on the top of the hardware to enable the development of the software applications. The Representational State Transfer application programming interfaces are used to enable the communication between the software components and the user interface, the web, and multimedia applications. The end users can be the owners of one or more VMs.

data abuse. In addition, the majority of the IaaS providers offer virtualization environments (i.e., VMs) that are part of their IaaS services catalog which can be easily adjusted in a PaaS layer so as to be used for the development of healthcare applications. Examples of popular IaaS cloud computing service providers that are currently involved in promoting a healthcare ecosystem include the Amazon AWS [73], the Microsoft Azzure [12], and the Google Compute Engine [74], all of which are addressed to end users with both low and high expertise and deploy high-performance, scalable components for secure networking, storage, disaster recovery, maintenance, etc. (Table 6.2).

6.4.2 Platform as a service

The PaaS model provides a complete development platform with tools and resources that enable the development of software applications by the end users with technical expertise and/or the acquisition of these applications by the typical end users [1−10]. The term "development" involves the programming of tools, schedulers, debuggers, and software testing, integration and maintenance which can be seen as part of the software provider's use-case diagram (Fig. 6.3). The PaaS model lies above the hardware layer and is the model which supports the "software lifecycle" process of a cloud platform according to which the user can directly develop and/or acquire applications that are part of the SaaS model that lies on top of the hierarchy (Fig. 6.1). The PaaS providers establish the operating systems (e.g., Linux) and the software frameworks (e.g., Java) that exist on the higher level of a typical VM in Fig. 6.5. Moreover, the PaaS model can offer cloud applications that are either ready or under development. Thus, the actor of a PaaS model can either be the software provider or the end user. In any case, the end user is not able to control the hardware layer but only deploy the cloud's software applications.

In healthcare, the PaaS providers offer the basis for the development of the healthcare applications. In fact, the PaaS providers offer software solutions that can be used for the development of different applications that vary from diagnostic imaging analysis and genomic analysis to mobile health applications for health monitoring and patient selection for clinical trials (Table 6.2). Popular PaaS providers, such as the Microsoft Azure [12], the Amazon AWS [73], the Google Cloud [35], and the IBM [40,45], offer high-performance environments for big data analytics (e.g., machine learning methods) that can be used for the development of prediction models to study the onset and progress of various types of diseases. These software environments can either be directly delivered to the end users or indirectly developed by their side. Moreover, a straightforward part of their overall strategy is to make use of blockchain technology [27] to enable distributed computing and thus provide computationally efficient solutions for the evaluation of prediction data models, especially in genomic analysis where the amount of collected data (genes) is vast.

6.4.3 Software as a service

The SaaS model delivers the applications to the end users and is probably the most straightforward model in terms of how the cloud service is delivered to the end users through the web [1–10]. These applications can be usually accessed through a web browser. In addition, these applications can be requested by thousands or even millions of users at the same time. For this case, the SaaS model is based on a single environment to handle multiple user requests and at the same time reduce the computational costs. In the SaaS, the end users are not able to control the underlying infrastructure of the cloud's platform (i.e., the hardware layer) or the software layer but only make use of these services without the need to install any special kind of software on their computer (except from a web browser).

Through the SaaS, the software providers can develop applications that can run remotely on multiple servers and be accessed from any computer that is connected to the cloud (through a secure network) instead of developing them locally on their computer. Furthermore, in the case of updates on web applications, the latter can be installed directly on the main cloud server, and from there the end users can receive the updates immediately through the web. This makes the SaaS highly scalable and easy to use which comprises one of the major benefits of the SaaS model during the delivery of services. In addition, the SaaS model is easy to be maintained and is highly customizable (e.g., uses a common programming language) [4,75]. Moreover, the fact that the cloud management process is centralized improves the (i) monitoring, (ii) control, (iii) maintenance, and (iv) distributed update and delivery of the web applications to the end users.

Examples of popular commercial SaaS models include the Google Apps (e.g., Google Mail, Google Docs) and Microsoft web applications. In healthcare, the SaaS providers offer services for data management of almost any type of clinical data (e.g., clinical images, genomic data, EHRs) through user-friendly interfaces, mobile health applications that vary from fitness and lifestyle indications to diabetes (like the GlucoMe [37]). These cloud computing services provide important health indicators not only to the end users but also to healthcare providers, through proper "smart" devices (e.g., smartphones), on-demand. Other popular SaaS providers, such as the IBM Clinical Development [45] and the Medidata Cloud [46], offer software tools for the design and evaluation of clinical trials using almost any type of medical data. The high-performance data analytics tools that are offered by Microsoft Azure [12], Amazon AWS [73], Google Cloud [35], and IBM [43,45] can also be considered as part of the SaaS model because these providers are both PaaS and SaaS. In general, the majority of the SaaS providers in healthcare offer services for medical data management to promote the sharing of medical and other related data (Table 6.2).

6.4.4 Data as a service

The DaaS model (often referred to as data storage as a service model) is a special type of a SaaS service that specifically aims to fulfill the needs for the on-demand delivery of virtualized data to the end users [4,76,77]. The DaaS model allows the end users to gain direct access to their personal data through a self-maintained personal cloud repository without the need to establish a whole IaaS model for simply accessing their personal data. Thus, a DaaS model greatly overcomes additional financial costs regarding the server maintenance and the expensive underlying hardware resources of a typical IaaS model. In addition, the DaaS model enables faster data integration, giving more emphasis to the importance of delivering the product of interest, i.e., the data that the user actually pays for, instead of providing a complete infrastructure of unnecessary hardware components. This is the main reason why more and more cloud providers have recently begun to adopt the DaaS model as part of their overall cloud architecture.

The DaaS models are often accompanied by appropriate web GUIs which offer a user-friendly environment for data management and thus promote the improvement of the data quality. The fact that the DaaS model delivers web user interfaces means that they deliver a SaaS type of cloud computing services, and for this reason the DaaS is sometimes referred to as a "cousin" of the SaaS. Furthermore, the fact that the DaaS model supports the execution of web services through REST APIs increases their overall scalability and interoperability. Moreover, the DaaS model can support database management systems using relational models (RDMS) which provide easier multiuser access to the data in a human-readable format (the data are presented in tabular forms), faster access to the data through special memory indices and reduced queries, and virtual tables for storing sensitive data [78]. Toward this direction, some DaaS providers have designed new methods for manipulating terabyte- or petabyte-scale amount of data using single-keyed rows that are accessed with very low latency (similar to data compression), such as the Google BigTable [79]. The DaaS model, however, allows the users to access their data by means of "renting" their data (i.e., virtualized data) and sometimes do not provide options for downloading the data.

6.5 Storage

To establish a data storage environment, the cloud provider must first define the specifications of the database management system (DBMS) [78,80–82]. The DBMS is a software component which is responsible for the data management operations that take place in electronic databases. The data management operations include data storage, data retrieval, data update, and any interactions that might occur between the end users, the cloud applications, and the database. A database can be any collection of data that is stored in physical storage units (e.g., hard disk drives) and can be accessible through a computer system. The data can be stored into a database according to a database model which describes the logical structure according to which the data are stored and organized in the database [78,80–82]. Examples of popular database models include the hierarchical model where the data are stored in the form of a tree structure (the nodes represent the data records and the links the relationships between them), the network model where the data are stored in the form of a graph (the object types are represented as nodes and the relationships are represented with arcs), and the relational model which is a declarative way to manage data for querying based on schemas that describe the relations among the data attributes [78,80].

Relational database management (RDBM) is the most prominent method for handling structured data in databases [78,80–83]. A popular query programming language is the Structured Query Language (SQL) [78,80] which is widely used in relational DBMS (RDBMS). In SQL, the structured data consist of the entities and the relations between them. An entity consists of a set of variables that can be used to describe a specific domain. For example, an entity might be a "patient" who has a set of variables, such as a name, an address, a social security number, etc. The entity "patient" can also be related with other entities, e.g., a "patient" has "laboratory tests" or a "patient" has "therapies," through appropriate object properties (e.g., "has"). The relational databases can be used to store similar semantic data structures that exist in the form of ontologies. The ontologies are widely used in the semantic web and knowledge-based systems [84,85] as hierarchical data models that represent complex data structures using entities (e.g., classes and subclasses) along with object properties that reflect the "relationship" between the entities (see Section 5.6.2 for more information).

In general, there are three major types of databases, namely the centralized databases, the decentralized databases, and the distributed databases [80,86–88] (Fig. 6.8). In a centralized database, the data are collected from multiple sites (nodes) and stored into a single (centralized) database where they exist in a unified form (Fig. 6.8A). In a distributed network, the data are stored in multiple locations (nodes) which are interconnected through intermediate nodes that end-up in a central administration node (Fig. 6.8B). In fact, in a distributed network, the centralized database is physically spread across different sites which are connected through a data communication network that supports the sharing of data through intermediate local DBMS. These intermediate DBMS nodes do not interact with each other, but only with the central administration node. In a decentralized network, however, the data are stored in multiple sites (e.g., geographically dispersed locations) with the absence of a central data storage node, where all the sites are interconnected through a data management network (Fig. 6.8C). In a decentralized network, however, all the nodes are equal, with a complete absence of a centralized node. This means that the nodes interact with each other without the existence of a master node with control over the nodes. In a decentralized network, the data are physically spread across multiple sites, where each site serves as a local DBMS node.

The decentralized topology can be met in the blockchain technology [27,87–89]. The blockchain is an emerging technology which is based on the Distributed Ledger Technology (DLT) [89]. The DLT uses a peer-to-peer (P2P) network, i.e., a network where the computational tasks are split between the nodes (i.e., peers) which share resources without the existence of a central authority (Fig. 6.8C), to validate cryptographically secured transactions (e.g., information sharing) without the existence

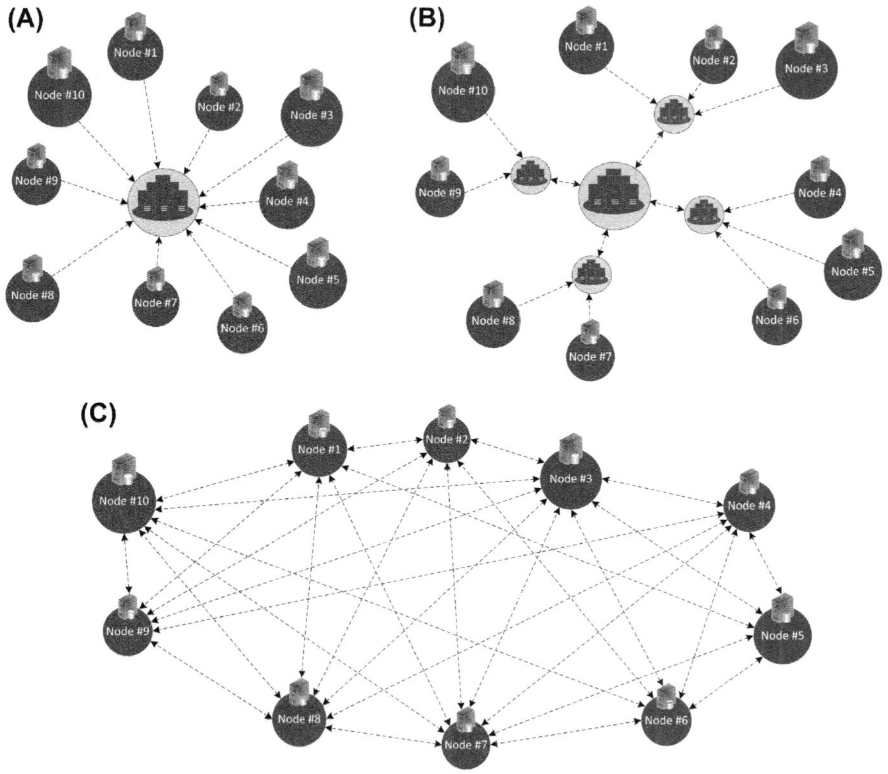

FIGURE 6.8

The three fundamental types of networks for data storage. (A) A centralized network where the data are stored in a centralized node. (B) A distributed network where the data are interlinked through the intermediate local nodes under the coordination of a central authorization node. (C) A distributed network where the local nodes communicate with each other without the existence of a central authorization node.

of a central supervised authority. The transactions that are transferred between the nodes of the P2P network are usually referred to as "smart contracts," i.e., records of information. In fact, the blockchain network can be seen as a state-machine, where the transactions are transferred from one node (state A) to another node (state B), in complete secrecy [87–89]. In a blockchain network, the transactions are invoked by specific nodes which are called "clients" and the validation is performed by the "peers," i.e., the nodes that are part of the P2P network. The "peers" maintain a copy of the chain and use computational power to validate the transactions. A "peer" that manages to successfully validate a type of transaction is usually rewarded with a financial amount. The validated transactions are finally added as a new "block" into an existing "chain" of validated transactions. Once a transaction becomes part of the blockchain, it cannot be changed or removed.

Significant attempts are being made toward the adoption of the blockchain technology in healthcare. The MedRec [90] is a blockchain architecture which realizes the transactions as data ownership and viewership requests that are distributed within a private P2P network. The Gem Health Network [27,89] is another initiative which envisages to establish a blockchain-based ecosystem that aims to improve the patient-centered care through a shared network infrastructure with the participation of healthcare practitioners toward an improved healthcare ecosystem. An example of a startup is the Healthbank [27,91] which envisages to promote the sharing of patient data through a blockchain network, so that the patients can anytime be informed about their personal data (e.g., blood pressure, sleep habits) through wearable devices [27,91]. The majority of these initiatives adopt the Ethereum blockchain technology [92], a distributed public blockchain network that provides services for the transactions transfer within the network's nodes.

The blockchain offers anonymity by ensuring the nodes that participate in the network remain in complete secrecy. In addition, the blockchain networks are not easily hacked because the information regarding the transactions is not stored in a central location but it is distributed across the nodes in a cryptographic and secure way. In a blockchain network, any update regarding the records is conducted in real time and in each node. The fact that in a blockchain network, the information in the blockchain cannot be removed or altered enhances its privacy and security. However, maintaining a record of transactions that is distributed through a blockchain can pose significant data privacy issues and often create problems with the existing data protection regulations. In addition, the level of trust in a blockchain network is small because a peer is not always aware of the information that might arrive to his or her premises. Money laundering cases can also exist in a financial blockchain network unless appropriate anti-money laundering measures are taken.

Centralized databases are less complex and convenient. The data management and maintenance process is easier because the data are gathered in a unified form. However, centralized databases are not reliable because all the data can be compromised and abused in the case of an attack and thus lacks of crucial security measures. This security issue is reduced in distributed databases, because an attack to a specific local DBMS node can cause the rest of the local DBMS nodes to lock down the access to their connected databases and, thus, provide a moderate data security level. This is also present in decentralized databases, where a malicious attack on a single node does not compromise any of the rest nodes and thus accomplishes high security levels. In addition, in the case of an error in a single node, the data can be removed from the problematic node to another node, for safety purposes. This can be also applied in distributed databases but at a higher level because the data from the single nodes can only be moved through their corresponding local DBMS nodes. Furthermore, the recovery rate in a centralized database is small due to the fact that a query (e.g., a search operation) has to be executed on the whole data, whereas in distributed networks, the query is distributed to smaller portions of data where the execution is faster. In addition, a centralized database cannot be easily expanded due to their low scalability, whereas the distributed and decentralized databases can be easily

Table 6.3 Comparison of different performance factors across the three types of database storage networks.

	Centralized	Distributed	Decentralized
Reliability	Low	High	High
Expandability	Low	High	High
Manageability	High	Moderate	Low
Complexity	Low	Moderate	High
Maintainability	High	Moderate	Low
Interoperability	Low	High	High
Level of Trust	High	Moderate	Low
Response	Low	High	High
Recovery	Low	Moderate	High
Security	Low	Moderate	High

expanded due to the high scalability they offer. On the other hand, distributed databases are more complex than centralized databases because the former require continuous communication with the local DBMS nodes to send the queries and receive the results. This complexity is largely increased in decentralized networks where the communication needs to grow exponentially. The aforementioned benefits and drawbacks of the three different types of database storage networks are summarized in Table 6.3.

6.6 Security protocols and guidelines

6.6.1 Cloud security alliance

The Cloud Security Alliance (CSA) is a global, nonprofit organization which was established in 2009 with the purpose of promoting the use of best practices, raise awareness, and provide education on the uses of cloud computing to help ensure a secure cloud computing environment [93]. Within its latest documentation [94], the CSA provides a complete security guidance for critical areas of focus in cloud computing varying from information governance and enterprise risk management to data security and encryption, as well as access management and popular SaaS categories. The CSA recommends the use of a cloud computing model which is based on the NIST standards for cloud computing [1,2]. More specifically, the CSA states that in a cloud computing model, (i) the cloud provider distributes the cloud sources to the end users, (ii) the end users can make use of the cloud resources using on-demand self-service, (iii) the end users can make use of all types of cloud resources, (iv) the end users are able to request an expansion of the cloud resources according to their needs, and (v) the end users can make use of what they pay for under appropriate measured service meters. The CSA defines the three standard types of cloud computing services, namely the IaaS, the PaaS, and the SaaS, and the four types of cloud deployment models, namely the public, the private, the hybrid, and the community (see Section 6.2).

The CSA classifies the security responsibilities on each cloud computing layer according to the activities that take place on the corresponding layer. In the SaaS layer, the cloud provider is responsible for ensuring the entire security of the SaaS layer because the end user has only access rights to the applications without being able to modify them at all. In the PaaS layer, however, both the cloud provider and the end user are responsible for the security because the end user is responsible for the implementation of the software services, and the cloud provider is responsible again for providing the security of the platform. Similarly, in the IaaS layer, the cloud provider is responsible for the foundational security, whereas the end user is responsible for whatever they establish on the infrastructure layer. The CSA recommends the use of Software Defined Networking [95] for the establishment of the underlying network in the IaaS layer instead of the conventional Virtual Local Area Networks (VLANs), because the former can offer security isolation and can support multiple end users under the same IP address (multiple overlapping IPs are segregated). VLANs do not support physical network segregation and are commonly used in single-tenant networks which are located in enterprise data centers instead of the cloud computing platforms which require multitenant networks [94]. In addition, the CSA promotes the use of security agents which can be adapted on the cloud environment and are able to manage the network workloads, and establish proper cloud firewalls to manage the network bottleneck and ensure a stable performance.

As far as the storage security is concerned, the CSA highlights the Storage Area Network and the Network-Attached Storage as the two common methods for storage virtualization to encrypt the physical storage units and prevent data exposure either during hardware maintenance and update or during an unauthorized data management operation [94]. Furthermore, the CSA promotes the use of dockers [96,97] for the development of the software services instead of the conventional VMs. The dockers are software containers which can be established directly on the top of the IaaS layer and can be executed on any virtual environment without the need to install a complete operating system or any software components at all [96,97]. A software container system consists of the container which offers the environment for the services execution, the scheduler which handles the software libraries and additional software components, and the repository which includes the program that is going to be executed [96,97].

As for the security of the cloud computing applications and vulnerability testing, the CSA recommends the use of both the Static Application Security Testing (SAST) and the Dynamic Application Security Testing (DAST). The former (SAST) [98] checks for the reliability of the cloud API calls and for the existence of any stored API credentials that could possibly harm the system, whereas the latter (DAST) [98] tests for web vulnerabilities during the execution of the APIs, i.e., during the running state of the platform. Regarding the data security and encryption, the CSA recommends the use of the Cloud Access and Security Blockers [98] that provide continuous monitoring of the cloud API connections that are established during the use of the cloud computing services from the end users and thus which services are used by the users and the distribution of the data within the platform (data lineage). Other options include the use of IP filtering and Data Loss Prevention [99] methods which, however, are less accurate because they may be affected by the data encryption protocols. Both API and data monitoring is of great importance because they can directly affect the compliance of the platform with the existing data protection regulations. As for user authentication, the CSA promotes the use of multifactor authentication (MFA) [100] in all its layers as it is the most effective security control system for user authentication. The OAuth (either the OAuth 1.0 or the OAuth 2.0) framework [101] is proposed as a standard method for the authorization of the web services over HTTP requests along with the standard Attribute-Based Access Control [102] which accounts for the role, the location, the authentication method, etc., of the user who wishes to gain access into the platform through the user authentication process instead of the conventional Role-Based Access Control model which considers only for the role of the user [103].

CSA has launched the CSA Security, Trust & Assurance Registry program [104], a security certification program that is addressed to cloud providers and consists of a three-stage process (self-assessment, third-party audit, and continuous monitoring) which certifies the security of a cloud computing platform. In addition, the CSA has launched the Certificate of Cloud Security Knowledge [104] and the Certificate of Cloud Security Professional [104] for professionals. The CSA has also initiated the first-ever baseline control framework for risk management in cloud computing environments, namely the Cloud Controls Matrix [104]. In 2011, the CSA was selected as the US federal cloud computing strategy. Undoubtedly, the CSA is currently the world's leading organization toward the establishment of guidelines toward a secure cloud computing environment with legal entities in Europe, Asia Pacific, and a large number of separate, USA-controlled, chapters. Currently, the CSA has 90,000 members worldwide with a large number of hosted educational seminars.

6.6.2 Institute of electrical and electronic engineers standards

The Institute of Electrical and Electronic Engineers Standards Association (IEEE-SA) is a promising initiative which aims to promote the global technologies and advances all over the world, through IEEE [105]. In 2011, the IEEE-SA launched two ongoing (active) drafts, namely the IEEE P2301 Draft Guide for Cloud Portability and Interoperability Profile (CPIP) [106] and the IEEE P2302 Data Standard for Intercloud Interoperability and Federation (SIIF) [107] toward a sustainable, interoperable, portable, and standardized cloud computing model. The purpose of the CPIP is to provide a set of guidelines regarding the existing cloud computing standards, whereas the SIIF aims at providing standards for cloud-to-cloud interoperability to ensure the transparency between the end users and the cloud applications. Both of these projects are under development with the scientific community being on hold for the initiative's outcomes.

The CPIP is addressed to cloud vendors, developers, and end users. Its main purpose is to formulate a guide to assist them toward the development of standardized cloud-based products and services [106]. A cloud computing environment consists of a variety of software and hardware components, each one with its own specifications and interactions with the services. For example, a software service might be written in a different programming language which requires specific data formats and individual visual interfaces, as well as have specific interactions with the rest of the services or the software components. For this reason, the guide will aim at providing guidelines for technologies and standards to be used for visual interfaces, data formats, and standards for data storage using semantic mechanisms in heterogeneous databases, among many others. This will enhance the portability, commonality, and interoperability of cloud computing.

The SIIF focuses on the federation of cloud computing environments [107]. A federated cloud computing environment is an environment where multiple cloud computing systems interact under a common purpose. The SIIF aims at providing a set of guidelines for standardizing between cloud operations within multiple cloud environments to accomplish a federated cloud platform. In fact, the SIIFs envisage to standardize the governance (e.g., audit and compliance with regulations), functional (e.g., the interactions between the cloud resources), and topological (e.g., roots and gateways) elements of disparate cloud environments to promote cloud federation. The standards that will be provided will not consider for the standardization of hybrid cloud models because the latter involve the combination of two or more public, private, or community cloud models. Moreover, the standards will not consider for the standardization of the cloud operations (in a single cloud environment) because the latter depends on the implementation specifications.

6.6.3 European network and information and security agency guidelines

The European Network and Information and Security Agency (ENISA) [108] is an EU agency which was formed in 2004 and consists of a group of experts on information security issues with the purpose of providing advice and high-quality recommendations toward the assurance of information security in cloud computing environments. Since 2009, the ENISA has an active role toward the substantial improvement of the information security in different areas of cloud computing. In 2009, the ENISA published a set of guidelines in cooperation with experts from the industry, academia, and governmental organizations, regarding the security risks and opportunities of cloud computing business models and technologies [108]. According to the ENISA's definition of cloud computing, the types of the cloud computing services are the IaaS, the PaaS, and the SaaS and the types of cloud computing deployment models is the public, the private, and the community (the hybrid cloud deployment model is not stated).

The ENISA has identified a list of 11 security risks which are of high importance, including [108] (i) the vulnerabilities in the software security of the cloud services which might lead to the identification of the end user, (ii) the network attacks through the web, such as Denial-of-Service (DoS) attacks,[1] (iii) the social engineering attacks which might lead to the identification of the user's credentials from social networks, (iv) the compromise of data management interfaces which might lead to data abuse and loss, (v) device theft or loss which might lead to compromised authentication credentials, (vi) any physical disaster which might harm the data centers, (vii) data overload due to DoS attacks,[1] (viii) the unexpected high cost of the cloud services, (ix) the vendor lock-in which disables the end user from changing a cloud provider, (x) the administrative and/or legal conflicts between the cloud services, and (xi) the foreign legal conflicts.

To deal with the aforementioned security risks, the ENISA provides a set of guidelines in the form of questionnaires followed by possible answers [108]. In fact, the ENISA urges the cloud providers to take appropriate security measures against such risks. More specifically, the cloud providers shall deploy software models and underlying network infrastructures with reduced likelihood of being affected from malicious attacks, as well as enhance the user access management and authentication system. Furthermore, the cloud providers must be well aware of data protection regulations and fulfill all the necessary ethical and legal requirements regarding the use of personal data and control the billing limitations and obtain security certifications. Independent auditors shall conduct the evaluation of the cloud computing services to ensure the reliability of the evaluation process. Moreover, the sensitive data must be handled in a secure way to avoid data abuse through standard data management interfaces. In addition, the platform developers must be trained for reducing the vulnerability of the software services under development and the cloud providers shall evaluate the vulnerability scan reports from third-party audits of software. The cloud providers shall also conduct disaster recovery plans in the case of physical disasters, by employing enforced perimeter defense units and support data mitigation in remote data centers.

In addition, the ENISA provides a list of 11 opportunities (benefits) that can arise from the use of cloud computing environments against traditional IT infrastructures, including [108] (i) the geographic spread of cloud computing data centers, (ii) the elasticity they offer in terms of the resource usage for big data, (iii) the distribution of services (on-demand) to a large number of end users, (iv) the reduced cost for physical security because the resources are collected in a common place, (v) the reduced surveillance costs because the latter are shared across the large number

---

[1] DoS: A Denial-of-Service attack is performed by flooding a host server or an underlying network infrastructure with continuous requests until it crashes, and thus preventing the legitimate users from connecting to the server host [127]. A Distributed Denial-of-Service (DDoS) attack occurs when multiple machines are operating together to perform the attack yielding significant computational power.

of end users, (vi) the secure software development, (vii) the fast patching and update of services to prevent attacks, (viii) the automated and scheduled maintenance of backups, (ix) the reduced amount of data that arrive on the devices of the end users, (x) the regular detection of malicious content on custom software, and (xi) the easy evaluation of the cloud services from the auditors regarding the organizational security, governance, and risk management. Each one of these opportunities is directly linked with various cloud security challenges including the cloud's interoperability, portability, continuous monitoring, software security, scaling, and sizing costs, among many others (see Section 6.7).

6.6.4 National institute of standards and technology guidelines

The NIST was initially found in 1901 as an entity which is responsible for the development of standards and guidelines for providing adequate information security under the US Department of Commerce. Since 2011, the NIST provided definitions for the cloud computing, under the Federal Information Security Management Act (FISMA) of 2002 and has been designated by the Federal Chief Information Officer to lead the implementation of secure cloud computing in the federal government, as well as define standards and guidelines toward secure cloud computing [1,2]. In November 2010, the NIST cloud computing program was launched which aimed to transform the federal government's traditional system into a cloud computing model, in coordination with other related federal government agencies. At the beginning of 2011, the program provided a technical strategy and defined standards for federal cloud computing both in national and international level.

The NIST provides definitions for three cloud computing services, namely the IaaS, the PaaS, and the SaaS, along with four cloud deployment models, namely the private cloud, the public cloud, the hybrid cloud, and the community cloud (the definitions and interactions between these models have been already presented in Section 6.2). According to the NIST's cloud computing reference architecture, there are five major types of cloud actors [2]: (i) the cloud provider (person or organization or entity) who makes the cloud computing services available to the cloud consumers, (ii) the cloud consumer (person or organization) who uses the services from the cloud provider, (iii) the cloud auditor (third-party) who evaluates the performance of the cloud platform and its services, (iv) the cloud broker (entity) who manages the use, the performance, and the delivery of the cloud services and serves as a mediator between the cloud providers and the cloud consumers, and (v) the cloud carrier (intermediary) who is responsible for the delivery of the services from the cloud providers to the cloud consumers. A typical IaaS consumer can manage the hardware resources and IT infrastructure operations, a typical PaaS consumer can develop and test software services, and a typical SaaS consumer can run the cloud applications (e.g., data management, social networking). The cloud provider is responsible for (i) the security of the platform, (ii) the privacy of the cloud consumers, (iii) the management of the cloud services, (iv) the deployment of the services, on-demand, and (v) the overall cloud orchestration. The cloud auditor ensures the compliance of the cloud operations with regulations and security policies and standards. The distinction among the cloud broker and the cloud provider is that the former can establish a connection among different cloud providers for technical/business support. Finally, the cloud carrier provides the access to the cloud services in terms of the underlying network infrastructure (e.g., an organization that supports the transfer of media).

At the IaaS level, the NIST suggests the adoption of two well-established and promising standards, namely the Open Cloud Computing Interface from the Open Grid Forum [109] and the Cloud Infrastructure Management Interface [110] from the Distributed Management Task Force. The former mainly deals with the application of REST services to control the interactions between the components of the infrastructure layer, whereas the latter provides a set of interfaces that support the provisioning and management of the cloud infrastructure in multiple clouds by standardizing the interactions between the cloud environments (federated cloud management). As far as the data management level is concerned, the NIST suggests the Cloud Data Management Interface [111] as functional interface that enables the user to manage his or her data from the cloud data repository, as well as to support the management of software containers. For the PaaS applications, the NIST recommends the use of the Cloud Application Management Protocol [112] which serves as a high-level system that provides all the necessary software components for the development of services and applications. As for the SaaS applications, there are different web-based standards that can be used for the transferring of web services between the cloud server and the web browser of the end user, including the HTTP, SSL/TLS, and IP, among others. For achieving interoperability between the cloud services, the NIST suggests the OpenID [113], a decentralized authentication protocol that enables the execution of multiple cloud services across different cloud systems without the need for the user to keep a separate account for each individual system.

For portability purposes, the NIST suggests the use of the Open Virtualization Format [114] as an open standard toward an efficient packaging and distribution of software components that are intended to be executed in VMs similar to the dockers [96]. Regarding the portability of hardware and overall cloud applications, the NIST identifies the Topology and Orchestration Specification for Cloud Applications [115] from the Organization for the Advancement of Structured Information Standards as a promising attempt to standardize the services and applications by considering for regulatory concerns and the relationships among them. For information exchange between or within cloud systems, the JSON or XML formats can be employed as the two standard data structures for information transfer. The standard REST and SOAP protocols can be used for access to the web services and finally the popular OAuth framework, and the Security Assertion Markup Language [116] framework can be used for exchanging information regarding the user's authentication and authorization. The widely adopted technologies and standards in cloud computing environments are summarized in Table 6.4.

Table 6.4 A summary of popular technologies/standards for different types of operations that take place in federated cloud computing environments.

Layer	Operation	Technologies/standards
PaaS	Connection to the cloud platform	SSL/TLS with private encryption and public decryption keys
SaaS	User authentication	OAuth (v1.0, v2.0), SAML
SaaS	User access control	ABAC, ARBAC
SaaS	Decentralized authentication	OpenID
ALL	User authentication among the cloud layers	MFA
PaaS	Access to the web services of the platform	REST, SOAP
PaaS	Implementation of an operating system	Linux, Windows
PaaS	Operating-system level virtualization	Software containers (dockers), VMs
PaaS	Development of software components	C#, Java, Python, PHP
SaaS	Data structures for information transfer	JSON, XML
IaaS	Application of an underlying network layer	VLANs, SDN
PaaS	Portability of software components	OVF
IaaS, PaaS	Portability of software and hardware components	TOSCA
IaaS	Management of hardware components	OCCI, CIMI
IaaS	Data and software containers management	CDMI
IaaS	Data security and encryption	CASB, URL filtering, DLP
IaaS	Storage virtualization	SAN, NAS
SaaS, PaaS	Vulnerability testing	SAST, DAST

6.7 Challenges

The technical challenges of a federated cloud platform are persistent and demanding [3–9,31,117–119]. Examples of such challenges that have been highlighted by the scientific community include the following:

Data sharing: The cloud computing platform must ensure the compliance of all the operations with the individual data protection regulations (harmonization of data protection regulations) to enable data sharing. Ethical issues for data collection introduced by different countries must be also taken into account. The standardization of national data protection laws is a crucial barrier toward the sharing of medical data in federated cloud computing environments. Furthermore, a cloud computing platform must account for the complexity of the hospital IT infrastructures during the sharing of medical data, as well as include semantic models for data standardization according to international clinical guidelines during the data acquisition process. In addition, a research cloud platform shall offer data analytics services for decentralized analysis including batch-based processing mechanisms (i.e., distributed methods for data analytics), especially when the sensitive data are stored in decentralized databases, where continuous communication is necessary.

Network security: All operations within the cloud must take place in secure virtual private networks for ensuring the confidentiality during the transfer of sensitive data. Data security comprises the biggest barrier of cloud computing. The lack of data security results to data leakage, data abuse, loss of data integrity, and control over the hosted data and the cloud applications. To deal with such issues, OAUTH-type authorization frameworks must be adopted to ensure secure user authentication and access management and secure access to the information and services of the platform. The flow of sensitive information outside the platform (e.g., user credentials) must be encrypted and decrypted through SSL/TLS protocols using public decryption keys and private encryption keys. The inner information flows are performed through secure firewalls and virtual private networks which enhance the reliability of the platform and ensure a highly secure information transfer.

Data security: The platform must support efficient web communication for faster data transfer. Most importantly, the platform must offer effective deidentification mechanisms through the construction of unique identifiers per patient (e.g., hash keys). In addition, the metadata must be followed by an expiration date for security reasons. The data must be stored in secure private repositories and the access should be controller through multiple-MFA systems. Informed consent forms must be requested in the case of pooled data analysis to overcome the fear for data abuse. Any data transfer within the platform must be tracked down by proper system monitoring and log mechanisms including audit tables which record the date, the time, and additional information regarding the user's access to the stored data.

Data management: The cloud platform must offer a user-friendly data management system that supports both retrospective and prospective data. Furthermore, the data management system shall include long-term data curation mechanisms for ensuring the constant quality of the prospective and retrospective data. The data must be accurate, up-to-date, relevant, adequate, complete, and in a readable form. Finally, the stored data must be always available to the users to support the reuse of health data for clinical research purposes.

Scalability: The cloud computing platform must be highly scalable in terms of an efficient resource management of both the hardware and the software components. A cloud computing platform must be also expandable in terms of the types of the hardware and the software components. The cloud infrastructure and especially the hardware components must be easily adjustable. Any changes within the hardware layer shall be easily tracked by the PaaS layer. The availability of the hardware and software components must be monitored and distributed according to the needs of the cloud computing services. The trade-off between cost scalability over security must be carefully considered.

Adaptability: A cloud platform must be easily scalable in terms of hardware expandability and software updates. A central system which handles the cloud resources according to the computational needs of the requested services is required for the development of efficient provisioning rules. These rules shall be able to fulfill the computational needs of the cloud computing services according to their computational demand. Moreover, the maintenance of the hardware and software components must be fast and precise.

Performance: The performance of a cloud platform is directly related with its hardware and software components and indirectly related with the development of efficient provisioning rules (i.e., schedules for efficient management of the cloud resources). The hardware layer shall include efficient hardware components, and the software components must be compatible with the hardware layer to fulfill the needs of services that require high computational needs (e.g., memory usage or processing speed).

Support: The platform must be able to offer advanced user support to the end users of the platform. Bring the analysis to the data instead of bringing the data to the analysis. Secure private data layers within the cloud for the storage of personal data (in remote private spaces). In addition, the servers that are hosted by the cloud platform must be easily accessible from the web through proper SSL/TLS protocols. All software components must be followed by Creative Common licenses. The cloud platform must also include automated mechanisms for reduced latency and fast error recovery in the case of high demand and account for efficient disaster recovery plans in the case of physical disasters.

Multidimensional interoperability: The diversity of offers combined with mainstream cloud services lead to scenarios of distributed SaaS applications whose parts are hosted on different cloud platforms and will therefore need to interoperate and cooperate through efficient and reliable protocols. Secure REST services need to be used to transfer any workflows outside the platform, e.g., for the execution of services in the premises of a company or organization and finally receive the results in appropriate JSON data structures so as to be parsed for visualization purposes. A federated cloud platform shall also be able to support the management of different types of data (e.g., imaging data, genetic data). The standardization of cloud standards (e.g., for user authentication, data sharing, data storage) is necessary to accomplish the interoperability among different cloud computing environments and enable the easier migration of the cloud computing services from one environment into another.

Usability: A cloud platform must provide user-friendly menus and high-quality visualization components. The visualization components must be easily scalable and coordinated by a central system. Furthermore, the visualization components must be properly adjusted based on the type of the cloud computing services. For example, a data management system must be friendly and enable the easy access to the stored data. The results from the data analytics services must be displayed in user-friendly graphical illustrations.

6.8 Conclusions

Undoubtedly, cloud computing is on the edge of a revolution which envisages to create a global network of computational resources that can be easily shared. Cloud computing provides a shared pool of flexible IT resources, such as hardware components, software components, web applications, etc., on-demand, and through a secure underlying network infrastructure, which is based on SLAs between the cloud provider and the end users. The impact toward the global adoption of cloud computing is estimated to be high, with 83% of the enterprise workloads (including machine learning and artificial intelligence applications, mobility, digital transformation, among others) being estimated to take place in the cloud by 2020, where the 41% of the workloads is estimated to run in public cloud computing models, followed by an estimated increase from 67 billion dollars (in 2015) to 162 billion dollars (19% compound annual growth rate) in 2020 [18,120]. In healthcare, cloud computing offers exciting opportunities for the healthcare stakeholders regarding medical data sharing, multimodal medical data management, medical data analytics, etc., where the annual growth rate had already been equal to 20.5% in 2017 [121].

The benefits of cloud computing are many. Cloud computing nullifies the need for enterprise companies and organizations to purchase and maintain expensive hardware and software resources, as well as reduces the high maintenance and implementation costs of the traditional infrastructure. The end users have access to a vast category of cloud computing services without the need to download anything on their computer except from a typical access through a web browser. Furthermore, a cloud computing environment offers a variety of cloud computing services, varying from the delivery of hardware resources (e.g., storage units, memory units, processing power, network components) and software components (e.g., operating systems, frameworks for software development) to software services (e.g., web applications, web tools) that are available on-demand, against the traditional licensed software components and obtained hardware. The cloud computing services take place on cloud computing environments (cloud models) which may be public, private, community, or hybrid, according to the technical needs (e.g., security and computing specifications) of each group of end users. The hybrid cloud models are in general preferred because they are able to combine the benefits that arise from the combination of the individual cloud computing models.

As we have already seen, the types of the cloud computing services can be classified into three large categories (models): (i) the IaaS which involves the distribution of hardware resources, (ii) the PaaS which involves the distribution of technology tools for the development of software applications, and (iii) the SaaS which involves the distribution of the applications to the end users. Each model is addressed to different types of end users and thus fulfills the demands of a wide range of end users with low and high technical expertise. For example, a typical end user can make use of SaaS services (e.g., manage his or her personal data through a data management interface), whereas a more expert user can make use of PaaS services to develop applications (e.g., use the software components to develop applications) or even make use of IaaS services to install software components (e.g., install an operating system) and establish VMs.

These types of cloud computing services can be considered as the layers of a cloud computing architecture where the IaaS layer lies on the basis because it provides the infrastructure for the establishment of the cloud services. The PaaS layer provides the technology tools that are used for the development of the software services which are finally distributed through the SaaS layer. In general, the majority of the offered services are SaaS applications, taking the form of web-based tools or web applications. The type and quality of the delivered cloud computing services are usually agreed on in an SLA between the cloud provider and the end users, a fact that provides flexibility in the delivery of the cloud services. Furthermore, the virtualization of the cloud resources (i.e., the establishment of VMs) followed by their efficient provisioning can reduce the latency and yield an increased performance. These layers can be interlinked with similar layers from other cloud computing environments to enable cloud federation.

In healthcare, the cloud computing vendors deliver a great number of cloud computing services for the sharing and management of almost all types and sources of medical data (e.g., omics, clinical images, biosignals), the preprocessing and the extensive analysis of medical big data through high-performance computing, and the design and execution of clinical trials. In addition, the majority of them have already taken into consideration measures for dealing with data governance ensuring the compliance of their services with existing data protection regulations and accounting for data quality assessment methods. The list of the cloud providers in healthcare is large with Amazon AWS, Microsoft Azure, Google Cloud, and IBM appearing to be at the top four leading international cloud computing vendors with the highest adoption rate [122], offering a variety of healthcare services including multimodal data sharing, data management, data storage, big data analytics, and mobile health monitoring, among others. All these services are always available on-demand, with reduced latency, under highly secure infrastructure, fulfilling all the necessary data protection and security standards. The latest advancements in data analytics include Google's high-quality custom machine learning models under the Cloud AutoML [123] and the Google BigTable [79] for managing petabytes of data with sub-10 ms latency, Azure's Machine Learning Service [124], and Amazon's Elastic Inference [125] for deep learning applications, among others. Furthermore, Amazon's Virtual Private Cloud (VPC) [126] has recently emerged as a promising solution for cloud federation through the construction of virtual networking environments that consist of reconfigurable cloud computing resources.

Apart from the significant benefits and the cost-efficient solutions that a cloud computing environment offers, there are many key challenges and risks that still need to be addressed. Information security is the most prominent, with the transferring of personal data to the cloud infrastructure raising several concerns regarding unauthorized data access and data abuse. Security risks may arise from weak encryption protocols and identity management systems, unsecure underlying network infrastructures, and from the lack of physical security and disaster recovery plans.

In addition, critical risks may also arise from weak and rare audits and the lack of compliance with data protection regulations and international cloud security guidelines (such as those posed by NIST, CSA, and ENISA). Other important risk factors include the vendor lock-in which disables the end user from transferring the existing services to another cloud provider, the lack of transparency in the cloud operations, and the lack of data monitoring and control. Risks may also arise during the migration of data and cloud services to shared cloud computing environments which are less secure, threatening the integrity of the data.

The heterogeneity of the SaaS and PaaS services in different cloud computing environments is a serious technical challenge which hampers the standardization of cloud service protocols especially during the collection of heterogeneous medical data. Additional challenges include the harmonization of national data protection laws for enabling the sharing of medical and other related data from different sources, the development of long-term data quality assessment methods for standardizing data formats, the development of efficient provisioning rules for high cloud computing performance, and other important indicators, such as the scalability, the agility, the flexibility, the multidimensional interoperability, and the resilience against malicious attacks. All these risks and challenges must be carefully taken into consideration by the cloud computing providers to enable cloud federation.

References

[1] Liu F, Tong J, Mao J, Bohn R, Messina J, Badger L, Leaf D. NIST cloud computing reference architecture. NIST Special Publ 2011;500:1–28.
[2] Mell P, Grance T. The NIST definition of cloud computing. Natl Inst Stand Technol 2009;vol. 53(6):50.
[3] Khan MA. A survey of security issues for cloud computing. J Netw Comput Appl 2016;71:11–29.
[4] Rittinghouse JW, Ransome JF. Cloud computing: implementation, management, and security. CRC Press; 2017.
[5] Singh S, Jeong YS, Park JH. A survey on cloud computing security: issues, threats, and solutions. J Netw Comput Appl 2016;75:200–22.
[6] Fernandes DA, Soares LF, Gomes JV, Freire MM, Inácio PR. Security issues in cloud environments: a survey. Int J Inf Secur 2014;13(2):113–70.
[7] Puthal D, Sahoo BPS, Mishra S, Swain S. Cloud computing features, issues, and challenges: a big picture. In: Proceedings of the 2015 international conference on computational intelligence and networks; 2015. p. 116–23.
[8] Ali O, Shrestha A, Soar J, Wamba SF. Cloud computing-enabled healthcare opportunities, issues, and applications: a systematic review. Int J Inf Manag 2018;43:146–58.
[9] Yang C, Huang Q, Li Z, Liu K, Hu F. Big data and cloud computing: innovation opportunities and challenges. Int J Digital Earth 2017;10(1):13–53.
[10] Marinescu DC. Cloud computing: theory and practice. Morgan Kaufmann; 2017.
[11] Amazon elastic compute cloud (Amazon EC2), http://aws.amazon.com/ec2/.

[12] Microsoft azzure cloud computing platform and services, http://azure.microsoft.com/en-us/.
[13] Kavis MJ. Architecting the cloud: design decisions for cloud computing service models (SaaS, PaaS, and IaaS). John Wiley & Sons; 2014.
[14] Al-Fuqaha A, Guizani M, Mohammadi M, Aledhari M, Ayyash M. Internet of things: a survey on enabling technologies, protocols, and applications. IEEE Commun Surveys Tutor 2015;17(4):2347–76.
[15] Lin J, Yu W, Zhang N, Yang X, Zhang H, Zhao W. A survey on internet of things: architecture, enabling technologies, security and privacy, and applications. IEEE Internet Things J 2017;4(5):1125–42.
[16] Botta A, De Donato W, Persico V, Pescapé A. Integration of cloud computing and internet of things: a survey. Future Gener Comput Syst 2016;56:684–700.
[17] Lee I, Lee K. The internet of things (IoT): applications, investments, and challenges for enterprises. Bus Horiz 2015;58(4):431–40.
[18] 83% of enterprise workloads will be on the cloud by 2020, Forbes, http://www.forbes.com/.
[19] Rose K, Eldridge S, Chapin L. The internet of things: an overview. The Internet Society (ISOC); 2015. p. 1–50.
[20] Vandelanotte C, Müller AM, Short CE, Hingle M, Nathan N, Williams SL, et al. Past, present, and future of eHealth and mHealth research to improve physical activity and dietary behaviors. J Nutr Educ Behav 2016;48(3):219–28.
[21] Barello S, Triberti S, Graffigna G, Libreri C, Serino S, Hibbard J, Riva G. eHealth for patient engagement: a systematic review. Front Psychol 2015;6:2013.
[22] Gagnon MP, Ngangue P, Payne-Gagnon J, Desmartis M. m-Health adoption by healthcare professionals: a systematic review. J Am Med Inform Assoc 2015;23(1):212–20.
[23] Langmead B, Nellore A. Cloud computing for genomic data analysis and collaboration. Nat Rev Genet 2018;19(4):208.
[24] Tan Z, Wei H, Chen Y, Du M, Ye S. Design for medical imaging services platform based on cloud computing. Int J Big Data Intell 2016;3(4):270–8.
[25] Abbas N, Zhang Y, Taherkordi A, Skeie T. Mobile edge computing: a survey. IEEE Internet Things J 2018;5(1):450–65.
[26] Cai H, Xu B, Jiang L, Vasilakos AV. IoT-based big data storage systems in cloud computing: perspectives and challenges. IEEE Internet Things J 2017;4(1):75–87.
[27] Mettler M. Blockchain technology in healthcare: the revolution starts here. In: Proceedings of the 18th IEEE international conference on e-health networking, applications and services. Healthcom); 2016. p. 1–3.
[28] Navale V, Bourne PE. Cloud computing applications for biomedical science: a perspective. PLoS Comput Biol 2018;14(6):e1006144.
[29] Schweitzer EJ. Reconciliation of the cloud computing model with US federal electronic health record regulations. J Am Med Inform Assoc 2011;19(2):161–5.
[30] Gao F, Thiebes S, Sunyaev A. Rethinking the meaning of cloud computing for health care: a taxonomic perspective and future research directions. J Med Internet Res 2018; 20(7):e10041.
[31] Darwish A, Hassanien AE, Elhoseny M, Sangaiah AK, Muhammad K. The impact of the hybrid platform of internet of things and cloud computing on healthcare systems: opportunities, challenges, and open problems. J Ambient Intell Hum Comput 2017: 1–16.

[32] Kshetri N, Fredriksson T, Torres DCR. Big data and cloud computing for development: lessons from key industries and economies in the global south. Routledge; 2017.
[33] Rumbaugh J, Booch G, Jacobson I. The unified modeling language reference manual. Addison Wesley; 2017.
[34] State of the cloud adoption and security. Forbes; 2017. http://www.forbes.com/.
[35] Google Cloud, http://cloud.google.com/.
[36] Microsoft healthVault, http://international.healthvault.com/.
[37] GlucoMe – Digital diabetes care platform, http://www.glucome.com/.
[38] Medicloud – Free software for medical clinics, http://medicloud.me/.
[39] Practise fusion – Cloud-based electronic health record, http://www.practicefusion.com/.
[40] IBM merge iConnect cloud archive, http://www.merge.com/Solutions/Interoperability/iConnect-Cloud-Archive.aspx.
[41] Merge eMix, http://www.merge.com/Solutions/Interoperability/Merge-eMix™.aspx.
[42] RxEye Cloud, http://www.rxeye.com.
[43] IBM Explorys, http://www.ibm.com/watson/health/explorys/.
[44] eClinical works, http://www.eclinicalworks.com/.
[45] IBM clinical development, http://www.ibm.com/us-en/marketplace/clinical-development.
[46] Medidata cloud, http://www.mdsol.com/en.
[47] Hou L, Zhao S, Li X, Chatzimisios P, Zheng K. Design and implementation of application programming interface for internet of things cloud. Int J Netw Manag 2017;27(3):e1936.
[48] Ren L, Zhang L, Wang L, Tao F, Chai X. Cloud manufacturing: key characteristics and applications. Int J Comput Integr Manuf 2017;30(6):501–15.
[49] Doukas C, Maglogiannis I. Bringing IoT and cloud computing towards pervasive healthcare. In: Proceedings of the 2012 6th international conference on innovative mobile and internet services in ubiquitous computing; 2012. p. 922–6.
[50] Petcu D, Macariu G, Panica S, Crăciun C. Portable cloud applications—from theory to practice. Future Gener Comput Syst 2013;29(6):1417–30.
[51] Assunção MD, Calheiros RN, Bianchi S, Netto MA, Buyya R. Big data computing and clouds: trends and future directions. J Parallel Distrib Comput 2015;79:3–15.
[52] Vitolo C, Elkhatib Y, Reusser D, Macleod CJ, Buytaert W. Web technologies for environmental big data. Environ Model Softw 2015;63:185–98.
[53] Paik HY, Lemos AL, Barukh MC, Benatallah B, Natarajan A. Web service implementation and composition techniques. Springer International Publishing; 2017.
[54] Di Martino B, Cretella G, Esposito A. Cross-platform cloud APIs. In: Cloud portability and interoperability. Cham: Springer; 2015. p. 45–57.
[55] Bray T. The javascript object notation (json) data interchange format (No. RFC 8259). 2017.
[56] Corbin JR. The art of distributed applications: programming techniques for remote procedure calls. New York: Springer Science & Business Media; 2012.
[57] Meloni JC, Kyrnin J. HTML, CSS, and JavaScript all in one: covering HTML5, CSS3, and ES6, Sams teach yourself. Sams Publishing; 2018.
[58] Alawar MW, Naser SSA. CSS-Tutor: an intelligent tutoring system for CSS and HTML. Int J Acad Res Dev 2017;2(1):94–8.
[59] Angular docs, http://angular.io/.
[60] Spring framework, http://spring.io/projects/spring-framework.

[61] Freier A, Karlton P, Kocher P. The secure sockets layer (SSL) protocol version 3.0 (No. RFC 6101). 2011.
[62] Stallings W. Cryptography and network security: principles and practice. Upper Saddle River: Pearson; 2017.
[63] Singh G. A study of encryption algorithms (RSA, DES, 3DES and AES) for information security. Int J Comput Appl 2013;67(19).
[64] Schneier B. Foundations. Applied cryptography. In: Protocols, algorithms, and source code in C. 2nd ed. John Wiley & Sons; 2015. p. 1−18.
[65] Igoe K, McGrew D, Salter M. Fundamental elliptic curve cryptography algorithms (No. RFC 6090). 2011.
[66] Karagiannis V, Chatzimisios P, Vazquez-Gallego F, Alonso-Zarate J. A survey on application layer protocols for the internet of things. Trans IoT Cloud Comput 2015;3(1):11−7.
[67] Rescorla E. The transport layer security (TLS) protocol version 1.3 (No. RFC 8446). 2018.
[68] Oppliger R. SSL and TLS: theory and practice. Artech House; 2016.
[69] Graniszewski W, Arciszewski A. Performance analysis of selected hypervisors (virtual machine monitors-VMMs). Int J Electron Telecommun 2016;62(3):231−6.
[70] Usmani Z, Singh S. A survey of virtual machine placement techniques in a cloud data center. Proc Comput Sci 2016;78:491−8.
[71] Masdari M, Nabavi SS, Ahmadi V. An overview of virtual machine placement schemes in cloud computing. J Netw Comput Appl 2016;66:106−27.
[72] Kivity A, Kamay Y, Laor D, Lublin U, Liguori A. Kvm: the Linux virtual machine monitor. Proc Linux Symp 2017;1:225−30.
[73] Amazon Web Services (AWS), http://aws.amazon.com/.
[74] Google cloud − Compute engine, http://cloud.google.com/compute/.
[75] Erl T, Cope R, Naserpour A. Cloud computing design patterns. Prentice Hall Press; 2015.
[76] Sarkar P. Data as a service: a framework for providing Re-useable enterprise data services. John Wiley & Sons; 2015.
[77] Terzo O, Ruiu P, Bucci E, Xhafa F. Data as a service (DaaS) for sharing and processing of large data collections in the cloud. In: Proceedings of the 7th IEEE international conference on complex, intelligent, and software intensive systems (CISIS); 2013. p. 475−80.
[78] Elmasri R, Navathe SB. Fundamentals of database systems. Pearson; 2017.
[79] Google cloud bigtable − Scalable NoSQL database service, http://cloud.google.com/bigtable/.
[80] Coronel C, Morris S. Database systems: design, implementation, & management. Cengage Learning; 2016.
[81] Hogan R. A practical guide to database design. Chapman and Hall/CRC; 2018.
[82] Laudon KC, Laudon JP. Management information systemsvol. 8. Prentice Hall; 2015.
[83] Harrington JL. Relational database design and implementation. Morgan Kaufmann; 2016.
[84] Becerra-Fernandez I, Sabherwal R. Knowledge management: systems and processes. Routledge; 2014.
[85] Rowley J, Hartley R. Organizing knowledge: an introduction to managing access to information. Routledge; 2017.
[86] Wu J. Distributed system design. CRC press; 2017.

[87] Wright A, De Filippi P. Decentralized blockchain technology and the rise of lex cryptographia. SSRN; 2015.
[88] Lakshman A, Malik P. Cassandra: a decentralized structured storage system. ACM SIGOPS – Oper Syst Rev 2010;44(2):35–40.
[89] Kuo TT, Kim HE, Ohno-Machado L. Blockchain distributed ledger technologies for biomedical and health care applications. J Am Med Inform Assoc 2017;24(6): 1211–20.
[90] Azaria A, Ekblaw A, Vieira T, Lippman A. Medrec: using blockchain for medical data access and permission management. In: Proceedings of the 2016 IEEE international conference on open and big data (OBD); 2016. p. 25–30.
[91] Chen Y, Ding S, Xu Z, Zheng H, Yang S. Blockchain-based medical records secure storage and medical service framework. J Med Syst 2018;43(1):5.
[92] Wood G. Ethereum: a secure decentralised generalised transaction ledger. Ethereum Project Yellow Paper 2014;151:1–32.
[93] Brunette G, Mogull R. Security guidance for critical areas of focus in cloud computing v2. 1. Cloud Security Alliance; 2009. p. 1–76.
[94] Cloud Security Alliance (CSA), http://cloudsecurityalliance.org/.
[95] Rawat DB, Reddy SR. Software defined networking architecture, security and energy efficiency: a survey. IEEE Commun Surveys Tutor 2017;19(1):325–46.
[96] Merkel D. Docker: lightweight linux containers for consistent development and deployment. Linux J 2014;2014(239):2.
[97] Chae M, Lee H, Lee K. A performance comparison of linux containers and virtual machines using Docker and KVM. Cluster Computing; 2017. p. 1–11.
[98] Parizi RM, Qian K, Shahriar H, Wu F, Tao L. Benchmark requirements for assessing software security vulnerability testing tools. In: Proceedings of the 2018 IEEE 42nd Annual Computer Software and Applications Conference (COMPSAC). vol. 1; 2018. p. 825–6.
[99] Jakimoski K. Security techniques for data protection in cloud computing. Int J Grid Distrib Comput 2016;9(1):49–56.
[100] Panse D, Haritha P. Multi-factor authentication in cloud computing for data storage security. Int J Adv Res Comput Sci Softw Eng 2014;4(8):629–34.
[101] Hardt D. The OAuth 2.0 authorization framework (No. RFC 6749). 2012.
[102] Hu CT, Ferraiolo DF, Kuhn DR, Schnitzer A, Sandlin K, Miller R, Scarfone K. Guide to attribute based access control (ABAC) definition and considerations. 2019. No. Special Publication (NIST SP)-800-162.
[103] Coyne E, Weil TR. ABAC and RBAC: scalable, flexible, and auditable access management. IT Professional 2013;vol. 15(3):14–6.
[104] Cloud security alliance (about), http://cloudsecurityalliance.org/about/.
[105] IEEE-SA – The IEEE standards association, http://standards.ieee.org/.
[106] IEEE P2301 – Guide for Cloud Protability and Interoperability Profiles (CPIP), http://standards.ieee.org/project/2301.html.
[107] IEEE P2302 – Standard for Intercloud Interoperability and Federation (SIIF), http://standards.ieee.org/project/2302.html.
[108] European Union Agency for Network and Information Security (ENISA), http://www.enisa.europa.eu/publications/cloud-computing-risk-assessment.
[109] Open Cloud Computing Interface (OCCI), http://occi-wg.org/.
[110] DMTF's Cloud Management Initiative – Cloud Infrastructure Management Interface (CIMI), http://www.dmtf.org/standards/cloud.

[111] SNIA's Cloud Data Management Interface (CDMI), http://www.snia.org/cdmi.
[112] http://www.oasis-open.org/committees/tc_home.php?wg_abbrev=camp.
[113] OpenID Foundation, http://openid.net/.
[114] DMTF's Open Virtualization Format (OVF), http://www.dmtf.org/standards/ovf.
[115] OASIS's Topology and Orchestration Specification for Cloud Applications (TOSCA), http://www.oasis-open.org/committees/tosca/.
[116] OASIS's Security Services Technical Committee – Security Assertion Markup Language (SAML), http://www.oasis-open.org/committees/tc_home.php?wg_abbrev=security.
[117] Dave D, Meruliya N, Gajjar TD, Ghoda GT, Parekh DH, Sridaran R. Cloud security issues and challenges. In: Big data analytics. Singapore: Springer; 2018. p. 499–514.
[118] Singh A, Chatterjee K. Cloud security issues and challenges: a survey. J Netw Comput Appl 2017;79:88–115.
[119] Díaz M, Martín C, Rubio B. State-of-the-art, challenges, and open issues in the integration of Internet of things and cloud computing. J Netw Comput Appl 2016;67:99–117.
[120] Roundup of cloud computing forecasts. Forbes; 2017. http://www.forbes.com/.
[121] O'Driscoll A, Daugelaite J, Sleator RD. 'Big data', Hadoop and cloud computing in genomics. J Biomed Inform 2013;46(5):774–81.
[122] The top 5 cloud computing vendors, Forbes, http://www.forbes.com/.
[123] Google Cloud AutoML – Custom machine learning models, http://cloud.google.com/automl/.
[124] Microsoft Azure Machine Learning Service, http://azure.microsoft.com/en-us/services/machine-learning-service/.
[125] Amazon Elastic Inference, http://aws.amazon.com/machine-learning/elastic-inference/.
[126] Amazon Virtual Private Cloud (VPC), http://aws.amazon.com/vpc/.
[127] Gupta BB, Badve OP. Taxonomy of DoS and DDoS attacks and desirable defense mechanism in a cloud computing environment. Neural Comput Appl 2017;28(12):3655–82.

Further reading

Django – The web framework for perfectionists with deadlines, http://www.djangoproject.com/.

CHAPTER

Machine learning and data analytics

7

Chapter outline

Medical Data Sharing, Harmonization and Analytics. https://doi.org/10.1016/B978-0-12-816507-2.00007-4

7.1 Overview

So, how can we enable the coanalysis of disparate sources to deal with the unmet needs for various medical diseases and conditions? The answer to this question has been the core of discussion from the first chapter of this book. As we have already mentioned in Chapter 3, everything begins from data sharing. The sources and types of big medical data are many and thus data sharing is a primary step to interlink disparate sources of medical data, such as cohorts and omics registries, to increase the population size and enhance the scientific impact of the clinical studies that make use of such data to provide outcomes with significant statistical power. The interlinked data might share a common medical domain but often exhibit structural heterogeneities due to the different data collection protocols and data recording schemas that are adopted by the data providers. So, once the interlinking mechanisms for data sharing are established, the next step is to overcome the structural heterogeneities that are present in the shared data. Toward this direction, data harmonization using lexical, semantic, and statistical matching based on reference schemas has been proven to be a powerful strategy that can enable the homogenization of the heterogeneously structured, interlinked medical data. Once the medical data are harmonized at the highest level of available information, the next step is to coanalyze the harmonized datasets using machine learning and data analytics.

Before the application of any data analytics strategy for distilling knowledge from medical data, it is first necessary to apply a preprocessing pipeline on the medical data. Conventional computational approaches for data preprocessing include (i) data curation [1] for enhancing the quality of the medical data, such as the methods that have been already presented in Section 3.3, and (ii) data discretization [2] for dealing with recording errors during the data collection process, where the continuous data are discretized into an equal number of bins or bins with equal frequency or into a specific number of bins that minimize the information entropy or maximize the overall information gain, such as the entropy-MDL (minimum description length) approach, among others. Feature selection and feature ranking approaches [3] are also useful for reducing the dimensionality of the input features by highlighting a subset of prominent features based on a target one, which is usually a disease outcome. Once the quality of the data is ensured, a data analytics pipeline can be applied to mine knowledge from the medical data. Regarding the application of machine learning in healthcare, a notable progress has been made over the past years toward the extensive analysis of large amounts of clinical, laboratory, histological, and omics data to develop machine learning models for (i) the prediction of disease outcomes, (ii) the detection of biomarkers, (iii) the effective treatment monitoring, and (iv) the development of patient stratification models, among others [4]. The capabilities of machine learning in medical data analytics are tremendous [5] with increased impact toward the clinical decision-making process.

Machine learning can shed light into complex data structures to reveal hidden patterns and associations between the variables that can lead to the detection of prominent variables with high contribution toward the prediction of a specific disease outcome [6]. Toward this direction, machine learning algorithms have been extensively applied in omics data and clinical data for the development of supervised learning models that are able to predict disease outcomes given a specific subset of annotated data for training [7]. Popular supervised learning algorithms, such as regression [8], support-vector machines (SVMs) [9], the decision trees [10], the Naïve Bayes [11], and the artificial neural networks (ANNs) [12], have been utilized on large subsets of clinical data to develop robust patient stratification models toward the identification of groups of individuals with high risk for the development of a disease outcome [13]. In addition, machine learning has been also applied on clinical and genetics data to develop unsupervised learning models for grouping (clustering) complex data structures with similar patterns [14]. Unsupervised learning algorithms, such as the hierarchical clustering and the *k*-means [15], along with more complex ones, such as the spectral clustering [16] and the hierarchical clustering [17], have been extensively employed for the categorization of highly associated features. The fundamental difference between a supervised and an unsupervised learning algorithm lies on the fact that the former is trained on a set of input features (variables), where one feature is set as the target feature, whereas the latter is directly applied on the set of features without any prior knowledge regarding the existence of a target feature. As far as the development of prediction models is concerned, a supervised learning algorithm is usually trained on a set of training features where the feature that represents the disease outcome is specified as the target feature. The performance of the supervised learning model is then evaluated using conventional methods, such as the *k*-fold cross-validation and train/test split ratio.

It is obvious now that the adoption of a proper machine learning algorithm depends on the type of application study and the definition of the scientific problem that needs to be addressed by the study. Apart from the variety of the existing machine learning algorithms for data analytics in healthcare, emphasis must be given on the data storage environment. As we have already described on Chapter 6, the most common way for storing medical data is through the adoption of a centralized database where all the medical data are stored into a common physical environment. Mining knowledge from large amounts of medical data requires the application of deep learning algorithms [18], such as multilayer neural networks with error propagation (e.g., the long short-term memory (LSTM) neural network [19]) and convolutional neural networks (CNNs) [20], which are capable of detecting hidden motifs within the complex big data structures and dealing with the development of supervised learning models for predicting disease outcomes with multiple applications in medical imaging segmentation [20] and biosignal analysis [19]. Several methods have been proposed toward the effective analysis of big medical data in centralized databases, such as the batch processing method according to which the data are divided into smaller subsets, i.e., batches, and the machine learning algorithm is sequentially applied on the set of batches until all the batches are parsed [21].

Keeping the data in a common, centralized database, however, poses significant security threats in the case of a data breach, as well as obscures the efficient analysis of big data, especially of omics data, where the amount of generated data is so huge that it significantly hampers the application of any machine learning algorithm due to the lack of sufficient memory units and the demand of high computational power [22,23]. A prominent solution to this is to use the distributed database schema. Distributed healthcare environments [22,24,25] have gained a lot of attention these days due to the need to process massive amounts of accumulated medical data. In distributed databases, the medical data are stored in multiple sites (or locations). An overwhelming scientific challenge, however, in distributed databases, is the need to develop prediction models across the data that are stored in multiple databases, without the data to leave these databases at all [26,27]. Another challenge lies on the fact that the application of the existing machine learning algorithms is not always feasible due to the nonconvex optimization problem that the majority of these algorithms try to solve [28].

Toward this direction, batch processing methods have been proposed to provide an adequate solution for the development machine learning models in distributed environments [28–30]. Online learning [29,31,32] is such an approach that updates an existing machine learning model according to a global cost function that is sequentially adapted on upcoming data streams. In fact, online learning uses stochastic gradient descent (SGD) optimization methods to update the existing machine learning model on upcoming training samples by minimizing a global cost function [29], where the model is continuously updated on new data points or on a series of accumulated data points, over time. Existing machine learning implementations that support online learning include linear SVMs [33], hybrid online learning attempts using nonlinear kernels [34], and gradient descent approaches for convex optimization [32,35].

A similar strategy that shares a common basis with the former one is the incremental learning [28,29,36]. In contrast to online learning, incremental learning tries to adapt an existing machine learning model on upcoming data streams without the burden of being applied only on upcoming data streams, i.e., in an "online" manner, but also on existing data streams or batches. More specifically, incremental learning uses a batch processing method to train a machine learning model on an initial batch and then adjust the model on a series of upcoming batches, by solving additive optimization tasks [29,36]. This makes incremental learning ideal in the case of out-of-core learning, where the large-scale data do not even fit into the memory and thus need to be processed sequentially, as well as when the batches are treated as harmonized data that are stored in multiple locations. Existing methods that support incremental learning include methods for convex optimization [36,37], gradient boosting trees [38], and Naïve Bayes [29,39]. Meanwhile, stacked generalization techniques [40,41] have also been proposed for combining individual classification outcomes. Such parallelized methods, however, suffer from biases introduced by the assembly stage [42].

In this chapter, computational methods to analyze clinical, laboratory, histological, and omics data will be presented along with methodological challenges and solutions toward the development of both supervised and unsupervised machine learning models for predicting disease outcomes, clustering similar features, and detecting biomarkers, as well as patient stratification and treatment monitoring in different medical areas. Pseudocode implementations for popular algorithms will be provided along with indicative illustrations of the decision-making process and clustering performance of different algorithms. In addition, widely used data visualization methods, such as heatmaps [43] and graphs [44], will be also discussed. Popular machine learning and deep learning frameworks will be presented, including the scikit-learn [45] and the keras [46] along with cloud-based solutions, such as the Amazon SageMaker [47].

7.2 Data preprocessing

7.2.1 Data curation

The fundamental basis of any data analytics strategy should first take into consideration the enhancement of the quality of the data in terms of outliers, duplicated fields, and missing values. These methods have been already presented in Chapter 3, Section 3.1. Data discretization methods for reducing data recording errors and feature extraction methods for feature selection and feature ranking are presented next.

7.2.2 Data discretization

Data discretization is a rigorous data preprocessing method that is often used to reduce recording errors [2]. It is a data binning method that discretizes the continuous features into bins, i.e., data instances, by either using a fixed number of bins or using approximately equal-sized bins. This can be achieved by splitting the data into data intervals until the desired number of bins or the desired width of data bins is met. Through this approach, the continuous data can be separated into data instances to reduce the errors that are posed during the data recording stage. Data discretization is also used during the construction of decision trees [48], where the continuous data are first discretized according to a given outcome (i.e., the target feature for classification) based on the instance that results to the highest information gain.

To demonstrate the discretization process, let us assume a 1D vector x with age values, say $x = \{71, 77, 64, 67, 68, 68, 56, 52, 66, 72\}$. The equal width approach for N-bins, assume $N = 2$, will try to divide the age values into two data intervals so that the difference between the maximum and the minimum value in each data interval will be smaller than (max−min)/N, which in this case is equal to 77−52/2 = 12.5. To do so, the algorithm finds a threshold 64.5 to create two data instances, namely the data instance A = {71, 77, 67, 68, 68, 66, 72} and the data instance B = {64, 56, 52}, so that the maximum difference between the values in each data instance is less than 12.5. The final form of the vector ages will then be $x = \{1, 1, 0, 1, 1, 1, 0, 0, 1, 1\}$. On the other hand, the equal frequency approach will set a threshold at 67.5 to create two data instances C = {71, 77, 68, 68, 72} and D = {64, 67, 56, 52, 66} with the same number of data instances. In that case, the final form of the vector ages will be $x = \{1, 1, 0, 0, 1, 1, 0, 0, 0, 1\}$.

To demonstrate the applicability of the discretization toward the detection of recording errors, assume an outlier, say 120, in the vector $x = \{71, 77, 64, 67, 68, 68, 56, 52, 66, \mathbf{120}\}$. In that case, the equal frequency approach would ignore the outlier and yield a discretized vector, $x = \{1, 1, 0, 0, 1, 1, 0, 0, 0, 1\}$, whereas the equal width approach would yield a discretized vector $x = \{0, 0, 0, 0, 0, 0, 0, 0, 0, 1\}$, thus giving one bin for only one value that indicates a recording error. In the case where $N = 3$, the equal frequency method would yield a discretized vector, $x = \{3, 3, 1, 1, 2, 2, 0, 0, 1, 3\}$, which would assign the new bin to the values 71, 77, and 120, whereas the equal width method would yield a discretized vector, $x = \{1, 1, 0, 0, 0, 0, 0, 0, 0, 3\}$, which again isolates the outlier from the rest of the values.

A more complex approach for data discretization is the entropy-MDL [49] approach, which is a top-down discretization method that recursively splits the attribute at a data instance that maximizes the information gain in entropy, until the gain is lower than the minimal description length of the cut. This discretization can result in an arbitrary number of intervals, including a single interval, in which case the attribute is discarded as useless (removed). To understand how the algorithm works, let us assume a new vector that denotes the output class, say $m = \{1, 3, 2, 1, 1, 2, 1, 1, 2\}$, for the age vector $x = \{71, 77, 64, 67, 68, 68, 56, 52, 66, 72\}$. The output vector m can be the disease outcome that corresponds to a person with the specific age in x. The entropy-MDL approach evaluates the information gain in entropy for all possible splitting values and picks the one that maximizes the information gain [50]:

$$IG(x|m) = H(x) - H(x|m), \tag{7.1}$$

where $H(x)$ is the information entropy of x (i.e., the average rate of existing information in x):

$$H(x) = -\sum_{i=1}^{n} p(x_i)\log(p(x_i)), \tag{7.2}$$

with $p(x_i)$ denoting the fraction of the data point x_i across the sample, and $H(x|m)$ is the conditional entropy of x given the output vector m:

$$H(x|m) = -\sum_{i=1}^{n} p(x_i)H(m\ x = x_i) = -\sum_{j=1}^{m}\sum_{i=1}^{n} p(x_i, m_j)\log\frac{p(x_i, m_j)}{p(x_i)}, \tag{7.3}$$

with $p(x_i|m_j)$ denoting the conditional probability of data point x_i given the data point m_j as the fraction of all the data points x that have the value m_j. According to the MDL approach, Eq. (7.1) is evaluated for all possible splitting values in x until the splitting value, assume t, that yields the highest $IG(x, m)$ is found. For demonstration purposes, Eq. (7.1) was evaluated for different thresholds t on the interval [51,52] for the age vector, x, using a step 0.5. To enhance the statistical power of the process, we generated 300 additional samples for x and m, on the same interval. The highest IG value was 0.0212 for a threshold value $t = 62.5$, and the second largest IG value (0.0155) was obtained for $t = 64.5$. Thus, the optimal splitting threshold for discretization of x into two bins, given the outcomes in m, is 62.5.

7.2.3 Feature extraction

Feature extraction is a straightforward method that aims to discriminate a set of predominant features within the data. This set of features is usually determined according to its relationship with a target feature. In the majority of the cases, this relationship is determined by the significance (importance) of each feature toward the prediction a target feature, which is usually a variable that represents a disease outcome (e.g., non-Hodgkin's lymphoma). The discrimination of this set of features can be accomplished by either using feature selection methods, such as filter methods such as the fast correlation-based filter (FCBF) method [53] and wrapper methods such as the exhaustive feature selection [51], which extract a specific subset of important features from the data, or by feature ranking methods, such as the gain ratio [54], the information gain [50], and the Gini index [55], which rank all the features based on their importance in predicting the target feature. Feature selection can reduce the dimensionality and computational complexity of processing large amounts of data and reduce the overfitting during the training process [3]. Feature extraction and feature ranking methods have been extensively used in the literature for the identification of predominant features, which are referred to as biomarkers [56], with increased contribution in predicting disease outcomes, such as lymphoma [57]. The biomarkers can enhance the overall performance of the disease prediction models, as well as reveal the underlying mechanisms of the disease's onset and progress.

An illustration of the difference between feature ranking and feature selection methods is depicted in Fig. 7.1. The features in the original dataset are displayed

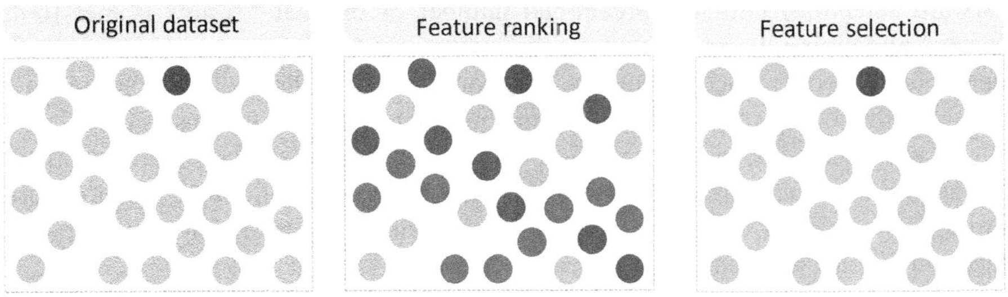

FIGURE 7.1

An illustration of the difference between feature ranking and feature selection.

as nodes in the 2D space for visualization purposes. The target feature is depicted with red color and the rest of the features with gray. A feature ranking algorithm ranks the importance of each feature using well-known measures from information theory, such as the gain in entropy. The maximum information (maximum gain in entropy or mutual information [MI]) that can be expressed by each feature regarding the target feature is calculated in the form of a score, and the features are ranked according to this score into different categories (like the blue and the green categories in Fig. 7.1). Examples of such methods include the information gain, gain ratio, and Gini index. Another rank-based approach is the recursive feature elimination (RFE) method, which uses a machine learning model to remove the weakest features, i.e., the less-informative features, by ranking the features according to their prediction performance into informative and redundant and finally by recursively eliminating the weakest features to reduce any inner dependencies within the data [58]. The RFE method requires a predefined number of optimal features before its execution, which can, however, be determined using the cross-validation strategy to evaluate the performance of each subset of features [58].

On the other hand, a feature selection algorithm can automatically extract a subset of important features without using any prior knowledge regarding the number of optimal features. Feature selection methods measure the similarity between each pair of features, as well as between each feature separately with the target feature, and keep only those features having the highest similarity with the target feature and the lowest similarity with the rest of the features. This family of methods is known as the filter-type method [59] and focus on measuring the relevance of the features. An example of a filter-type method for feature selection is the FCBF algorithm [53], which uses the correlation as a similarity measure, as well as the variance thresholding approach, which is an arbitrary approach that ignores features having less variance than a specific threshold. Another approach for feature selection is the wrapper method [51], according to which a machine learning model is evaluated on all possible combinations of features through a greedy strategy [51], which exhaustively searches for the optimal combination that achieves the highest prediction performance. The optimal combination of features can be also predefined like in the sequential feature selection approach where a recursive process is applied by either discarding redundant features until the predefined number of optimal features is met (sequential backward selection) [60] or by starting with one feature and adding a new feature that increases the overall prediction performance (sequential forward selection) until the predefined number of optimal features is met [60]. Instead of defining the predefined number of optimal features, one can alternatively determine a performance evaluation criterion, such as a threshold on the model's accuracy. In contrast to the filter-type method that quantifies the "usefulness" of a subset of features, the wrapper method is more robust as it solves a feature optimization problem that is much closer to the real problem of predicting a target feature through an "optimal" subset of predictors that additionally takes into consideration the interactions between them.

The gain ratio and the information gain is entropy-based measures [50,54]. The information gain [50], which is also referred to as MI, measures the amount of information that is shared between a target feature and one or more training features (i.e., the mutual dependence). The mathematical basis of information gain was presented in Section 7.2.1. Given a set of training features, assume T, the information gain of a value, assume a, is given as

$$IG(T) = -\sum_{i=1}^{n} p(a_i, T)log(p(a_i, T)), \tag{7.4}$$

where $p(a_i, T)$ is the frequency of a_i in $T, f(a_i, T)$, divided by the total number of observations in T, $|T|$:

$$p(a_i, T) = \frac{f(a_i, T)}{|T|}. \tag{7.5}$$

The gain ratio (information gain ratio) is a normalized variant of the information gain, which is used to reduce any biases that might be introduced during the calculation of the mutual dependence between the features [50,54], where the normalization is achieved by dividing the amount of shared information with the entropy of the target feature so that the ratio values will lie in the range [0, 1]. The gain ratio of a split subset, assume F, is defined as

$$GR(F) = \frac{IG(F)}{SI(F)}, \tag{7.6}$$

where $IG(F)$ is the information gain of the split F, and $SI(F)$ is the split information that acts as a weight of the split and is given as

$$SI(F) = -\sum_{j=1}^{m} \frac{|F_j|}{|F|} \log\left(\frac{|F_j|}{|F|}\right), \tag{7.7}$$

where $|F_j|$ is the number of observations that include the feature j in the split F, $|F|$ is the total number of observations in the split, and m is the number of participating features in the split.

The Gini impurity index is widely used in the construction of the decision trees, as a measure of impurity, where each feature is ranked according to how often it is incorrectly labeled during the splitting-node process, toward the establishment of an efficient decision tree structure [55]. The Gini index quantifies the frequency of misclassification for a randomly selected element that belongs to class j with probability p_j, where $j \in \{1, 2, \ldots, M\}$, weighted by the misclassification probability, $1 - p_j$, as follows:

$$GI(p) = \sum_{j=1}^{M} p_j(1 - p_j) = 1 - \sum_{j=1}^{M} p_j^2, \tag{7.8}$$

where M is the number of classes (labels) and $1-p_j$ is the probability that the selected item does not fall in class j. The probability p_j can be seen as the probability of a random element with label j being classified into class j by taking into consideration the overall distribution of classes in the subset. The subset of features that manages to minimize the Gini index is finally selected as the one that minimizes the overall misclassification error. In the case of a binary classification problem, the Gini impurity index is reduced to

$$GI(p_0, p_1) = p_0(1-p_o) + p_1(1-p_1), \tag{7.9}$$

which is a linear combination of the probabilities p_0 and p_1 of correctly classifying a random element in classes 0 and 1, respectively, weighted by the misclassification probabilities. In the ideal case where the GI is 0.5, the classes are equally weighted. Let us assume a binary classification problem where the probability of an element being classified in class 0 is 0.6, i.e., $p_o = 0.6$, and the probability of an element being classified in class 1 is 0.4, i.e., $p_1 = 0.4$. The Gini index is then $GI(0.6, 0.4) = 0.48$. To demonstrate how the Gini index is used in the decision tree making for measuring the homogeneity of a random split, the gain in Gini index is defined as a measure that evaluates a branch during the decision-making process:

$$GGI = GI(T) - \frac{N_{left}}{N_T} GI(T_{left}) - \frac{N_{right}}{N_T} GI(T_{right}), \tag{7.10}$$

where $GI(T)$ is the Gini index of the subset T, i.e., 0.48 in our case, N_T is the total number of elements in T, N_{left} is the number of elements at the left child node, N_{right} is the number of elements at the right child node, and GI_{left} and GI_{right} are the corresponding Gini index in each branch. Two classification instances for two random splitting values, assume A and B, are depicted in Fig. 7.2. In the original subset T, there are 100 elements, 60 with class label 0 and 40 with class label 1. Let us assume that for the splitting value A ($T < A$), the algorithm correctly classifies 40 out of 60 elements in class 0 and 20 out of 40 elements in class 1. In that case, (7.9) is used to compute the Gini index in the left and right branches, $GI(T_{left}) = 0.444$ and $GI(T_{right}) = 0$, respectively. Then, (7.10) yields the overall gain in the Gini index, $GGI = 0.48 - 0.6 * 0.444 = 0.2136$. For the splitting value

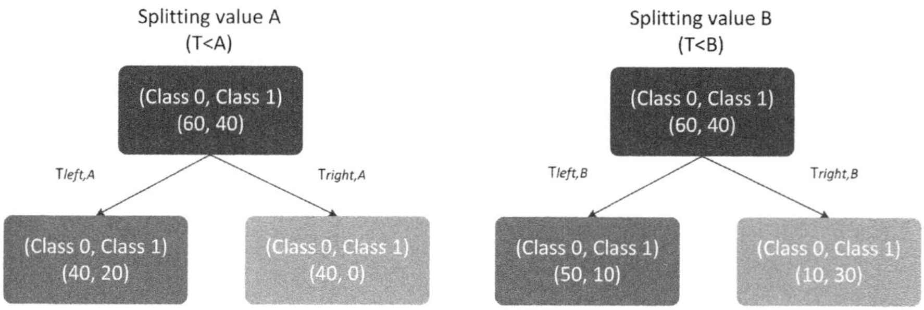

FIGURE 7.2

An illustration of a simple decision-making process for two different splitting values.

B, the algorithm correctly classifies 50 out of 60 elements in class 0 and 30 out of 40 elements in class 1. In that case, the Gini index in the left and right branches are $GI(T_{left}) = 0.277$ and $GI(T_{right}) = 0.5$, respectively. Then, (7.10) yields the overall gain in the Gini index, $GGI = 0.48 - 0.6 * 0.277 - 0.4 * 0.375 = 0.1637$. The process can be repeated for different splitting values, and the one that achieves the smallest GGI is selected. In our case, the subset with the splitting value B is selected as it achieves the best homogeneity.

Alternatively, the impurity in (7.10) can be measured using the entropy:

$$E(p) = -\sum_{j=1}^{M} p_j log(p_j). \tag{7.11}$$

In that case, (7.11) is used to compute the total entropy, $E(T) = 0.673$, and the entropy in the left and right branches, $E(T_{left}) = 0.636$ and $E(T_{right}) = 0$, respectively. Then, (7.10) yields the overall gain in the entropy, $GE = 0.673 - 0.6 * 0.636 = 0.2914$. For the splitting value B, the entropy in the left and right branches are $E(T_{left}) = 0.45$ and $E(T_{right}) = 0.5623$, respectively. Then, (7.10) yields the overall gain in the entropy, $GE = 0.673 - 0.6 * 0.45 - 0.4 * 0.5623 = 0.178$. The splitting value that maximizes the entropy is again B. Finally, the classification error can also be used in (7.10) as the impurity:

$$e = 1 - max(p_j), \tag{7.12}$$

which is large when the probability of correctly classifying a random element in class j is small. In that case, (7.12) is used to compute the total classification error, $e(T) = 0.4$, and the classification error in the left and right branches, $e(T_{left}) = 0.333$ and $e(T_{right}) = 0$, respectively. Then, (7.10) yields the overall gain in the entropy, $Ge = 0.4 - 0.6 * 0.333 = 0.2002$. For the splitting value B, the entropy in the left and right branches are $e(T_{left}) = 0.166$ and $e(T_{right}) = 0.25$, respectively. Then, (7.10) yields the overall gain in the classification error, $Ge = 0.4 - 0.6 * 0.166 - 0.4 * 0.25 = 0.2004$. In this case, the splitting values A and B yield splits with similar gains.

Regarding the filter-type method, the FCBF algorithm [53] uses a "predominance" correlation strategy according to which the features that are both highly correlated with the target feature and less correlated with the rest of the features are automatically isolated by the rest of the features as predominant ones. This subset is usually extracted by computing the correlation of each feature with the target feature, as well as between each pair of features, and then determines the subset of important features by keeping only those features that achieve the maximum correlation with the target feature and at the same time the minimum correlation with the rest of the features. As far as the wrapper method is concerned, the exhaustive feature selection approach focuses on the evaluation of all possible feature combinations to seek for the one that achieves the best performance. The sequential feature selection methods use a RFE (in the case of the sequential backward selection) or feature inclusion strategy (in the case of the sequential forward selection) to seek for a predefined number of optimal features that achieve the highest performance. Another emerging family of feature selection methods is the embedded method, which is a hybrid method one that tries to combine the advantages of the filter-type and Wrapper methods by introducing a misclassification penalty or a regularization term (e.g., the *l*1 or the *l*2 penalty) during the training of the machine learning model to improve both the performance and the training process. Such types of methods include regression models and decision trees that will be further described in Section 7.4.

7.3 Machine learning and data mining

7.3.1 Overview

In this section, fundamental methods for machine learning and data mining will be presented for solving prediction, regression, and clustering problems, including supervised learning and unsupervised learning models, methods for association analysis, and methods for evaluating the performance of the classifiers.

7.3.2 Supervised learning algorithms

In this subsection, supervised learning algorithms will be presented including regression analysis, SVMs, Bayesian models, decision trees and random forests (RF), ensembles, and neural networks.

7.3.2.1 Regression analysis

7.3.2.1.1 Linear and multiple regression

Regression analysis [8] is a classic approach for modeling the linear dependence between a controlled (or dependent) variable and one or more independent variables. In the case of a single independent variable, the regression analysis is referred to as simple regression analysis. In the case of more than one independent variables, the regression analysis is referred to as multiple regression analysis [61], which is the most widely used method for linear dependence modeling as the number of explanatory factors is usually more than one. Assuming a set of n-independent variables (features), say $\{x_1, x_2, \ldots, x_n\}$, and a dependent variable, say y, the linear regression analysis models the association between a controlled variable and a set of independent variables as a linear function:

$$y = w_0 1 + w_1 x_1 + w_2 x_2 + \ldots + w_n x_n + \varepsilon = x^T w + \varepsilon, \tag{7.13}$$

where $w_1, w_2, \ldots, w_n$ are the regression coefficients (weights or effects) of the independent variables, w_0 is the interception parameter x^T and w is the inner product of the weight vector $w = [w_1, w_2, \ldots, w_n]$, which includes the regression coefficients of each independent variable in vector $x = [x_1, x_2, \ldots, x_n]$ and ε is the error (or disturbance) parameter of the model. Eq. (7.13) can be generalized for a sample i as follows:

$$y_i = w_0 1 + w_1 x_{i,1} + w_2 x_{i,2} + \ldots + w_n x_{i,n} + \varepsilon_i = \boldsymbol{x}_i^T \boldsymbol{w} + \varepsilon_i, \tag{7.14}$$

where $i = 1, 2, \ldots, m$. Using a matrix representation, we can rewrite (7.13)

$$y_i = \begin{bmatrix} 1 & x_{i,1} & x_{i,2} & \cdots & x_{i,n} \end{bmatrix} \begin{bmatrix} w_0 \\ w_1 \\ \vdots \\ w_n \end{bmatrix} + \varepsilon_i, \tag{7.15}$$

which can be generalized as follows:

$$\boldsymbol{y} = \begin{bmatrix} y_1 \\ y_2 \\ \vdots \\ y_m \end{bmatrix} = \begin{bmatrix} 1 & x_{1,1} & x_{1,2} & \cdots & x_{1,n} \\ 1 & x_{2,1} & x_{2,2} & \cdots & x_{2,1} \\ \vdots & \vdots & \vdots & \ddots & \vdots \\ 1 & x_{m,1} & x_{m,1} & \cdots & x_{m,n} \end{bmatrix} \begin{bmatrix} w_0 \\ w_1 \\ \vdots \\ w_n \end{bmatrix} + \begin{bmatrix} \varepsilon_0 \\ \varepsilon_1 \\ \vdots \\ \varepsilon_m \end{bmatrix}$$

$$= \begin{bmatrix} \boldsymbol{x}_1^T \\ \boldsymbol{x}_2^T \\ \vdots \\ \boldsymbol{x}_n^T \end{bmatrix} \begin{bmatrix} w_0 \\ w_1 \\ \vdots \\ w_n \end{bmatrix} + \begin{bmatrix} \varepsilon_0 \\ \varepsilon_1 \\ \vdots \\ \varepsilon_m \end{bmatrix} = \boldsymbol{X}\boldsymbol{w} + \boldsymbol{\varepsilon}, \tag{7.16}$$

where $\boldsymbol{X}$ is an mxn observation matrix, where n is the number of features and m is the number of samples per feature, $\boldsymbol{w}$ is an $nx1$ weight vector that includes the regression coefficients for each feature, and $\boldsymbol{\varepsilon}$ is an $mx1$ error vector that includes the error parameters. It is obvious from (7.16) that one can make predictions for the dependent variable $\boldsymbol{y}$ using a set of continuous independent variables, $\boldsymbol{X}$, as long as the regression coefficients are known. The overall idea of regression analysis is to solve (7.16) to estimate the weight vector, $\boldsymbol{w}$, that minimizes a specific cost function, assume $\boldsymbol{E}(\boldsymbol{w})$:

$$\widetilde{\boldsymbol{w}} = argmin_{\boldsymbol{w}}(E(\boldsymbol{w})). \tag{7.17}$$

According to the literature, several approximations have been proposed to estimate the weight vector, including the ordinary least squares (OLS) [62], generalized least squares (GLS) [63], weighted least squares (WLS) [64], and recursive least squares (RLS) [65], among others. OLS method [62] seeks for the estimate $\widetilde{\boldsymbol{w}}$ that minimizes the following quadratic cost function:

$$E(\boldsymbol{w}) = \sum_{i=1}^{m} \left| y_i - \sum_{j=1}^{n} X_{i,j} w_j \right|^2 = \sum_{i=1}^{n} \left(y_i - x_i^T w\right)^2 = \boldsymbol{y} - \boldsymbol{X}\boldsymbol{w}^2 = \boldsymbol{\varepsilon}^2, \tag{7.18}$$

which minimizes the sum of squares of the difference between the estimations and the outcomes, where $X_{i,j}$ is the ith observation of the jth feature (independent variable). Note that the term $y_i - x_i^T w$ in (7.18) is called the error residual of the ith observation and measures the distance between the point (x_i, y_i) and the hyperplane $x_i^T w$ in the two-dimensional space. Thus, the OLS minimizes the sum of squared residuals to find a model that minimizes the distance between the predicted outcomes and the actual outcomes. Replacing (7.18) in (7.17) yields an estimation in the following closed matrix form:

$$\widetilde{w} = \left(X^T X\right)^{-1} X^T y. \tag{7.19}$$

Indeed, if (7.19) is substituted in (7.18), we get

$$\varepsilon = y - \widetilde{y} = y - X\left(X^T X\right)^{-1} X^T y = y - XX^{-1}X^{-T}X^T y = y - Iy = 0. \tag{7.20}$$

GLS method [63] is a generalized version of the OLS when correlation exists between the error residuals of the model. To account for these correlations, the GLS method minimizes (7.16) by taking into consideration the covariance matrix C of the error residuals as follows:

$$\widetilde{w} = \left(X^T C^{-1} X\right)^{-1} X^T C^{-1} y, \tag{7.21}$$

where the element $c_{i,j}$ of C is the covariance between the error residual of the ith and jth estimates. Finally, the WLS method is a special case of the GLS method where the covariance matrix C is diagonal [64]. The family of the regularized least squares methods [65] accounts for an additional regularization parameter that constrains the least square solutions in the case where the number of independent variables (features) is larger than the number of observations (i.e., $m > n$). This is a serious drawback of the OLS method, which results to an overfitting tendency. Regularization is also useful in the case where sparse solutions need to be obtained from the minimization of the cost function to improve the prediction performance of the regression model. The RLS minimizes an extension of (7.18), where an additional regularization function, assume $R(w)$, is added:

$$E(w) + \lambda R(w) = \frac{1}{n}\|y - Xw\|^2 + \lambda R(w), \lambda > 0, \tag{7.22}$$

where λ is a positive scalar value and $1/n$ is a normalization coefficient. An example of a widely used regularization method is the Tikhonov regularization or Ridge regression [66], where an $l2$-norm penalty is introduced as the regularization function in (7.22):

$$\widetilde{w} = \left(X^T X + aI\right)^{-1} X^T y, \tag{7.23}$$

which yields the following estimation:

$$\widetilde{w} = \left(X^T X + aI\right)^{-1} X^T y, \tag{7.24}$$

where the term aI ensures that the eigenvalues of the covariance matrix $X^T X$ are larger than zero to provide unique solutions. Note that in the case where $a = 0$, the OLS method is obtained. The least absolute selection and shrinkage operator

(LASSO) [67] is another method in the RLS family where an $l1$-norm penalty is introduced as the regularization function in (7.22):

$$E(w) + \lambda R(w) = \frac{1}{n} y - Xw^2 + \lambda \sum_{j=1}^{p} |w_j|. \tag{7.25}$$

Unlike the Ridge method, the LASSO method can yield zero coefficient and thus provide sparse solutions by performing some kind of feature selection (when $n < p$). In addition, it can yield more accurate predictions in the case where the number of independent variables is small. The Ridge regression, however, is preferred in the case where $n > p$, as well as when the input features are correlated [66]. In an attempt to combine the advantages of the Ridge and LASSO regression methods, into a hybrid method, the elastic net regularization has been proposed [68]:

$$E(w) + \lambda R(w) = \frac{1}{n} \|y - Xw\|^2 + \lambda_1 \sum_{j=1}^{p} |w_j| + \lambda_2 \sum_{j=1}^{p} |w_j|^2, \tag{7.26}$$

where λ_1 and λ_2 are nonnegative scalars. In fact, the elastic net combines the $l1$- and $l2$-norm penalties that were introduced in (7.23) and (7.25) to combine the resilience of the Ridge regression in the case of highly correlated features when $n > p$ along with the shrinkage (feature elimination) that is offered by the LASSO.

7.3.2.1.2 Logistic regression

Logistic regression (or logit regression or log-linear classifier) is an alternative regression analysis strategy that provides class probabilities in a direct way and allows for the presence of a categorical dependent variable. Logistic regression is generally preferred in the case where the target feature (or the dependent variable) is categorical, i.e., takes two possible values (e.g., 0 and 1). This is similar to the solution of the binary supervised learning problem (e.g., when we want to predict a disease presence or absence), which is the most frequent supervised learning problem in the medical domain. Logistic regression can be also extended to the multiclass classification problem where the number of classes is more than one. Given a set of n-features, $\{x_1, x_2, \ldots, x_n\}$ and a dependent variable with two possible outcomes, say y, logistic regression tries to estimate the log-odds (or logarithm of the odds) as a linear combination of the input features:

$$logit(p) = ln\left(\frac{p}{1-p}\right) = w_0 + w_1x_1 + w_2x_2 + \ldots + w_nx_n, \tag{7.27}$$

where p is the probability that the dependent variable takes the value 1, i.e., $p = P(y = 1)$. Solving (7.27) for p, we get the probability of predicting an outcome $y = 1$ given the set of input features, $\boldsymbol{x}$:

$$p = P(y = 1|\boldsymbol{x}) = \frac{1}{1 + exp(-(w_0 + w_1x_1 + w_2x_2 + \ldots + w_nx_n))}, \tag{7.28}$$

where w_0 is the y-intercept, i.e., the point where the distribution intersects the y-axis. Eq. (7.28) is the formal expression of the standard logistic function (sigmoid):

$$\sigma(t) = \frac{1}{1+e^{-t}}, \tag{7.29}$$

where $t = w_0 + w_1x_1 + w_2x_2 + \ldots + w_nx_n$. For more information regarding the sigmoid function, see Chapter 5, Section 5.5. The logit of (7.29) can then be calculated as follows:

$$logit(\sigma(t)) = ln\left(\frac{1}{1+e^{-t}}\right) = t, \tag{7.30}$$

which yields the odds as a linear combination of the input features.

In logistic regression, the most common way of estimating the weight vector $w = (w_0, w_1, w_2, \ldots, w_n)$ is to estimate the vector w that maximizes the log-likelihood, i.e., the probability of getting an outcome $y = 1$. In the case where the outcomes are two, we can use the Bernoulli distribution [69] to model the outcomes, where the likelihood, assume $L(y, w)$, is defined as follows:

$$L(y, w) = \prod_{i=1}^{n} p^{y_i}(1-p)^{1-y_i}, \tag{7.31}$$

where

$$p = P(y_i = 1|x_i) = \frac{1}{1+exp(-x_iw)}, \tag{7.32}$$

and thus

$$P(y_i = 0|x_i) = 1 - P(y_i = 1|x_i) = 1 - \left(\frac{1}{1+\exp(-x_iw)}\right). \tag{7.33}$$

Then, (7.31) can be rewritten using the log of the likelihood:

$$ln(L(y, w, X)) = ln\left(\prod_{i=1}^{n}\left[\left(\frac{1}{1+exp(-x_iw)}\right)^{y_i}\left(1-\frac{1}{1+exp(-x_iw)}\right)^{1-y_i}\right]\right), \tag{7.34}$$

$$=> ln(L(w, y, X))$$

$$= \prod_{i=1}^{n}\left[y_i ln\left(\frac{1}{1+exp(-x_iw)}\right) + (1-y_i)\ln\left(1-\frac{1}{1+exp(-x_iw)}\right)\right], \tag{7.35}$$

which can be reduced to Ref. [70]:

$$ln(L(w, y, X)) = \sum_{i=1}^{n}[-ln(1+exp(x_iw)) + y_ix_iw]. \tag{7.36}$$

A solution to (7.36) can be given by the Newton–Raphson (or simply Newton's) method [71] by minimizing the gradient of the likelihood with respect to the weights, i.e., $\nabla_w ln(L(w, y, X))$. The $l1$- and $l2$-regularization terms from (7.23) and (7.25) can also be introduced to penalize the cost function depending on the conditions that were previously described.

7.3.2.1.3 Differences

An example that demonstrates the differences between linear and logistic regression is depicted in Fig. 7.3. In Fig. 7.3A a linear regression model was fitted on 1000 randomly generated samples using Python's scikit-learn [72]. The red line represents the decision boundary of the model. On the other hand, Fig. 7.3B presents a weakness of the linear regression model in the case of a binary classification problem. For simulation purposes, 1000 samples were again randomly generated, and the Newton–Raphson method was used as a solver. The results demonstrate the superiority of the logistic regression model toward the correct discrimination of the actual classes by parsing the linear weights through a sigmoid function that somehow "squash" the predicted values in the range [0, 1] against the linear method that simply fits a line.

The main drawback of regression analysis lies on the fact that it assumes a linear independence between the independent variables along with the additional assumption that the relationship between the target variable and the independent variables is linear, which often simplifies the analysis and introduces large variance in the predictions (i.e., overfitting). For this reason, linear and logistic regression models are sensitive to outliers and the levels of noise between the independent variables in the model.

7.3.2.2 Support vector machines

SVMs can be used for solving high-dimensional regression and classification problems [9]. The SVM method seeks for a hyperplane in the Nth dimensional space that can effectively separate the data points according to the given labels. The hyperplanes serve as decision boundary for solving regression and classification problems. An example of a hyperplane can be seen in Fig. 7.3A, where a linear regression line is used to predict data points in the 2D space ($N = 2$). In the case where $N = 3$, the hyperplane is a plane in the 3D space and so on. For illustration purposes, we will present the hyperplanes in the 2D space. An example of two hyperplanes with small and large margins for a binary classification problem is depicted in Fig. 7.4. In each case, the data points that are close to the hyperplane are called support vectors and are depicted in red color. So, the overall idea of the SVM is to find the support vectors that maximize the margin of the hyperplane.

In Fig. 7.4, the hyperplane can be defined as follows:

$$x_1 w_1 + x_2 w_2 + \ldots + x_n w_n = b. \tag{7.37}$$

Using a matrix notation, (7.37) can be rewritten as follows:

$$\mathbf{x}^T \mathbf{w} - b = 0, \tag{7.38}$$

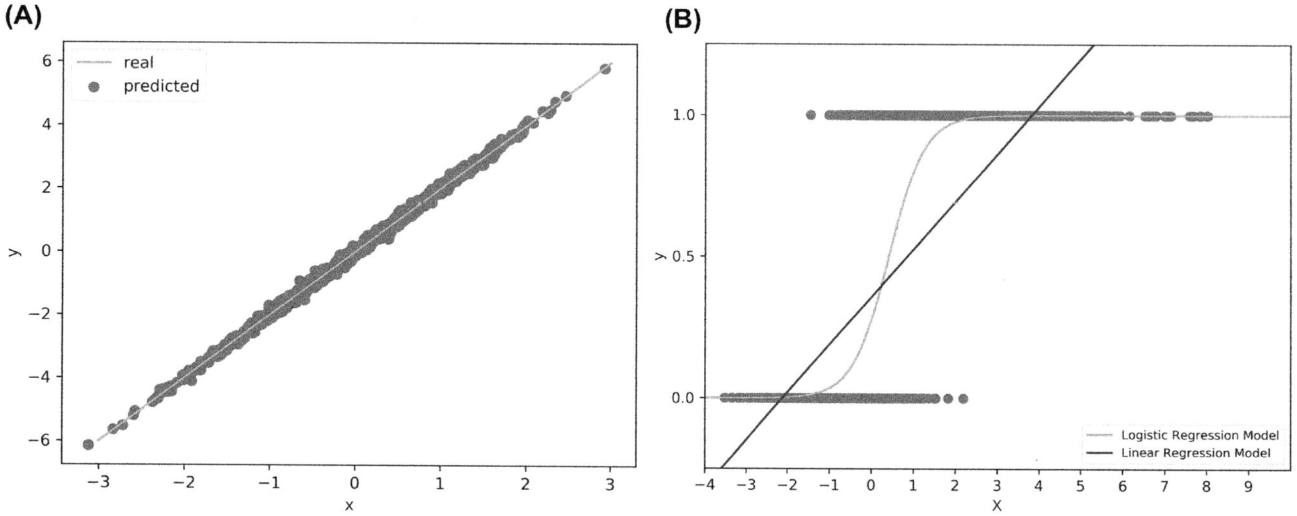

FIGURE 7.3

Differences between linear and logistic regression: (A) A linear regression line and (B) a logistic regression for solving a binary classification problem.

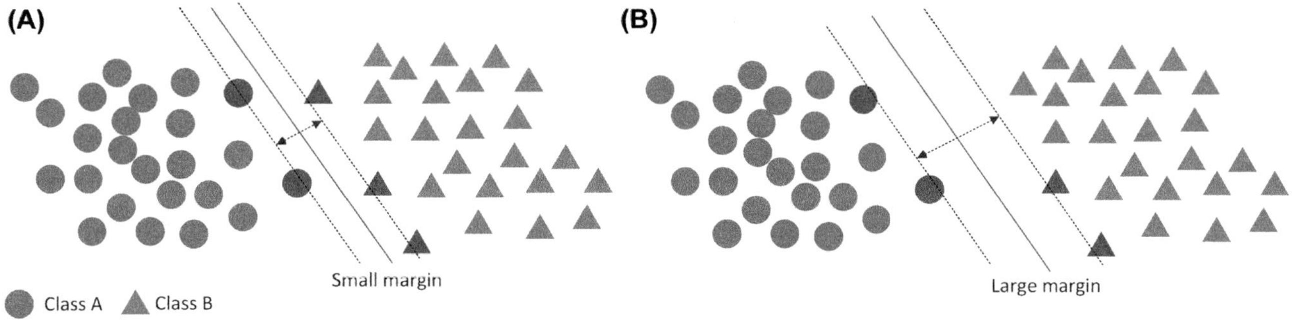

FIGURE 7.4

Hyperplanes with small and large margins. The support vectors are depicted in red color (dark gray in print version).

where b is a scalar that determines the offset of the hyperplane in the 2D space. In the case where the data points are linearly separable like in Fig. 7.4, the idea is to find the region between the upper and lower hyperplanes (i.e., the dashed lines in Fig. 7.4), or the maximum margin hyperplanes, and then define the hyperplane in the halfway between them like the red line in Fig. 7.4. Assuming that the data points are normalized and assuming that we have two class labels, say -1 and 1, for the target y_i, we can easily define the boundaries for the maximum margin hyperplanes, as in

$$rule = \begin{cases} \boldsymbol{x}^T\boldsymbol{w} - b = 1, & upper\ margin \\ \boldsymbol{x}^T\boldsymbol{w} - b = 0, & hyperplane \\ \boldsymbol{x}^T\boldsymbol{w} - b = -1, & lower\ margin \end{cases} . \tag{7.39}$$

where the upper margins and the lower margins correspond to the black dashed lines in Fig. 7.4 and the hyperplane is the red line. Using the above rule, a data point is classified with label 1 when

$$\boldsymbol{x}^T\boldsymbol{w} - b \geq 1, \tag{7.40}$$

i.e., when it exceeds the upper margin, and with label -1 when

$$\boldsymbol{x}^T\boldsymbol{w} - b \leq -1. \tag{7.41}$$

Therefore, the goal is to find a weight vector, $\boldsymbol{w}$, that minimizes the following objective function:

$$y_i\left(\boldsymbol{x}_i^T\boldsymbol{w} - b\right) \geq 1, \tag{7.42}$$

where $y_i = \{-1,\ 1\}$ and $i = 1, 2, \ldots, n$. After estimating the weight vector $\boldsymbol{w}$ and the parameter b, we can get the support vectors, assume $\widetilde{\boldsymbol{x}}$, as follows:

$$\widetilde{\boldsymbol{x}} = sign\left(\boldsymbol{x}^T\boldsymbol{w} - b\right), \tag{7.43}$$

where $sign()$ is a function that returns 1 in the case of a positive value and -1 otherwise. The importance of the support vectors is high as they determine the location of the marginal hyperplanes and thus the location of the hyperplane. This type of SVM is referred to as hard-margin classification [73]. In the case where the data points are not linearly separable, the soft-margin classification can be applied, where the hinge loss function [74] is introduced:

$$max\left(0, 1 - y_i\left(\boldsymbol{x}_i^T\boldsymbol{w} - b\right)\right) = 0. \tag{7.44}$$

In that case, the objective function can be written as follows:

$$max\left(0, 1 - y_i\left(\boldsymbol{x}_i^T\boldsymbol{w} - b\right)\right) + \lambda\left|\boldsymbol{w}\right|^2, \tag{7.45}$$

where $\lambda\left|\boldsymbol{w}\right|^2$ is similar to the $l2$-norm penalty in the Ridge regression case. The difference between the hard and the soft margin is depicted in Fig. 7.5. In the hard margin case (Fig. 7.5(B)), the data points that lie on the vertical lines are the support vectors that determine the maximum marginal hyperplanes so that no other

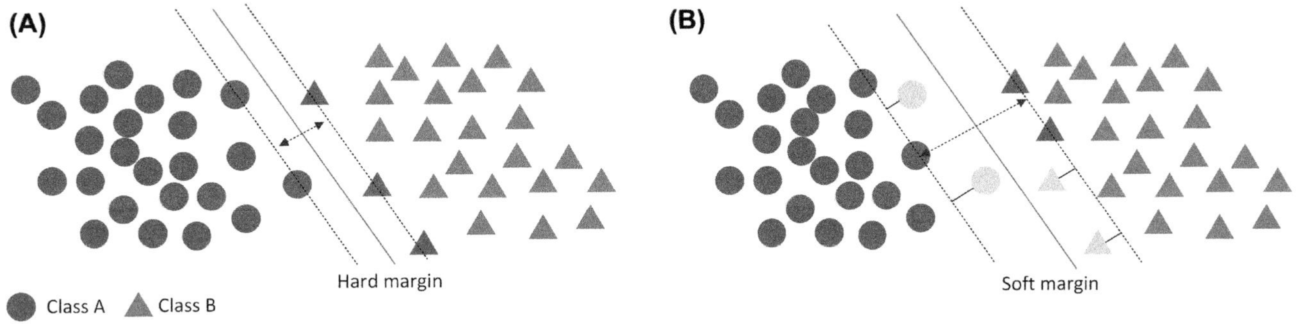

FIGURE 7.5

Differences between the soft and hard margin support-vector machine (SVM) methods.

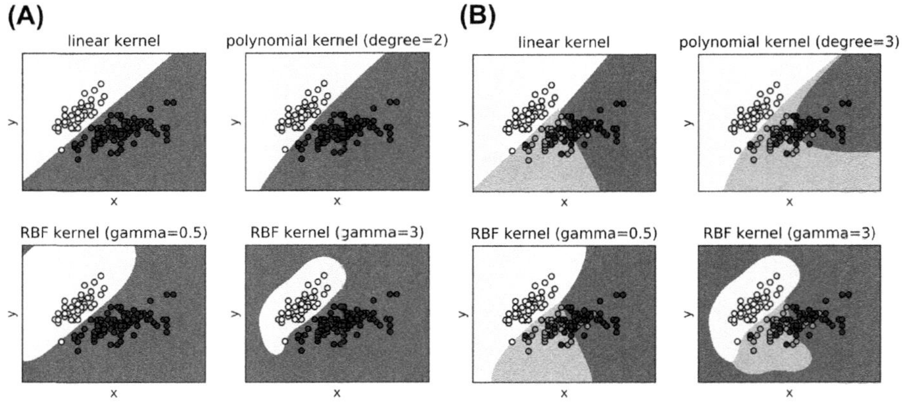

FIGURE 7.6

Evaluation of the support-vector machine (SVM) method with four different kernels. (A) A binary classification problem with one hyperplane that divides the 2D space into two sections. (B) A three-class classification problem with two hyperplanes that divide the 2D space into three sections. In each case, the decision boundaries are depicted in white and magenta color for the classes with labels 1 and 2 and with an additional aqua color (light gray in print version) in the multiclass problem for the additional class with label 3.

data point lies inside the region between the marginal hyperplanes and thus is strict. On the other hand, the soft margin case allows for the existence of data points that fall into the region between the marginal hyperplanes. These data points are usually referred to as violating data points and are depicted in orange color. Thus, the soft margin methods provides a more "flexible" hyperplane.

The impact of the soft margin method is not obvious in the previous case, but it is extremely valuable in the case where the data points are not linearly separable (see Fig. 7.6). Indeed, the SVM can be extended to deal with nonlinear classification problems, where (7.38) can be rewritten as follows:

$$\boldsymbol{k}\left(\boldsymbol{x},\boldsymbol{x}^{\boldsymbol{T}}\right)\boldsymbol{w}-b=0, \tag{7.46}$$

where the dot product in (7.38) is replaced by the kernel function $\boldsymbol{k}\left(\boldsymbol{x},\boldsymbol{x}^{\boldsymbol{T}}\right)$ to adapt the SVM model into transformed feature spaces. The nonlinear kernel function is defined as follows:

$$\boldsymbol{k}\left(\boldsymbol{x},\boldsymbol{x}^{\boldsymbol{T}}\right)=\left\langle\boldsymbol{\varphi}(\boldsymbol{x}),\boldsymbol{\varphi}\left(\boldsymbol{x}^{\boldsymbol{T}}\right)\right\rangle, \tag{7.47}$$

where $\boldsymbol{\varphi}(\boldsymbol{x})$ is a transformation function.

Examples of popular kernels include the polynomial kernel [75]:

$$\boldsymbol{k}\left(\boldsymbol{x},\boldsymbol{x}^{\boldsymbol{T}}\right)=\left(\boldsymbol{x}^{\boldsymbol{T}}\boldsymbol{x}+\boldsymbol{c}\right)^{\boldsymbol{d}}, \tag{7.48}$$

where d is the polynomic degree, and the Gaussian radial basis function (RBF) [76]:

$$\boldsymbol{k}\left(\boldsymbol{x},\boldsymbol{x}^{\boldsymbol{T}}\right)=\boldsymbol{e}^{\left(-\gamma\left|\left|x-x^{T}\right|\right|^{2}\right)}, \tag{7.49}$$

where the parameter γ is equal to $1/2\sigma^2$, and σ is the variance of the Gaussian distribution. In this case, the minimization of the objective function in (7.45) can be found through its Lagrangian dual, which introduces the additional variables, assume c_i, so that [77]

$$w = \sum_{i=1}^{n} c_i y_i x_i, \tag{7.50}$$

where $0 \leq c_i \leq 1/(2n\lambda)$ and

$$\sum_{i=1}^{n} c_i y_i = 0. \tag{7.51}$$

The effect of different types of kernels on the hyperplanes can be better illustrated in the case where the data points are not linearly separable like in Fig. 7.6. In Fig. 7.6A, different types of kernels are evaluated on Python's iris dataset [52] for binary classification. Four kernels were used, namely the linear kernel, the polynomial kernel for degree 2, the RBF kernel for gamma 0.5, and the RBF kernel for gamma 3. For the three class classification problem in Fig. 7.6B, the same kernels were used except from the polynomial kernel, the degree of which was set to 3. As it is shown in Fig. 7.6A, the linear kernel misses one data point in the white class, whereas the polynomial kernel, which works with the soft margin, uses a nonlinear curve to correctly capture the data point. The same occurs in the case of the RBF kernel for gamma equal to 0.5, which is not able to capture the same data point due to the increased variance. Using a smaller variance (i.e., gamma value 3), the SVM hyperplane is constrained to a smaller region and thus captures the data point. Regarding the three-class classification problem, the additional class is depicted in light aqua color. In all the cases, the hyperplanes divide the 2D space in three sections.

7.3.2.3 Naïve Bayes

The Naïve Bayes family [11] is a wide family of probabilistic supervised learning algorithms, which focus on estimating the conditional probability that a random data point belongs to a class, assume j, where $j \geq 2$, given a set of input features. More specifically, given a set of n-training features, $\{x_1, x_2, \ldots, x_n\}$, and a target feature y, the Bayes theorem is first applied [78]:

$$P(y|x_1, x_2, \ldots, x_n) = \frac{P(y)P(x_1, x_2, \ldots, x_n|y)}{P(x_1, x_2, \ldots, x_n)}, \tag{7.52}$$

where $P(y|x_1, x_2, \ldots, x_n)$ is the conditional probability of y given the set of input features $\{x_1, x_2, \ldots, x_n\}$, which is called the posterior probability, $P(x_1, x_2, \ldots, x_n|y)$ is the likelihood, $P(x_1, x_2, \ldots, x_n)$ is the joint probability of the input features, and $P(y)$ is the class probability. To derive the objective function here, the trick is to simplify the likelihood expression in (7.52). One way to do this is to assume that the conditional probabilities of each feature given the target feature y is independent through the conditional independence assumption [78]:

$$P(x_1, x_2, \ldots, x_n|y) = P(x_1|y)P(x_2|y)\cdots P(x_n|y). \tag{7.53}$$

Based on this assumption, (7.52) can be reduced to a product of conditional probabilities:

$$P(y|x_1, x_2, \ldots, x_n) \propto P(y) \prod_{i=1}^{n} P(x_i|y), \tag{7.54}$$

where the joint probability $P(x_1, x_2, \ldots, x_n)$ has been ignored as it is a predefined constant value. The assumption in (7.52) is meaningful as the prior probabilities $P(x_1), P(x_2), \ldots, P(x_n)$ might not be always independent, but providing a conditional knowledge regarding the target feature enhances the independence. Then, an estimation can be derived by solving the following objective function:

$$\tilde{y} = argmax_y \left(P(y) \prod_{i=1}^{n} P(x_i|y) \right), \tag{7.55}$$

where the maximum a posteriori (MAP) rule [79] can be used to find an estimate that maximizes the product of the likelihood and the prior probabilities. This makes sense as finding an estimation that maximizes the likelihood $P(x_i|y)$ yields the maximum posterior probability in (7.52). To demonstrate the MAP rule, let us assume that we have a vector of input features, $\boldsymbol{x} = \{\boldsymbol{x}_1, \boldsymbol{x}_2, \ldots, \boldsymbol{x}_n\}$, and a vector of unknown parameter vectors, assume $\tilde{\boldsymbol{y}} = \{\tilde{\boldsymbol{y}}_1, \tilde{\boldsymbol{y}}_2, \ldots, \tilde{\boldsymbol{y}}_n\}$, so that

$$P(\boldsymbol{x}|\tilde{\boldsymbol{y}}) = \prod_{i=1}^{n} P(\boldsymbol{x}_i|y). \tag{7.56}$$

The goal of the MAP rule is to find the parameter vector $\tilde{\boldsymbol{y}}$, that maximizes the following objective:

$$\tilde{\boldsymbol{y}} = argmax_{\tilde{\boldsymbol{y}}}(l(\tilde{\boldsymbol{y}})P(\tilde{\boldsymbol{y}})), \tag{7.57}$$

where

$$\boldsymbol{l}(\tilde{\boldsymbol{y}}) = \sum_{i=1}^{n} ln(P(x_i|\tilde{\boldsymbol{y}})). \tag{7.58}$$

An estimate for $\tilde{\boldsymbol{y}}$ can be found by computing the derivative:

$$\nabla_{\tilde{\boldsymbol{y}}}\boldsymbol{l}(\tilde{\boldsymbol{y}}) = \sum_{i=1}^{n} \nabla_{\tilde{\boldsymbol{y}}} ln(P(x_i|\tilde{\boldsymbol{y}})), \tag{7.59}$$

and solve for $\tilde{\boldsymbol{y}}$ by setting the derivative equal to zero. Note that if we ignore the $P(\tilde{\boldsymbol{y}})$ expression in (7.56), we obtain the maximum likelihood estimation [80]. An example of the likelihood maximization can be seen in Section 5.7, where the value that maximizes the likelihood of the model is located on the maximum peak of the density. The different variations of the Naïve Bayes classifier depend on the conditional probability $P(\boldsymbol{x}_i|y)$ that is used in (7.56). For example, in the case of the Gaussian Naïve Bayes, we set

$$P(\boldsymbol{x}_i|\boldsymbol{y}) = \frac{1}{\sqrt{2\pi\sigma_y^2}} exp\left(-\frac{\left(x_i - \mu_y\right)^2}{2\sigma_y^2} \right), \tag{7.60}$$

where the mean vector μ_y and the variance vector σ_y^2 can be estimated by solving (7.59).

In that case, (7.58) becomes (for the univariate case):

$$l(\tilde{y}) = -\frac{1}{2} ln\left(2\pi\mu_{\tilde{y}}\right) - \frac{1}{2\sigma_y^2}\left(x_i - \mu_{\tilde{y}}\right)^2, \tag{7.61}$$

which can be extended to the multivariate case as in

$$l(\tilde{\boldsymbol{y}}) = -\frac{1}{2} ln\left(2\pi\boldsymbol{\mu}_{\tilde{y}}\right) - \frac{1}{2\sigma_y^2}\left(\boldsymbol{x}_i - \mu_{\tilde{y}}\right)^{\boldsymbol{T}} \boldsymbol{\Sigma}_{\tilde{y}}^{-1}\left(\boldsymbol{x}_i - \boldsymbol{\mu}_{\tilde{y}}\right), \tag{7.62}$$

where $\boldsymbol{\Sigma}_{\tilde{\boldsymbol{y}}}$ is the covariance matrix. Solving (7.59) yields the following estimation for the mean vector:

$$\boldsymbol{\mu}_{\tilde{y}} = \frac{1}{n} \sum_{i=1}^{n} \boldsymbol{x}_i, \tag{7.63}$$

which is equal to the mean of the training samples, and the following estimation for the covariance matrix:

$$\boldsymbol{\Sigma}_{\tilde{y}} = \frac{1}{n} \sum_{i=1}^{n} \left(\boldsymbol{x}_i - \boldsymbol{\mu}_{\tilde{y}}\right)\left(\boldsymbol{x}_i - \mu_{\tilde{y}}\right)^{\boldsymbol{T}}. \tag{7.64}$$

To demonstrate the classification capabilities of the Gaussian Naïve Bayes classifier, we have applied the classifier on a slightly modified version of the Python's iris dataset [52] for solving a binary classification problem. The decision boundaries based on the estimations of (7.63) and (7.64) are depicted in Fig. 7.7 for both classes.

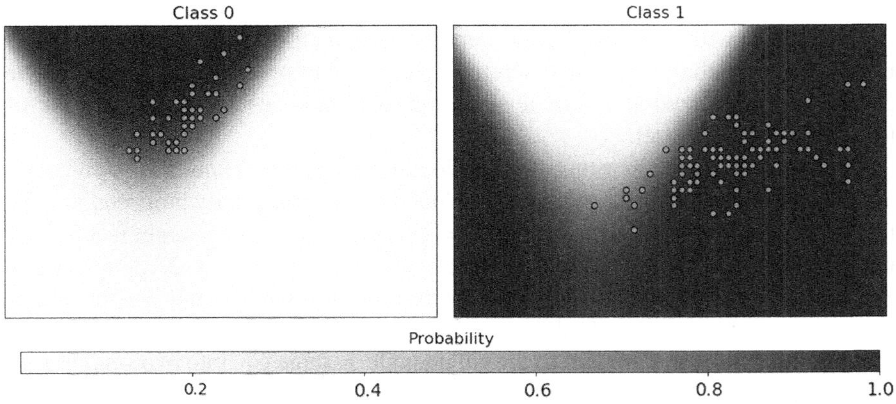

FIGURE 7.7

The discrimination capability of the Naïve Bayes classifier.

The blue-shaded areas correspond to the regions where the classifier has an increased classification probability (density), and the white areas to the regions where the density is small.

Other variants of the Naïve Bayes classifier include the multinomial Naïve Bayes classifier, which is preferred in the case where the input features are discrete and is widely used in text classification [81]. In the Multinomial Naïve Bayes, (7.56) becomes

$$P\left(\boldsymbol{x}|\tilde{y}\right) = \frac{\left(\sum_{i=1}^{n}\boldsymbol{x}_i\right)!}{\prod_{i=1}^{n}\boldsymbol{x}_i!}\prod_{i=1}^{n}p_i^{\boldsymbol{x}_i}, \tag{7.65}$$

where $p_i^{\boldsymbol{x}_i}$ is the probability of event x_i, which is equal to the frequency of that event in the histogram, e.g., the number of occurrences of a word in a text file. Another variant of the Naïve Bayes classifier is the Bernoulli classifier, where (7.56) becomes

$$P(\boldsymbol{x}|\tilde{y}) = \prod_{i=1}^{n}p_i^{\boldsymbol{x}_i}\left(1-p_i^{\boldsymbol{x}_i}\right), \tag{7.66}$$

where $p_i^{\boldsymbol{x}_i}$ is the conditional probability of the event $\boldsymbol{x}_i$ given the class $\boldsymbol{y}$, $P(\boldsymbol{x}_i|\boldsymbol{y})$.

7.3.2.4 Decision tree induction and ensembles

7.3.2.4.1 Classification and Regression Trees

An intuitive supervised learning method that has been widely used for solving both classification and regression problems in machine learning and data mining is the decision tree induction strategy [10]. Decision tree induction focuses on the construction of a tree that consists of a set of nodes that are recursively partitioned according to a splitting attribute to yield nodes with increased impurity toward the development of robust prediction and regression models that are much easier to interpret with reduced computational cost (memory efficient) and increased prediction performance [10]. The splitting attribute is usually determined by an impurity function, such as the information entropy and the Gini impurity index, which quantifies the amount of information that is provided by that node given the target. Decision trees can handle both numerical and categorical features (mixed types) with embedded data preprocessing methods, such as data imputation, data normalization, and discretization. In addition, decision trees can be graphically visualized at a node level, a fact that enhances the interpretability of the classification model.

One of the pioneer algorithms for developing decision trees was the classification and regression trees (CART) algorithm, which had been originally proposed in 1969 [10]. For a given set of n-features, say $\boldsymbol{x} = \{\boldsymbol{x}_1, \boldsymbol{x}_2, \ldots, \boldsymbol{x}_n\}$, and a target feature, say $\boldsymbol{y}$, a training set is first defined as $(\boldsymbol{x}_i, y_i)$, $i = 1, 2, \ldots, n$, where $\boldsymbol{x}_i = \left(x_{i,1}, x_{i,2}, \ldots, x_{i,m}\right)$ is the ith input vector (with m-samples) and y_i is the corresponding target value. The CART algorithm tries to partition a large subset of the training set, which is referred to as the parent node, assume C_P, into two smaller subsets, which are referred to as the left and right nodes (child nodes), say C_L and C_R, respectively, according to a splitting criterion set, say S_c, that is defined based on a splitting value s_c:

$$S_c = \{n \in C_P : x_{nm} \leq s_c\}. \tag{7.67}$$

The splitting value s_c is the threshold value that was defined in Section 7.2.3 as the value that maximizes the information entropy of the information gain given the target feature $\mathbf{y}$. The splitting criterion splits a parent node into left and right nodes according to a decision rule, i.e., when a value is less than (or equal to) c, follow the left child node or the right node otherwise. In fact, the splitting criterion is the one that provides the best value (i.e., the splitting value) so that the node splitting process yields a pure split, where a pure split is defined as a homogeneous split like in the example of Section 7.2.3. The node splitting process recursively builds a set of decision rules that are defined according to the value s_c, which in turn splits the child nodes into more left and right nodes (child nodes). The node splitting process is a greedy process that solves a recursive binary partitioning problem that is repeated until a stopping criterion is met, e.g., until the depth of the tree reaches a specific value or until the desired number of leaf nodes is met, where the leaf node is defined as a child node that is no longer partitioned into left and right nodes.

For a continuous splitting attribute s_c, that corresponds to a criterion subset S_c, the left child node (left subset) is defined as

$$C_L(s_c) = \{n \in C_P : x_{nm} \leq s_c\}, \tag{7.68}$$

whereas the right child node (right subset) is defined as

$$C_R(s_c) = \{n \in C_P : x_{nm} > s_c\}. \tag{7.69}$$

In the case of a discrete-valued splitting attribute, with two possible values (e.g., "yes" or "no," "true" or "false"), (7.68) becomes

$$C_L(s_c) = \{n \in C_P : x_{nm} \in S_c\}, \tag{7.70}$$

and (7.69) becomes

$$C_R(s_c) = \{n \in C_P : x_{nm} \in S'_c\}, \tag{7.71}$$

where S'_c is the complement of S_c.

In the regression tree problem, the CART's goal is to find the optimal s_c^* that minimizes an error loss function, such as the squared error (function):

$$s_c^* = argmin_{s_c}\left(\sum_n |y_n - e_L(s_c)|^2 + \sum_n |y_n - e_R(s_c)|^2\right), \tag{7.72}$$

where $e_L(s_c)$ is the error function $f()$ on the left subset (7.68):

$$e_L(s_c) = f(C_L(s_c)), \tag{7.73}$$

and $e_R(s_c)$ is the error function $f()$ on the right subset (7.69):

$$e_R(s_c) = f(C_R(s_c)). \tag{7.74}$$

In the K-class classification problem, we first define the probability of a training observation that belongs to the node C, with a class label k, where $k = 1, 2, \ldots, K$, as follows:

$$p_k(C) = \frac{1}{|C|} \sum_{i=1}^{n} I(y_i = k). \tag{7.75}$$

where $|C|$ is the number of observations in node C, and $I()$ is the indicator function that is equal to 1 if the condition is met and 0 otherwise. The Gini index from Section 7.2.3 can be redefined as follows (7.8):

$$G(C) = \sum_{i=1}^{n} p_k(C)(1 - p_k(C)), \tag{7.76}$$

which measures the impurity in node C, where $p_k(C)$ is calculated from (7.73).

The gain in the Gini index can then be used to evaluate a decision as it was already presented in (7.10). The optimal splitting attribute can then be defined as follows [82]:

$$s_c^* = argmin_{s_c} \left(\frac{|C_L(s_c)| G(C_L(s_c)) + |C_R(s_c)| G(C_R(s_c))}{|C_L(s_c)| + |C_R(s_c)|} \right), \tag{7.77}$$

where $G(C_L(s_c))$ is the Gini index in the left subset, $C_L(s_c)$, and $G(C_R(s_c))$ is the Gini index in the right subset, $C_R(s_c)$. The fundamental limitation of the CART algorithm is the fact that it does not permit a multiway branch construction on each node but allows only for a binary, i.e., only two branches are grown from each node. In contrast, the Iterative Dichotomiser 3 (ID3) and C4.5 algorithms (see next) allow for multiway tree construction, where two or more branches can be grown from each node during the tree growing (induction) process [83,84].

7.3.2.4.2 ID3, C4.5, and C5.0 implementations

Instead of the Gini impurity index, one can use the information entropy and the information gain measures that were defined in Section 7.2.3. The ID3 algorithm [83] and its predecessor, the C4.5 algorithm, [84] (and its updated version the C5.0 for memory-efficient workflow management [85]) use the information entropy and the information gain scores that were presented in Section 7.2.3 to determine the splitting attributes according to the levels of uncertainty in each subset based on the target feature. The C4.5 algorithm is an improved version of the ID3 where data imputation and pruning methods are embedded to prevent overfitting [84]. Tree pruning [86] is widely used to reduce the overfitting effects that are posed during the training process and increase the generalizability of the tree by removing subtrees that reduce the global performance (accuracy) and thus reducing the size of the decision tree either in a top-down or a bottom-up manner. Examples of tree pruning methods include [86] (i) the reduced error pruning, or fast pruning (supported by the C4.5), where all the nodes are first considered as candidates for pruning and each subtree that is connected to each candidate node is removed if no change in the global performance is observed, (ii) the error complexity pruning, where an error function is computed for each node and the node(s) with small values is removed through a recursive way until the tree with the highest performance is selected, and (iii) cost-complexity pruning, where an additional cost function is considered apart from the error rate that yields the subtree for pruning as the one with the lowest cost (supported by the CART).

In the ID3, C4.5, and C5.0 implementations, the optimal value s_c^* is then defined as the one that minimizes the information entropy (level of uncertainty) or maximizes the information gain according to the target feature, on each node. Both the CART and the ID3 and C4.5 algorithms follow a greedy, top-down approach, where the tree construction process starts from the root node (i.e., from the top) and goes down to the leaf nodes (i.e., to the bottom), following a node-splitting process according to which the training set is recursively partitioned into smaller subsets based on a splitting value that is derived by a score. In the CART case, the score is set to the Gini index, whereas in the ID3 and C4.5, the scores are set to the information entropy or the information gain. Using the generalized impurity function, assume $H()$, the impurity of a node m can be defined according to Section 7.3.2 as follows:

$$G(s_c) = \frac{n_{left}}{n_m} H(C_L(s_c)) + \frac{n_{right}}{n_m} H(C_R(s_c)), \tag{7.78}$$

where the best splitting attribute is given by minimizing the following objective function:

$$s_c^* = argmin_{s_c}(G(s_c)). \tag{7.79}$$

In the case of the CART, the impurity function is set equal to (7.76) and thus $H(C) = G(C)$, where C is the subset of the training data in node m. In the case of the ID3 and C4.5 implementations, the impurity function $H()$ is either the entropy

$$H(C) = -\sum_{i=1}^{n} p_k(C)\log(p_k(C)), \tag{7.80}$$

or the information gain

$$IG(C) = H(C) - H(C|C'), \tag{7.81}$$

where $H(C)$ is the entropy in (7.79) and $H(C|C')$ is the conditional entropy that quantifies the difference between the entropy levels before and after the split or even the misclassification gain:

$$H(C) = 1 - max(p_k(C)), \tag{7.82}$$

where $max(p_k(C))$ is the class label with the highest occurrence in node C.

A generalized pseudocode for decision tree learning is presented in Algorithm 7.1. The algorithm uses as input a training set of n-features along with the associated labels and a set of attributes A. The algorithm first creates a parent node C and then applies some conditions to test whether the class labels are the same, as well as

Algorithm 7.1 A PSEUDOCODE IMPLEMENTATION FOR THE CONSTRUCTION OF A DECISION TREE.

```
0  def decision_tree(x = {x1,x2,...,xN|y}, A):
1  initialize a node C;
2  if all values in y are equal to K:
3     return C as a leaf node with class K;
4  if A is empty then:
5     return C as a leaf node with the majority class k in x, where k = 1,2,...,K;
6  [sc, Sc] = findBestSplittingAttribute(X, A); #an impurity function is applied
7  for each j in Sc:
8     create the partition xj;
9     if xj is empty then:
10       add a leaf with a label that is equal to the majority class k in x, where k = 1,2,...,K;
11    else:
12       C = C∪ decision_tree(xj, A − {sc});
13 end
```

whether the set of attributes is empty. In the first case, the algorithm returns an empty node as a leaf node with the class label that is the same across the training samples, whereas in the latter case, the algorithm returns the node as a leaf node with the class label being equal to the majority class. Then, the algorithm proceeds with the identification of the best splitting attribute using either the Gini impurity index for the CART case or the entropy/information gain for the ID3 and C4.5 cases.

Once the splitting value s_c is found along with the corresponding set S_c, the algorithm iterates on all possible values and partitions the training samples in $\boldsymbol{x}$ accordingly. If the partition is empty, the algorithm adds a leaf with a label that is equal to the majority class, otherwise the algorithm repeats the whole process by adding the partition in a recursive manner (ignoring every time the best splitting attribute) until a stopping criterion is met. The stopping criterion is met either when the partition x_j has samples with the same class label or when there are no remaining attributes to be partitioned in line 8, i.e., when $A = \varnothing$. In that case, node C is returned as a leaf node with the majority class as its label.

An illustration of a decision tree for a binary classification problem is depicted in Fig. 7.8. The decision tree predicts binary outcomes that correspond to a disease outcome with two possible values, i.e., "yes," which denotes the presence of the disease and "no" otherwise. In this example, the depth of the tree (number of levels) is equal to three, and the features (both numerical and categorical) are presented as nodes, where the root node corresponds to the attribute "salivary gland enlargement." The paths in the tree lead to the leaf nodes that correspond to the disease outcome. Each decision rule is derived in a top-down manner, starting from the root note (level 1) to the leaf nodes (level 3). To understand how to interpret the decision rules from this tree, the following four paths were defined:

Path 1: IF the patient has a positive "salivary gland enlargement" (true), a positive "Raynaud's phenomenon" (true), and a positive "Anti-Ro/Anti-La" (true), THEN "disease outcome" = yes.

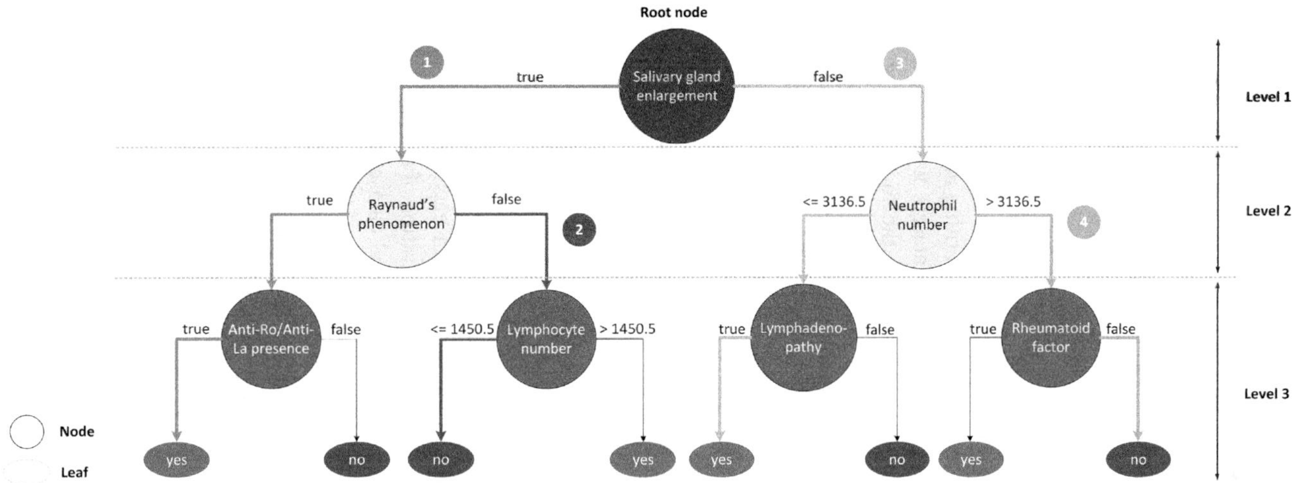

FIGURE 7.8

An illustration of a decision tree for classifying binary outcomes.

Path 2: IF the patient has a positive "salivary gland enlargement" (true) and is not subject to "Raynaud's phenomenon" (false) and has a "number of lymphocytes" $\leq$ 1450.5, THEN "disease outcome" = no.
Path 3: IF the patient does not have a positive "salivary gland enlargement" (false) and has a "neutrophil number" $\leq$ 3136.5 but is positive to lymphadenopathy, THEN "disease outcome" = yes.
Path 4: IF the patient does not have a positive salivary gland enlargement (true) and has more than 3136.5 neutrophils but is positive to lymphadenopathy, THEN "disease outcome" = yes.

7.3.2.4.3 Ensemble classifiers

There is no doubt that the decision trees can yield robust supervised learning models that can be used for a variety of machine learning applications from simple regression to supervised learning problems in datasets with mixed data types. A main drawback of the decision tree, however, is the fact that a single decision tree is not always able to capture the whole picture in a large and complex dataset and thus introduces a form of a bias during the decision-making process. A solution to this is to use ensemble methods that combine multiple decision trees to reduce the variance and enhance the performance of a single decision tree. Ensemble methods are separated into two large categories, namely the bagging and the boosting methods.

The bagging (or bootstrap aggregation) method selects random subsets of the original data with replacement and uses each subset to train a decision tree [87]. An example of a popular bagging ensemble method is the RF algorithm [88], which builds N-decision trees, where each decision tree is trained on random instances of the training set (with replacement) and on random feature subsets as well. The prediction outcomes of each tree are finally combined (aggregated) by either selecting the class with the highest occurrence across the individual trees (majority voting) or by weighting (averaging) the outcomes of each decision tree according to the individual tree's performance [88]. An illustration of the bagging process is depicted in Fig. 7.9. The orange circles on the original dataset represent the features and the red circle the target one. For N-stages, a random subset with replacement is constructed on each stage and a decision tree model is fitted on each subset. This produces a set of N-decision trees (RF), the outcomes of which are combined through a voting process yielding the final outcome.

RF algorithm works well with high-dimensional data and reduces the variance effect during the decision-making process as it combines single trees that capture different patterns. The main problem with the bagging methods, however, is the fact that the data aggregation process introduces a bias when combining multiple prediction outcomes that obscures the accuracy of the supervised learning model. In addition, the number of decision trees directly affects the computational complexity of the algorithm although parallel implementations have been proposed to reduce the execution time of training process. A pseudocode for the RF algorithm is presented in Algorithm 7.2. The algorithm receives as input the training set x, the

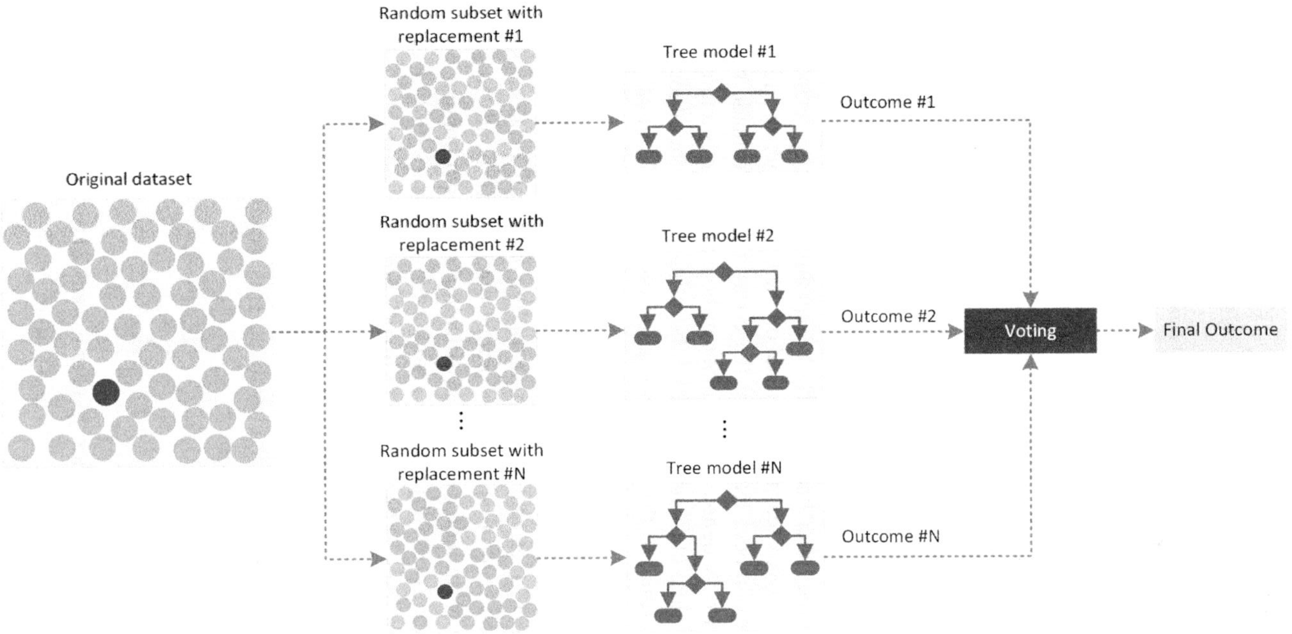

FIGURE 7.9

An illustration of the bagging process (the target feature is depicted in red (dark gray in print version)).

Algorithm 7.2 A PSEUDOCODE IMPLEMENTATION FOR THE RANDOM FORESTS CLASSIFIER.

```
def RF(x = {x1,x2,...,xN|y}, A, N):
T' = {}; #initialize a set of nodes that represents the random forest
for i = 1 : N do:
   define a random subset xi from the original dataset with replacement;
   T = decision_tree(xi,A); #fit a decision tree on the subset from Algorithm 7.1
   T' = T'∪T; #add the decision tree to the forest
end
Use the random forest T' for classification by majority voting or by weighted averaging;
end
```

set of attributes, and the number of decision trees, assume N. For each round, the algorithm builds a decision tree on a random subset from the training set, assume $\boldsymbol{x}_i$, with replacement and adds each tree on the forest. Then, the RF classifier combines the outcomes of each individual tree either by majority voting or by weighted voting where the outcomes are averaged across the models.

Boosting, on the other hand, adopts a sequential strategy, instead of a parallel one, where a set of weak learners is trained on the training subset and on each boosting round the next model learns from the errors that were made by the previous model. In this case, N denotes the number of boosting rounds and differs from the number of individual classifiers. On each boosting round, the algorithm reweights the features according to the misclassification rate. Thus, features that misclassify the target receive a larger weight than the features with small misclassification rate. Then, the next model focuses on the features with the higher weights to improve the overall classification performance. The procedure is repeated until the number of boosting rounds is met. An illustration of the boosting process is depicted in Fig. 7.10. In each random subset, the orange circle corresponds to a feature with high weight due to an increased misclassification rate. Two popular boosting algorithms are the AdaBoost [89] and the gradient boosting trees [38] (and the extreme gradient boosting trees [90]), which are briefly presented next.

The AdaBoost (Adaptive Boosting) classifier [89], originally proposed by Freund and Schapire, is an ensemble classifier that combines a set of N-weak learners in a sequential error reduction fashion, where the final output of the classifier is a weighted sum of the weak classifiers. This means that the final classifier can be expressed in the following form:

$$F_N(x) = \sum_{i=1}^{N} f_i(x), \tag{7.83}$$

where $F_N(x)$ is the final classifier, $f_i(x)$ is a weak classifier (learner), and N is the number of boosting rounds. The previous equation can be rewritten in a sequential form as in

$$F_i(x) = F_{i-1}(x) + a_i h_i(x) = F_{i-1}(x) + f_i(x), \tag{7.84}$$

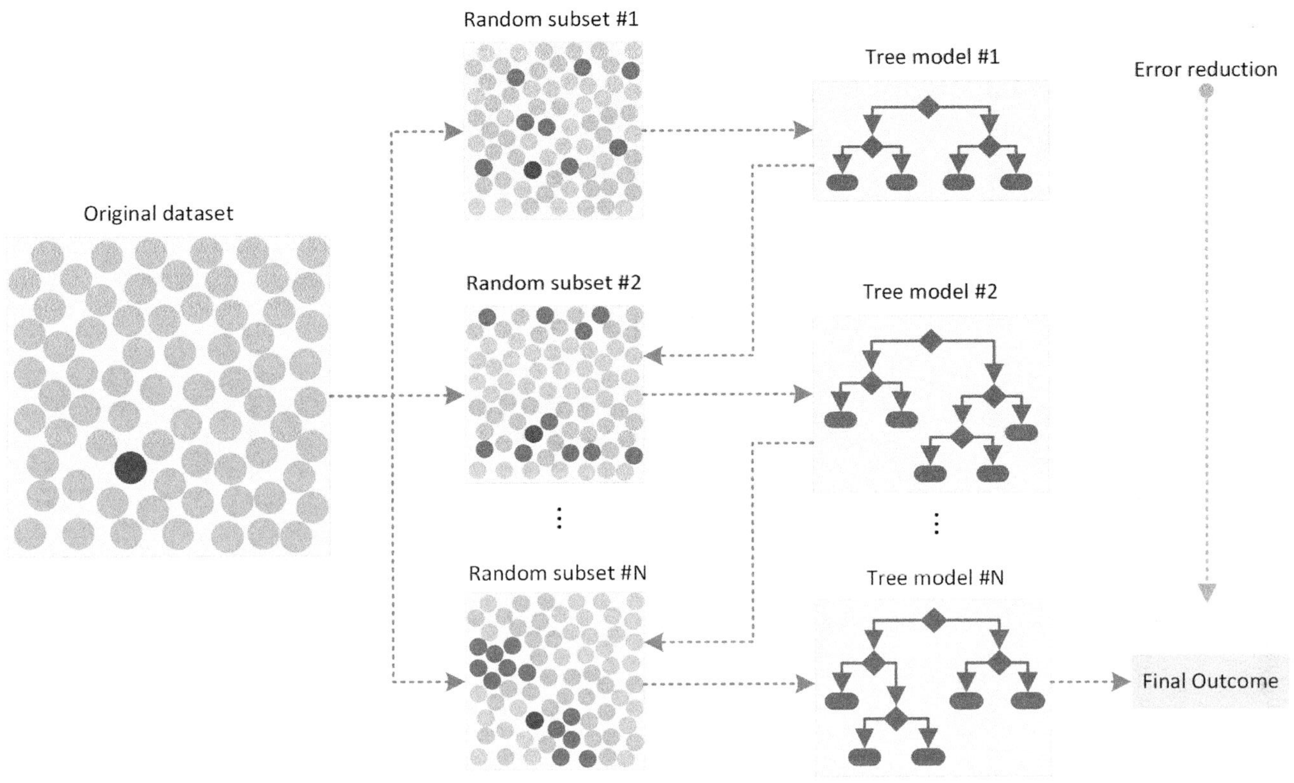

FIGURE 7.10

An illustration of the boosting process (the target feature is depicted in red (dark gray in print version)).

where $F_{i-1}(x)$ is the ensemble at step $i-1$, a_i is the weight that is given to the classifier at step i, and $h_i(x)$ is the outcome of the weak classifier at step i. It is proven [89] that the weights a_i are given by

$$a_i = \frac{1}{2} ln\left(\frac{1-\varepsilon_i}{\varepsilon_i}\right), \tag{7.85}$$

where ε_i is the weighted sum of the misclassified samples w_i at step i:

$$\varepsilon_i = \sum_{j f_i(x_j) \neq y_j}^{n} w_{j,i}, \tag{7.86}$$

where n is the number of features. The weights $w_{j,i}$ that are assigned on each feature fulfill the following:

$$\sum_{j=1}^{n} w_{j,i+1} = 1, \tag{7.87}$$

where $w_{j,i+1}$ is the weight of the feature j at step $i+1$. It is proven that to fulfill the condition in (7.87), the following must stand (in general):

$$w_i = \frac{w_i e^{-f_i(x) y_i}}{2\sqrt{\varepsilon_i(1-\varepsilon_i)}}, \tag{7.88}$$

where $y_i \in \{-1, +1\}$ and the term $e^{-f_i(x) y_i}$ is computed by the gradient of an exponential loss function over the weak learner $\nabla_{h_i(x)} l(h_i(x))$, where $l(h_i(x)) = \sum_{j=1}^{n} e^{-f_i(x_j) y_j}$. A pseudocode for the AdaBoost classifier is presented in Algorithm 7.3.

Algorithm 7.3 A PSEUDOCODE IMPLEMENTATION FOR THE ADABOOST CLASSIFIER.

```
define AdaBoost(x = {x1,x2,...,xn|y}, N)
for i = 1 : N do:
  define a weak learner h_i(x);
  define the weighted classification error ε_i as in (7.86);
  if (ε_i < 0.5):
    Compute the weights a_i as in (7.85);
    define f_i(x) = a_i h_i(x);
    update using (7.84) as F_{i+1}(x) = F_i(x) + f_i(x);
    update w_i as in (7.88);
  else
     return F_i(x);
   end
   return F_i(x);
 end
```

The gradient boosting algorithm is based on the idea that the boosting process can be used to sequentially optimize a differentiable error loss function [38,90]. The gradient boosting algorithm is an ensemble classifier that combines a set of weak learners into a stronger classifier, where on each boosting round, the algorithm minimizes the gradient of a loss function to optimize the overall performance of the classifier. At step i, the gradient boosting classifier seeks for a weak learner, say $f_i(x)$, so that

$$F_{i+1}(x) = F_i(x) + f_i(x). \tag{7.89}$$

Assuming that $\widetilde{y}$ is the predicted value at step i, we wish to minimize the following error:

$$y = f_i(x) - F_i(x). \tag{7.90}$$

To do so, the algorithm minimizes the following cost function at step i:

$$F_i(x) = F_{i-1}(x) + argmin_f\left(\sum_{j=1}^{n} L\left(\widetilde{y}_j, F_{i-1}(x_j) + f_i(x_j)\right)\right), \tag{7.91}$$

where $L(y, F(x))$ is the error loss function, e.g., the mean squared error, and n is the number of samples. The previous equation can be rewritten as follows:

$$F_i(x) = F_{i-1}(x) + \gamma_i f_i(x), \tag{7.92}$$

where

$$\gamma_i = argmin_\gamma\left(\sum_{j=1}^{n} L(y_j, F_{i-1}(x_j) + \gamma f_i(x_j))\right), \tag{7.93}$$

which can be rewritten using the steepest gradient descent as in

$$\gamma_i = argmin_\gamma\left(\sum_{j=1}^{n} L(y_j, F_{i-1}(x_j) - \gamma\nabla_{F_{i-1}} L(y_j, F_{i-1}(x_j))\right). \tag{7.94}$$

Then (7.93) becomes

$$F_i(x) = F_{i-1}(x) - \gamma_i \sum_{j=1}^{n} \nabla_{F_{i-1}} L(y_j, F_{i-1}(x_j)). \tag{7.95}$$

A pseudocode for the gradient boosting classifier is presented in Algorithm 7.4. The algorithm initializes a learner from the sequential update in (7.92). Then, for each boosting round, the algorithm fits a weak learner on the error residuals of the previous round, say $ps_i = \nabla_{F_i(x)} l(y, F_i(x))$, and computes the weights γ_i from (7.94). The process is updated through (7.95). The algorithm finally returns the boosted classifier.

Algorithm 7.4 A PSEUDOCODE IMPLEMENTATION FOR THE GRADIENT BOOSTING CLASSIFIER.

```
define GradientBoosting(x = {x_1, x_2, ..., x_n | y}, N)
Initialize the learner F_0(x);
for i = 1 : N do:
    define a weak learner h_i(x);
    define the pseudoresiduals, ps_i = ∇_{F_i(x)} l(y, F_i(x)), and fit the weak learner on the set {x, ps_i};
    compute the weight γ_i from (7.94);
    update using (7.95) as F_i(x) = F_{i-1}(x) − γ_i Σ_{j=1}^{n} ∇_{F_{i-1}} L(y_j, F_{i-1}(x_j));
end
return F_i(x);
```

The gradient boosting algorithm is most commonly appeared using CART trees as a base learner. In that case, (7.92) becomes [38,90]

$$F_i(\boldsymbol{x}) = F_{i-1}(\boldsymbol{x}) + \gamma_i \sum_{l=1}^{L} b_{li} 1(\boldsymbol{x} \in R_{li}), \tag{7.96}$$

where at each step i, a regression tree partitions the original space into L-disjoint regions $\{R_{lm}\}_1^L$, where R_{lm} is the L-terminal node, b_{li} is the value predicted in each region R_{lm}, and $1()$ is an indicator function (1:true, 0:otherwise). A popular and more enhanced implementation of the gradient boosting algorithm is the extreme gradient boosting algorithm [90], which adopts a regularized approach to further reduce the overfitting on the gradient and enhance the overall performance.

The gradient boosting approach is much more flexible than the AdaBoost as it is not only constrained on a specific loss function (the exponential loss) but on a generalized gradient loss. In addition, the gradient boosting approach uses the gradient direction to identify the samples with high misclassification rate (on the negative gradient direction) and reduces the overfitting of the next learner instead of the AdaBoost, which weights the samples according to the misclassification rate, where the samples that are correctly classified receive a small weight and a large otherwise. The gradient boosting approach does not alter the original training subset but trains each weak learner on the pseudoresiduals of the previous learner. The fact that it solves a gradient descent optimization problem instead of simply reweighting the samples according to the misclassification rate like in the AdaBoost enhances its performance. The AdaBoost, on the other hand, can be also adapted on any base classifier, apart from the decision trees, a fact that enhances its applicability.

7.3.3 Artificial neural networks and deep learning

The scientific challenge of mining useful knowledge from big medical data, such as omics data and multislice medical images, has led to the rise of deep learning, a modern field in data science that involves the application of deep neural networks (DNNs) for the detection of hidden patterns in complex big data structures [18]. Before the description of a DNN, it is first necessary to begin with a brief description of the ANNs [12]. The main difference between a simple neural network and a deep learning neural network lies in the number of hidden layers. An ANN consists of three layers, the input layer, the hidden layer and the output layer, where each layer includes a set of nodes, i.e., the input nodes, the hidden nodes and the output nodes. A deep learning neural network, on the other hand, consists of more than one hidden layers. Therefore, the number of hidden nodes and the number of hidden layers determines the computational complexity of the ANN.

We will now begin with a brief introduction into the simple ANNs. An example of an ANN with one hidden layer is depicted in Fig. 7.11. According to Fig. 7.11, the neural network consists of four input nodes, namely the x_1, x_2, x_3, and x_4. Each input node corresponds to an input feature observation. A hidden layer consists of additional hidden nodes, assume h_1, h_2, h_3, h_4, h_5, and h_6, where each hidden node is related to each input node through a specific weight value, $w_{i,j}$, $i = 1, 2, \ldots, 6$, and $j = 1, 2, 3, 4$. In a similar manner, each output node is related to each hidden node through a specific weight value $wh_{i,p}$, $i = 1, 2, \ldots, 6$, and $p = 1, 2$.

Based on Fig. 7.11, we can express the hidden node h_i as a linear combination of the four input nodes:

$$h_i = w_{i,1}x_1 + w_{i,2}x_2 + w_{i,3}x_3 + w_{i,4}x_4, \tag{7.97}$$

where $w_{i,j}$ are the weight values related to the input node i and the hidden node j. In a similar manner, the output node, assume y_p, $p = 1, 2$, can be written as in:

$$y_p = wh_{i,1}h_1 + wh_{i,2}h_2 + wh_{i,3}h_3 + wh_{i,4}h_4 + wh_{i,5}h_5 + wh_{i,6}h_6, \tag{7.98}$$

where $wh_{i,p}$ is the weight value between the hidden node i and the output node p. For m-input nodes and n-hidden nodes, (7.97) can be generalized as follows:

$$\boldsymbol{h} = \begin{bmatrix} h_1 \\ h_2 \\ \vdots \\ h_n \end{bmatrix} = \sum_{i=1}^{n} w_i^T \boldsymbol{x} = \begin{bmatrix} w_{1,1} & w_{1,2} & w_{1,3} & \cdots & w_{1,m} \\ w_{2,1} & w_{2,2} & w_{2,3} & \cdots & w_{2,m} \\ \vdots & \vdots & \vdots & \ddots & \vdots \\ w_{n,1} & w_{n,2} & w_{n,3} & \cdots & w_{n,m} \end{bmatrix} \begin{bmatrix} x_1 \\ x_2 \\ \vdots \\ x_m \end{bmatrix} = \boldsymbol{W_I x}, \tag{7.99}$$

where $\boldsymbol{W_I}$ is the weight matrix of the input nodes. For p-output nodes, (7.98) can be rewritten as follows:

$$\boldsymbol{y} = \begin{bmatrix} y_1 \\ y_2 \\ \vdots \\ y_p \end{bmatrix} = \sum_{i=1}^{p} wh_i^T \boldsymbol{h} = \begin{bmatrix} wh_{1,1} & wh_{1,2} & wh_{1,3} & \cdots & wh_{1,n} \\ wh_{2,1} & wh_{2,2} & wh_{2,3} & \cdots & wh_{2,n} \\ \vdots & \vdots & \vdots & \ddots & \vdots \\ wh_{p,1} & wh_{p,2} & wh_{p,3} & \cdots & wh_{p,n} \end{bmatrix} \begin{bmatrix} h_1 \\ h_2 \\ \vdots \\ h_n \end{bmatrix} = \boldsymbol{W_H h}, \tag{7.100}$$

where $\boldsymbol{W_I}$ is the weight matrix of the hidden nodes.

In neural network theory, each weighted sum in (7.97) and (7.98) is parsed through a specific function, which is referred to as the activation function [91]. The idea behind the activation function is inspired by the biology, where a neuronal activation is the result of the activation of multiple neurons, where the neuronal

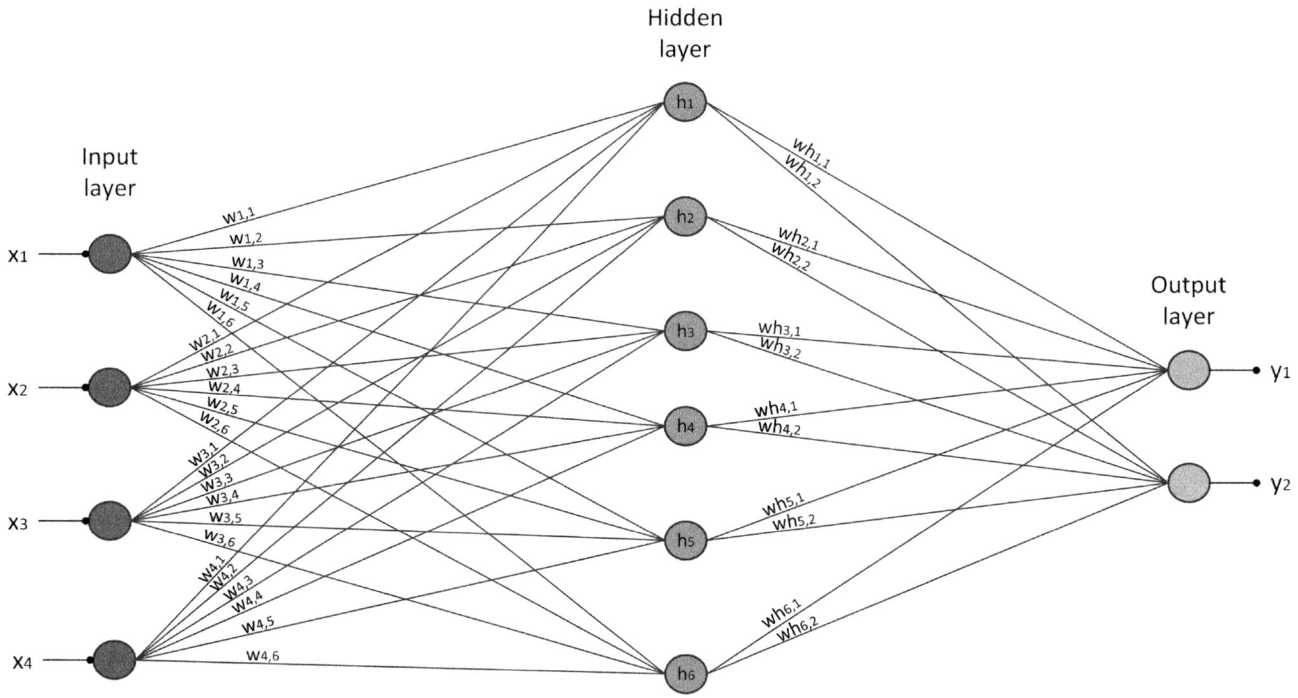

FIGURE 7.11

An illustration of an artificial neural network architecture with one hidden layer.

activation can be simulated by the weighted sum of the input nodes with a step function, say $\Phi()$, which is applied on the weighted sum. The equations of the system are then transformed accordingly as in

$$h_i = \Phi\left(w_{i,1}x_1 + w_{i,2}x_2 + w_{i,3}x_3 + w_{i,4}x_4\right), \left(w_{i,1}x_1 + w_{i,2}x_2 + w_{i,3}x_3 + w_{i,4}x_4\right), \tag{7.101}$$

and in

$$y_k = \Phi\left(wh_{k,1}h_1 + wh_{k,2}h_2 + wh_{k,3}h_3 + wh_{k,4}h_4 + wh_{k,5}h_5 + wh_{k,6}h_6\right). \tag{7.102}$$

In a similar way, (7.99) and (7.98) can be updated accordingly. An illustration of the neural network with the activation functions between the input-hidden nodes and the hidden-output nodes is depicted in Fig. 7.12. An example of a neuronal activation is depicted in Fig. 7.13. In this example, the weighted sum of the second hidden node, x_2, is activated yielding a subsequent activation of the second node in the output layer, h_2, which in turns yields an outcome y_1. By setting the input nodes to the input features and the output node to the target feature, we can construct a neural network for classification purposes. Because the number of output nodes can be larger than two, the ANN can be used for solving multiclassification tasks. The difficult part here is to initialize the weight values between the nodes of the input layer and the hidden layer. In practice, the weights are randomly initialized according to a distribution, such as the normal distribution.

One of the first activation functions was the step function [91], which activates the output node that achieves a weighted sum larger than a predefined threshold:

$$\boldsymbol{\Phi}\left(\boldsymbol{w}^T\boldsymbol{x} + b\right) = \begin{cases} 1, & \text{if } \boldsymbol{w}^T x + b > t \\ 0 & o.w. \end{cases}, \tag{7.103}$$

where $\boldsymbol{w}^T\boldsymbol{x} = \sum_{i=1}^{n} w_i^T\boldsymbol{x}$ is the weighted sum (dot product) of the feature vector $\boldsymbol{x}$ and t is a threshold. This activation function is used in the single-layer perceptron classifier [92], which yields a binary classifier that classifies an input observation as 1 if the weighted sum is larger than a predefined threshold t and 0 otherwise (usually $t = 0$). The single-layer perceptron is the simplest form of classifier that does not include a hidden layer but only the input and the output layer of Fig. 7.12. In the perceptron classifier, the weight update process can be expressed in a sequential way:

$$w(t+1) = w(t) + re(t)\boldsymbol{x}, \tag{7.104}$$

where $w(t)$ is the weight at step t, r is the learning rate, and $e(t)$ is the prediction error in the node of the perceptron. The error is the difference between the target value y and the perceptron output $\tilde{y}(t)$:

$$e(t) = |y - \tilde{y}(t)| \tag{7.105}$$

The convergence is then achieved when the error e is less than the threshold t, where $\tilde{y}$ is the perceptron's output. The objective function is solved by the classic

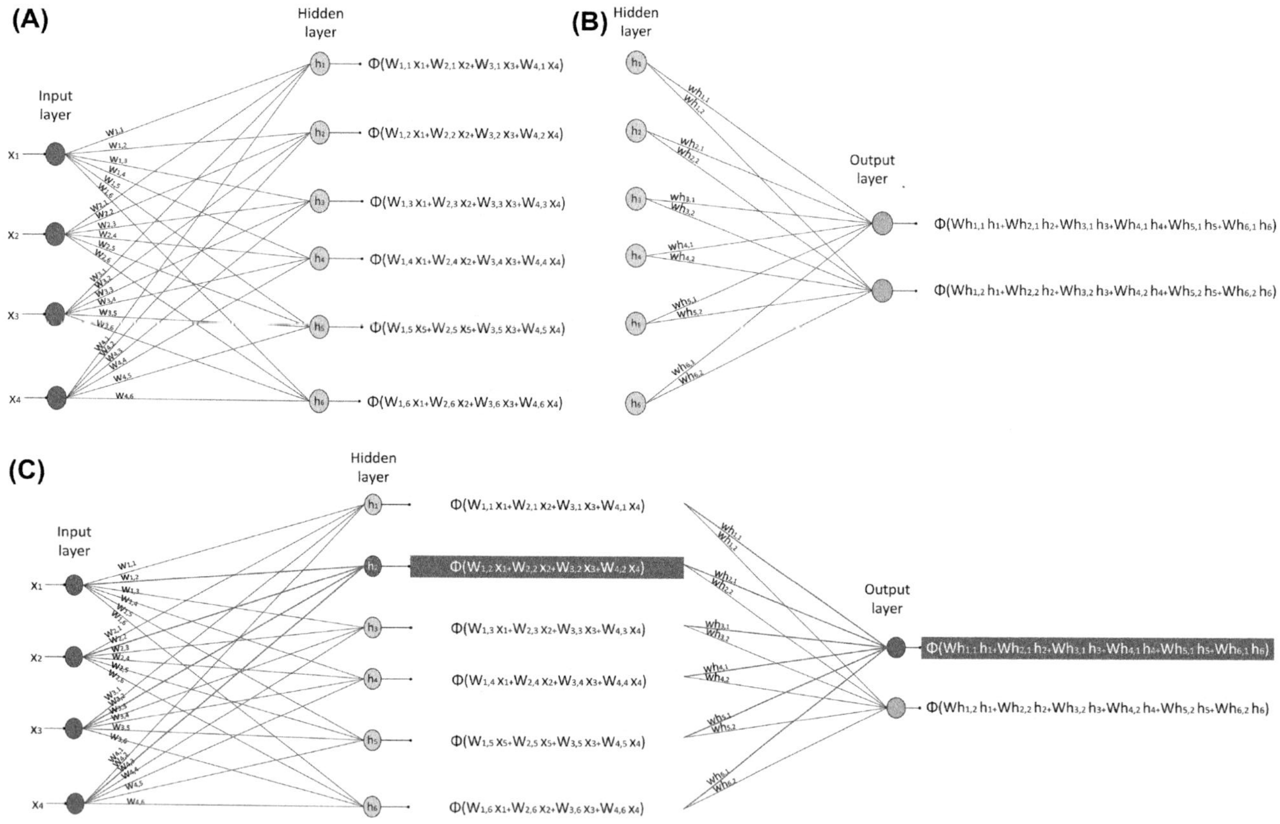

FIGURE 7.12

A deeper look into the activation functions between (A) the nodes between the input layer and the hidden layer and (B) the nodes between the hidden layer and the output layer. (C) An example of the information flow during a neuronal activation.

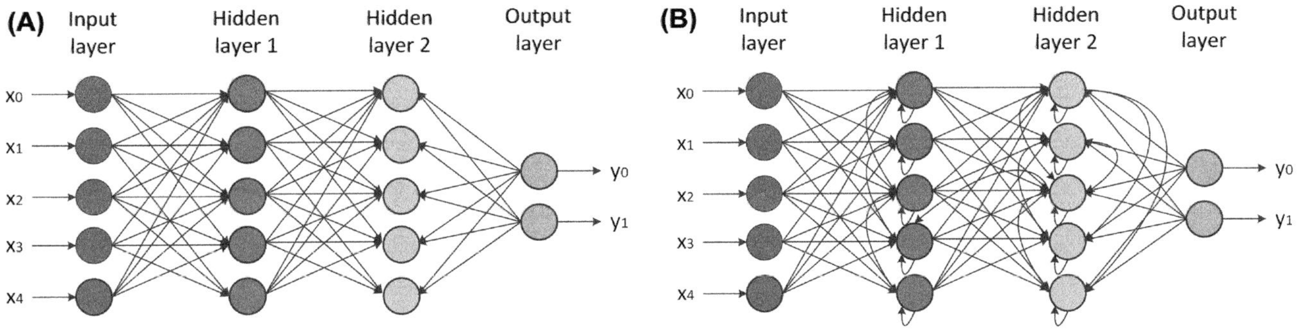

FIGURE 7.13

An illustration of the main differences between (A) a feedforward neural network architecture and (B) a recurrent neural network architecture.

way of minimizing the derivative of the error function, where the change in each weight is defined as

$$\nabla w(t) = -r\nabla e(t)y(t). \tag{7.106}$$

Popular activation functions that are widely used in practice include the sigmoid function [91]

$$\boldsymbol{\Phi}\left(\mathbf{w}^{T}\mathbf{x}+b\right) = \frac{1}{1+e^{-(\mathbf{w}^{T}\mathbf{x}+b)}}, \tag{7.107}$$

the rectified linear unit (ReLu) activation function [91]

$$\boldsymbol{\Phi}\left(\mathbf{w}^{T}x+b\right) = \begin{cases} 0, & \text{if } \mathbf{w}^{T}x+b<0 \\ 1, & o.w. \end{cases}, \tag{7.108}$$

and the *tanh* activation function [91]

$$\boldsymbol{\Phi}\left(\mathbf{w}^{T}\mathbf{x}+b\right) = \frac{e^{\mathbf{w}^{T}\mathbf{x}+b} - e^{-(\mathbf{w}^{T}\mathbf{x}+b)}}{e^{\mathbf{w}^{T}\mathbf{x}+b} + e^{-(\mathbf{w}^{T}\mathbf{x}+b)}}, \tag{7.109}$$

among many others. Each activation function has its own impact depending on whether someone wishes to solve nonlinear optimization problems with monotonic and/or differentiable error surfaces, as well as when the output of the classifier is continuous instead of categorical. For example, a sigmoid function can produce a continuous output instead of the step function that produces a binary output. In fact, Fig. 7.12 depicts a multilayer perceptron (MLP) network [93], which uses a sigmoid activation function as in (7.107). Despite each name, the MLP is not a perceptron ANN with multiple hidden layers but an ANN that combines multiple perceptron units to formulate a multiple hidden layer ANN. In addition, the MLP is not limited to threshold-based activations functions like in (7.103) but can support any type of activation functions, like those in (7.108) and (7.109). MLP has many applications in disease management, including EEG and EMG signal classification for BCI applications [94] and tumor diagnosis [95].

Depending on the direction of the information flow within the neural network, the neural networks can be divided into two main categories, namely the feedforward neural networks (FNNs) [96] and the recurrent neural networks (RNNs) [97]. In a FNN, the connections between the nodes of the input and the hidden layers, as well as between the hidden and the output layers, do not form any loops. This means that the information flows in a forward manner just like in Fig. 7.11. An example of a popular FNN is the MLP network that we have already presented in Fig. 7.11. On the other hand, in a RNN, directed pathways can exist between the nodes of the network. These directed pathways allow for the existence of loops between the nodes of the network. The differences between an FNN architecture and an RNN architecture are depicted in Fig. 7.14, where the connections in the RNN formulate a more "chaotic" information flow within the ANN. The RNN introduces a temporal factor during the construction of the ANN, where the information

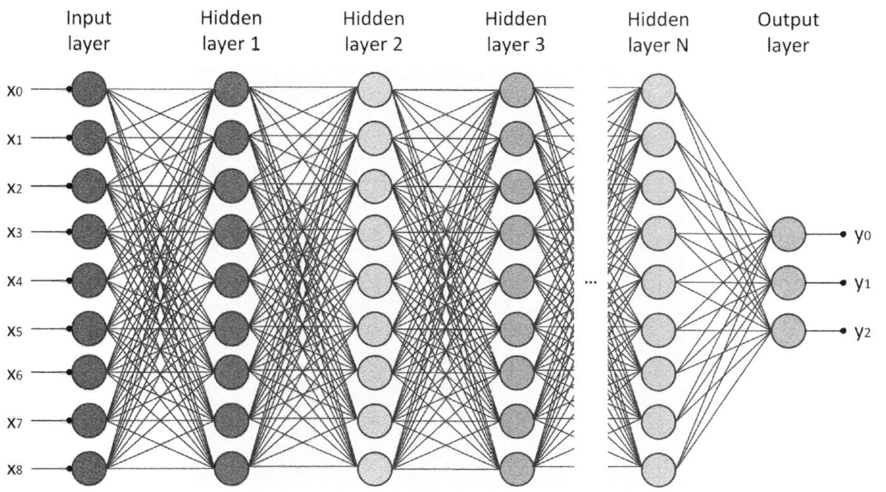

FIGURE 7.14

An illustration of a deep neural network architecture with N-hidden layers.

between the nodes of the hidden layer is exchanged in a temporal fashion (i.e., unrolled). This means that the gradients are backpropagated in the hidden layers to optimize the gradient of the error loss function during the training stage.

A typical architecture of a deep learning neural network (DNN) is depicted in Fig. 7.15. As we mentioned at the beginning of this section, a DNN architecture consists of $N-$hidden layers, where $N > 1$. Examples of popular types of DNNs include the CNNs and RNNs. The CNNs [20] are improved versions of the MLPs where a regularization process is applied to reduce the overfitting in an hierarchical manner using tensors and are widely used in medical imaging for classification and object detection due to their ability to simulate the biological processes that are invoked by the visual cortex. CNNs have been used to solve classification problems across large-scale image datasets including the detection of skin cancer [98] and diabetic retinopathy [99], as well as in omics studies for the prediction of chromatin marks in DNA sequences [100]. A widely used RNN is the LSTM [19], which includes "cell" components, where each cell consists of an input layer, an output layer, and a "forget" layer that controls the information flow within the RNN. The purpose of the "forget" gate is to eliminate the gradient "vanishing" problem [101] in RNNs. According to the gradient "vanishing" problem, the backpropagated gradients from one hidden layer to another "vanishes" (i.e., it approximates zero or infinity). The LSTM uses the cells to control the gradient backpropagation so that the backpropagated information will remain unchanged. The LSTM has numerous applications in biomedical signal analysis for the prediction of ictal states in epileptic seizures [102], the early detection of heart failure onset [103], and the freezing of gait in Parkinson's disease [104].

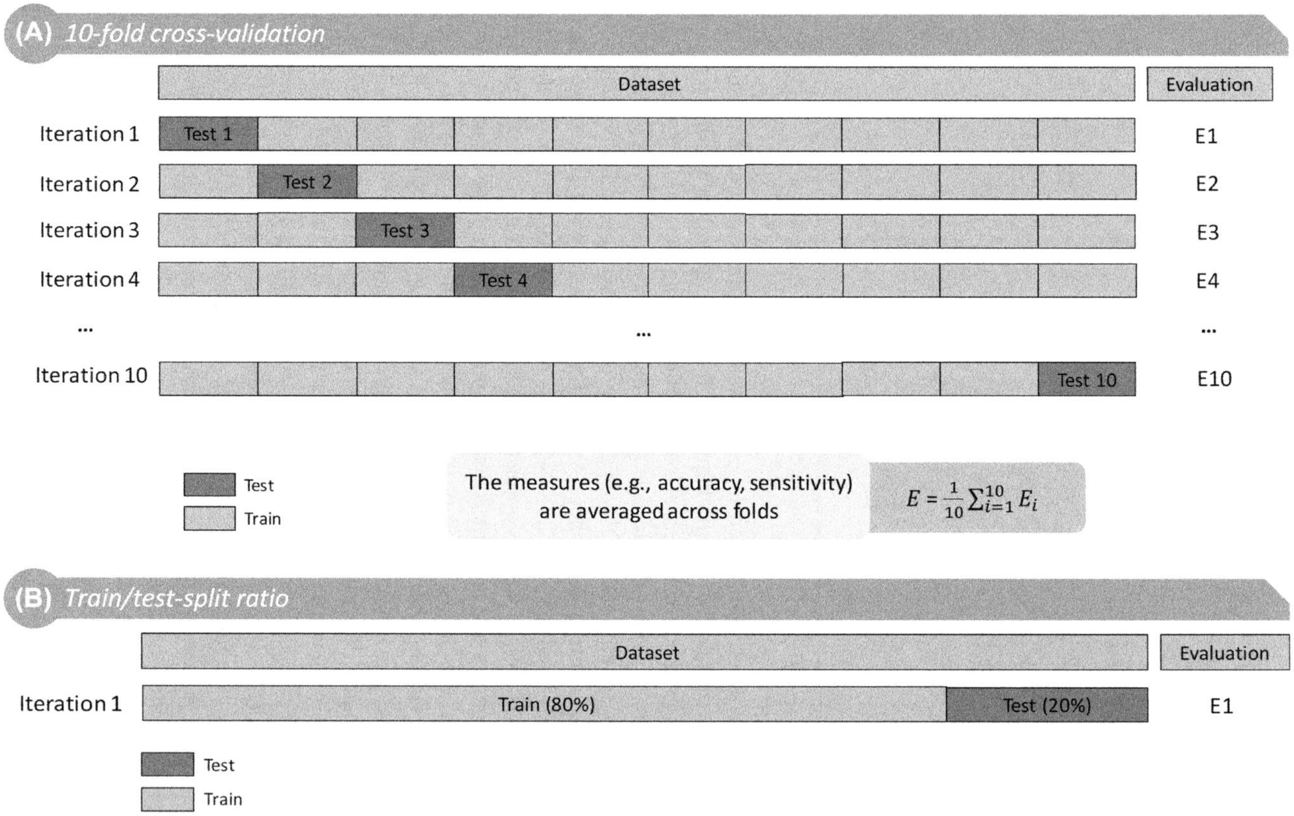

FIGURE 7.15

The two most common methods for evaluating the performance of a supervised learning model. (A) The k-fold cross-validation strategy for $k = 10$ and (B) the train/test split ratio for a 0.8 ratio.

7.3.4 Performance evaluation

The most widely used methods for performance evaluation include the cross-validation and the train/test split strategy. According to the literature, the variety of cross-validation methods is large, including the k-fold cross-validation [105], the repeated k-fold cross-validation [106], the leave-one-out (LOO) cross-validation [107], the stratified k-fold cross-validation [108], and the leave-P-out (LPO) cross-validation [109], among other variations.

The train/test split is the simplest method for performance evaluation (Fig. 7.3). It uses a train-over-test split ratio to divide the dataset into a training and a testing subset. For example, a train/test split ratio of 0.8 would use 80% of the dataset for training the supervised learning model and the rest 20% as a testing subset for evaluating the model (Fig. 7.3). On the other hand, the k-fold cross-validation procedure divides the original dataset into k-subsets or folds, with equal size (equal number of samples on each fold). Then, the supervised learning model (e.g., prediction model) is trained using the $k-1$ folds, where the remaining fold is used for testing purposes (Fig. 7.3). The process is repeated until all folds are used for testing (Fig. 7.3). The performance of the supervised learning model is finally evaluated on each fold and the performance scores are averaged across the k-folds. In practice, the number of folds may vary from 3- and 5- to 10-folds.

The stratified k-fold [108] is a variation of the k-fold cross-validation procedure that ensures that each fold (both training and testing) will preserve the ratio of the target class in the whole dataset. For example, if the target feature in the original dataset consists of 1000 samples where 400 are labeled as "1" (e.g., positive) and 600 are labeled as "0" (e.g., negative), the stratified k-fold will ensure that each fold will preserve the 4:6 ratio of positive over negative elements. In the case where the target feature in the original dataset consists of 1000 samples where 300 are labeled as "0", 200 are labeled as "1," and 500 are labeled as "2," the stratified k-fold will ensure that each fold will preserve the 3:2:5 ratio. The repeated k-fold cross-validation procedure [106] is another approach that repeats the k-fold cross-validation procedure n-times, where the value n is predefined. The same occurs in the case of the repeated stratified k-fold cross-validation procedure. LOO cross-validation procedure [107] creates learning subsets by discarding one sample at a time, until each sample is used for testing. This means that in the case of a dataset with n-samples, the LOO yields n-training subsets and n-testing subsets, where the supervised learning model is trained on $n-1$ samples instead of $n(k-1)/k$ samples. However, the LOO is a much more computationally complex approach than the k-fold cross-validation process especially in the case where $n \gg k$. In addition, as the LOO uses a slightly modified training subset every time, it is more prone to high variance in predicting outcomes. Finally, LPO cross-validation procedure [109] removes p-samples from the dataset and thus produces $\binom{n}{p}$ possible pairs of training and testing subsets with overlapping samples ($p > 1$). In practice, the 5-fold or the 10-fold cross-validation procedure is preferred as the most robust performance evaluation strategy (Fig. 7.15).

So, what metrics can be used to evaluate the performance of the classifier either in each fold during the cross-validation process or on the test subset during the train-test splitting process? Typical performance evaluation measures that are widely used in the literature include the [110] (i) accuracy, (ii) area under the curve (AUC), (iii) F1 score, (iv) sensitivity (or true positive rate [TPR]), (v) specificity (or true negative [TN] rate), (vi) positive predictive value or precision, and (vii) negative predictive value. The accuracy is defined as the proportion of the true positive (TP) and TN predicted values across the total population size. The higher the accuracy, the better the performance. However, the accuracy is not a robust measure as a classifier could perfectly classify not only the positive events but also the actual negative events as positive (i.e., have a high false negative [FN] rate). An alternative and more robust approach is the receiver operating characteristic (ROC) curve, which provides a graphical way to visualize the performance of a classifier in terms of TPR against the false positive rate (FPR). The TPR is also referred to as sensitivity and measures how well the classifier can detect all the actual positive events. In the binary classification problem, the positive event denotes a sample with a class label "1" (e.g., positive disease outcome), whereas the negative event denotes a sample with class label "0."

On the other hand, the FPR measures how many negative events are classified as positives (i.e., false positives [FPs]) by the classifier. Therefore, a good classifier is the one that has a high TPR (to accurately discriminate the actual positive events) and a small FPR to avoid FP events. AUC score measures how well the supervised learning model correctly classifies a positive random sample as positive (high TPR) and a negative random sample as negative (small FPR). The AUC evaluates the probability that a classifier will classify a random positive sample "higher" than a random negative one. This can be done by calculating the area under the ROC curve (Fig. 7.16) as follows [111]:

$$AUC = \int_0^1 TPR\left(FPR^{-1}(x)\right)dx = P(x_1 > x_o), \tag{7.110}$$

where x_1 is the predicted positive event, and x_o is the predicted negative event. An AUC score 1 denotes a perfect performance, whereas an AUC score of 0.5 denotes a poor prediction performance. The F1 score is a weighted average of the precision and recall scores with a value 1 denoting a perfect performance in terms of precision and recall and a value 0 otherwise.

In the binary classification problem, the performance of the classifier can be summarized in the form of a 2 × 2 confusion matrix (Table I), which depicts the performance of a classifier through a 2 × 2 table, where each row represents the predicted outcomes and each column represents the actual outcomes. According to Table 7.1, one can derive useful performance indicators, such as TP, which is the number of actual positive events that were predicted as positive events, FP, which is the number of actual negative events that were predicted as positive, FN, which is the number of actual positive events that were predicted as negative, and TN, which is the number of actual negative events that were predicted as negative. The FP is also referred to as

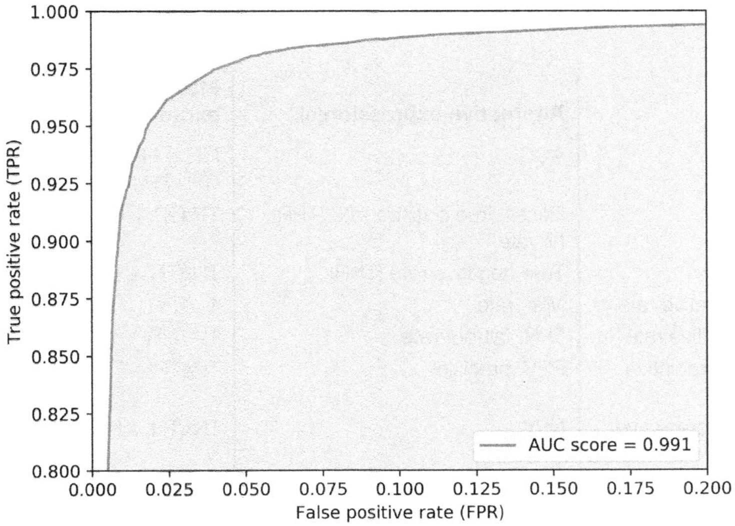

FIGURE 7.16

An illustration of a typical receiver operating characteristic (ROC) curve. The shaded region in gray corresponds to the area under the curve (AUC).

Table 7.1 The structure of the confusion matrix for a binary classification problem.

Population		True events	
		Actual positive event	**Actual negative event**
Predicted events	Predicted positive event	True positive (TP)	False positive (FP)
	Predicted negative event	False negative (FN)	True negative (TN)

a type I error, and the FN is referred to as a type II error. Moreover, the total number of actual positive events is TP + FN, and the total number of actual negative events is FP + TN. In a similar manner, the total number of predicted positive events is TP + FN, and the total number of predicted negative events is FP + TN. The total population is equal to TP + FP + TN + FN. It is obvious that a good classifier shall have high TP and TN values and small FP and FN values.

Using the four different notions of the confusion matrix in Table 7.1, namely the TP, TN, FP, and FN, one can easily define all the above performance evaluation metrics as shown in Table 7.2.

Table 7.2 A summary of the most common performance evaluation measures.

Metric	Alternative expression(s)	Mathematical expression
Accuracy	ACC	(TP + FP)/ (TP+FP+TN + FN)
Sensitivity	Recall, true positive rate (TPR), hit rate	TP/(TP + FN)
Specificity	True negative rate (TNR)	TN/(TN + FP)
False negative rate	Miss rate	1−TPR
False positive rate	FPR, fall-out rate	1−TNR
Positive predictive value	PPV, precision	TP/(TP + FP)
Negative predictive value	NPV	TN/(TN + FN)
F1 score	F-score, F-measure	2 * (TPR * PPV)/ (TPR + PPV)
Descriptive	**Alternative expression(s)**	**Mathematical expression**
Actual positive events	P	TP + FN
Actual negative events	N	FP + TN
Predicted positive events	–	TP + FP
Predicted negative events	–	FN + TN
Total population	–	TP + FP + TN + FN

7.3.5 Unsupervised learning algorithms

Unlike the supervised learning algorithms that use prior knowledge for the development of classifiers and predictors that combine the set of input features to predict or classify outcomes according to a target feature, the unsupervised learning algorithms are not based on labeled (annotated) data to make predictions but focus on grouping (clustering) the input features according to their similarity into a predefined number of clusters. In this section, three fundamental clustering algorithms will be presented with related examples.

7.3.5.1 K-means

A popular clustering algorithm that groups the input samples into k-predefined clusters by minimizing the variance within the clusters is the k-means algorithm [15]. For a given set of n-input features, say $\boldsymbol{x}$, and a predefined number of clusters, say k, the k-means algorithms tries to find k-disjoint sets (partitions), say $C_{opt} = \{C_1, C_2, \ldots, C_k\}$, where $k \leq n$, that minimize the following objective function:

$$C_{opt} = argmin_C \sum_{i=1}^{k} \sum_{\boldsymbol{x} \in C_j} min\left(\|\boldsymbol{x} - \mu_i\|^2\right), \tag{7.111}$$

Algorithm 7.5 A PSEUDOCODE IMPLEMENTATION OF THE k-MEANS.

```
0  define k_means(x, k):
1  select a set of k-random observations from x and set them as centroids,
   assume C = {c_1, c_2, ..., c_k}
2  compute the euclidean distance between and each clustering centroid. j = 1, 2, ...k.
3  assign x_i to the cluster that minimizes (7.110).
4  repeat lines 1–3 until the clustering assignment process does not alter (i.e., convergence) or until
   the maximum number of iterations is met.
5  return C;
```

where μ_i is the mean vector of cluster i and is referred to as the cluster centroid. The inner sum in (7.111) is referred to as the within-cluster sum of squares, which is also referred to as inertia, for cluster i, which implies the total variance of the observation in that cluster. For each cluster, the k-means algorithm tries to find a clustering centroid that is defined as the observation that has the minimum distance from the rest of the observations in the cluster. Then, the algorithm computes the distance of the observation from the clustering centroids and clusters the observation to the cluster whose centroid has the minimum distance from the observation. In the standard implementation of the k-means, the distance metric is usually set to the Euclidean distance. As a result, the k-means algorithm tries to seek for partitions (clusters) with a large variance between them (i.e., a large between-the-cluster variance [BCV]) or alternatively with a small variance within them (i.e., a small within-the-cluster variance [WCV]).

A pseudocode implementation of the k-means algorithm is presented in Algorithm 7.5. The algorithm first selects a random set of k-input samples and initializes them as the clustering centroids. Then, it computes the Euclidean distance of an observation x_i, where $i = 1, 2, \ldots, n$, from each clustering centroid, say c_j, where $j = 1, 2, \ldots, k$, and finally assigns the observation to the cluster whose centroid has the minimum Euclidean distance. The process is repeated by recomputing the clustering centroids and reassigning the observations to the appropriate clusters until no alteration is observed in the clustering assignment process.

An illustration of the k-means convergence is depicted in Fig. 7.17. For illustration purposes, 500 isotropic Gaussian samples were generated on the 2D space for clustering purposes using Python's make_blobs(). The k-means algorithm was applied by setting $k = 3$. On each iteration, the centroid locations are depicted using the symbol "X." A popular variant of the k-means algorithm is the minibatch k-means [112], which reduces the overall computational complexity in the case where the number of clusters is vast by processing the data in the form minibatches, i.e., random subsets of the original data, which are randomly selected on each iteration. The centroid update process is then applied on each minibatch by computing the global average of the samples in the minibatch that were assigned to cluster i including the previous samples that were assigned to the same cluster. Although the minibatch approach reduces the computational cost of the k-means algorithm and the overall convergence time of the clustering procedure, its clustering performance is much worse than the k-means due to the randomness that affects the inertia.

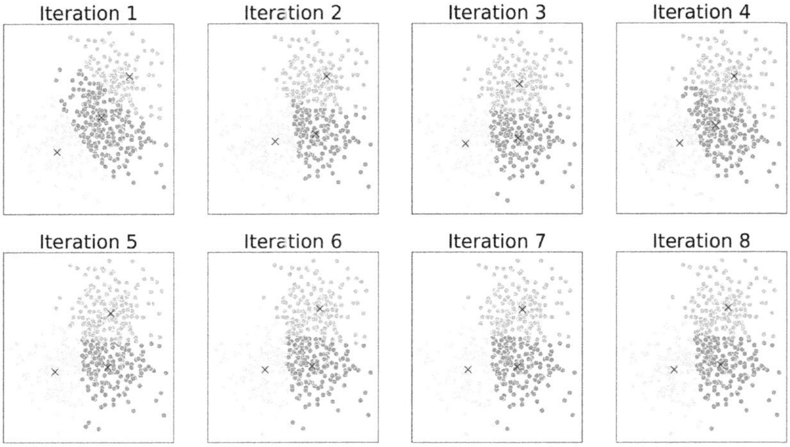

FIGURE 7.17

An illustration of the clustering centroids across eight iterations of the k-means algorithm, where $k = 3$. On each iteration, the centroids are depicted through the symbol "X."

7.3.5.2 Spectral clustering

Despite its name, spectral clustering [16] uses concepts from graph theory to cluster the input samples by minimizing the similarity between the clusters and maximizing the similarity within the clusters. To do so, spectral clustering does not apply the clustering process directly on the original data but first projects the original data into a different dimensional space, where the definition of clusters is easier. This is done by first computing the affinity matrix of the original data and then transforming the affinity matrix into its Laplacian form. Then, eigenvalue decomposition (EVD) is applied on the Laplacian matrix to lower the dimensional space and make the separation process easier. Given a set of n-input features, assume $\boldsymbol{x}$, the affinity matrix, say$\boldsymbol{W} \in R^{nxn}$ is a symmetric matrix:

$$W = (w_{ij}), i,j = 1,2,...,n, \tag{7.112}$$

where the element w_{ij} is the similarity between the input features i and j. The similarity measure that is used to calculate each element in the affinity matrix can be the Euclidean distance:

$$w_{ij} = \sum_{i=1}^{n} \sum_{j=1}^{n} |x_i - x_j|^2, \tag{7.113}$$

or other related measures such as the RBF kernel [76] that we have already presented in Section 7.3.2.2. The similarity matrix can be defined as , so that points with less distance between each other can be considered similar. Once the affinity matrix is constructed, its Laplacian matrix is defined as in Ref. [16]:

$$\boldsymbol{L} = \boldsymbol{D} - \boldsymbol{W}, \tag{7.114}$$

where $\boldsymbol{D}$ is an nxn diagonal matrix, where the ith diagonal element is given by

$$d_{ii} = (D)_{ii} = \sum_{j=1}^{n} w_{ij}, \tag{7.115}$$

yielding an alternative mathematical formulation of (7.114):

$$L = (L_{ij}) = \begin{cases} d_i, & i = j \\ -w_{ij}, & i \neq j \\ 0, & o.w. \end{cases} \tag{7.116}$$

The Laplacian matrix is a positive semidefinite matrix that offers a quadratic way of representing the original space. The keyfactor of spectral clustering is to apply EVD on L and cluster its eigenvectors into a predefined number of clusters. The eigenvectors of the Laplacian matrix have the interesting characteristic that they are smooth in areas where the data points are dense (small distance, large similarity) and vary in areas where the data points are less dense (large distance, small similarity).

Given a Laplacian matrix $\boldsymbol{L}$, EVD of $\boldsymbol{L}$ is given by

$$\boldsymbol{L} = \boldsymbol{UPU}^{-1}, \tag{7.117}$$

where $\boldsymbol{U}$ is an nxm matrix where each column represents an eigenvector, $\boldsymbol{P}$ is a mxm diagonal matrix with the eigenvalues in its main diagonal, and$\boldsymbol{U}^{-1}$ is the Moore–Penrose pseudoinverse of $\boldsymbol{U}$. Once the matrix $\boldsymbol{U}$ is estimated, the k-means algorithm is applied on the k-largest eigenvectors, say $\boldsymbol{u}_1, \boldsymbol{u}_2, \ldots, \boldsymbol{u}_k$, to assign each row of the k-largest eigenvectors, say $\boldsymbol{u}_1^k, \boldsymbol{u}_2^k, \ldots, \boldsymbol{u}_m^k$ into a cluster C_j, $j = 1, 2, \ldots, k$. Then, turning back into the original space, each observation $\boldsymbol{x}_i$ is assigned to the cluster C'_j if the corresponding row $\boldsymbol{u}_i^k$ is assigned to cluster C_j as follows [16]:

$$C'_j = \{\boldsymbol{x}_i | \boldsymbol{u}_i^k \in \mathrm{C_j}\} \tag{7.118}$$

An example of the difference between the k-means and the spectral clustering is depicted in Fig. 7.18. For this experiment, 5000 samples were randomly generated on a 2D space using Python's make_blobs() function [113] with additional noise levels. The purpose of this experiment is to cluster the two concentric patterns into two clusters, where the concentric pattern must be retained. It is obvious that the spectral clustering is able to identify the two patterns through the gradient information that is given by the Laplacian matrix, whereas the simple k-means implementation fails to identify them due to the projection in the Euclidean distance space, which is not able to discriminate the concentric pattern.

Other variants of the spectral clustering include the N-cut algorithm [114], which solves a graph partitioning problem based on the Laplacian matrix by minimizing the N-cut cost, and the spatially constrained spectral clustering, which introduces a spatial threshold to reduce the generation of noncontingent clusters [115], both having numerous applications in the field of neuroimaging toward the identification

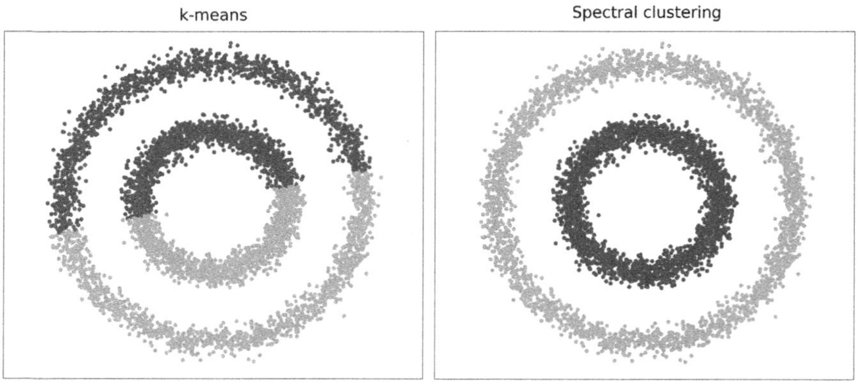

FIGURE 7.18

Detecting a concentric pattern using the k-means (left) and spectral clustering (right).

of brain regions with common activation patterns under resting state or different tasks (e.g., motor movement) [115] and the automated segmentation of tumor areas in the brain [116].

7.3.5.3 Hierarchical clustering

Hierarchical clustering [17] follows a nested strategy that uses either a bottom-up or a top-down approach to construct a hierarchical data clustering schema that is referred to as a dendrogram, where the root is a cluster with all the observations and the leaves are clusters consisting of only one observation. In the top-down approach (or the divisive approach), all data points are first assigned into one cluster, i.e., the root of the dendrogram, and then the cluster is split into pairs of similar clusters until the predefined number of clusters is met or until clusters with only one sample are created. On the other hand, in the bottom-up approach (or the agglomerative approach), each data point is assigned into its own cluster, and then similar pairs of clusters are merged (linked) until the predefined number of clusters is met or until a cluster with all the samples is created. The depth of the dendrogram is related to the number of predefined clusters.

In practice, the agglomerative approach is used where a linkage function is used to link similar clusters in a bottom-up direction. Examples of simple linkage functions that are widely used include (i) the Euclidean (or squared Euclidean) distance, (ii) the Mahalanobis distance, and (iii) the Manhattan distance. For two n-dimensional input vectors, assume $\boldsymbol{x}$ and $\boldsymbol{y}$, the Mahalanobis distance [117] is given by

$$M1(\boldsymbol{x},\boldsymbol{y}) = \sqrt{(\boldsymbol{x}-\boldsymbol{y})^T \boldsymbol{\Sigma}^{-1}(\boldsymbol{x}-\boldsymbol{y})}, \tag{7.119}$$

where $\boldsymbol{\Sigma}$ is the covariance matrix of $\boldsymbol{x}$ and $\boldsymbol{y}$, and the Manhattan distance [118] is given by:

$$M2(\boldsymbol{x},\boldsymbol{y}) = \sum_{i=1}^{n} |x_i - y_i| \tag{7.120}$$

Once a linkage function is specified, the linkage criterion is defined to determine the linkage strategy that will be used to merge similar pairs of clusters. The most common method is to either set a distance threshold on the previous scores to merge a cluster with the one that has the smallest distance (i.e., the nearest point algorithm) or the smallest average distance across all observations (average linkage) or even adopt probabilistic approaches, such as the v-linkage to merge clusters that tend to follow the same distribution [119]. A popular linkage criterion in agglomerative clustering is the Ward's method, which determines the pairs of clusters to be merged by solving the minimum variance problem through the Lance–Williams method [120]. Given a pair of candidate clusters to be merged, say (C_a, C_b), against a cluster C_d, the cluster distance is updated by recursively solving the following objective function:

$$d((C_a, C_b), C_d) = \frac{n_a + n_d}{n_a + n_b + n_d} d(C_a, C_d) + \frac{n_b + n_d}{n_a + n_b + n_d} d(C_b, C_d) - \frac{n_d}{n_a + n_b + n_d} d(C_a, C_b), \tag{7.121}$$

where n_a, n_b, n_d are the total number of observations in clusters C_a, C_b, C_d, respectively, and d_{ad}, d_{bd}, d_{ab} are the Euclidean distances of the observations between the clustering pairs (C_a, C_d), (C_b, C_b), and (C_a, C_b), respectively. An example of a dendrogram is depicted in Fig. 7.18. For this experiment, 20 random samples were generated Python's make_blobs() [113]. The k-means and the hierarchical clustering (using the Ward linkage method) algorithms were then applied by setting a number of 4 predefined clusters. The clustering regions that were identified by the two algorithms were the same (Fig. 7.18A and B). The resulting dendrogram of hierarchical clustering is depicted in Fig. 7.19C and presents the four different clusters in aqua, blue, red, and green colors, where the horizontal axis corresponds to the distance scores and the vertical axis to the data points in Fig. 7.19B.

7.3.5.4 Other data clustering approaches

Several other data clustering attempts have been made to enhance the computational efficiency of clustering including the DBSCAN (density-based spatial clustering of applications with noise) algorithm [121], which focuses on grouping data points that fall within a radius (neighboring data points) in terms of high density. The DBSCAN algorithm uses this density-based condition to also detect outliers. Other data clustering attempts include the Affinity Propagation algorithm [122], which establishes a "communication network" between the data points, where the data points "communicate" with each other, and then identifies those data points having the highest participation in the communication network as exemplars. The communication process is based on whether each data point will choose another data point as its exemplar based on a similarity score. Because this communication process requires a large number of communication links, its computational complexity is large. Several attempts have been made to reduce this complexity using sparse density matrices making it more suitable to small- or medium-scale datasets [123].

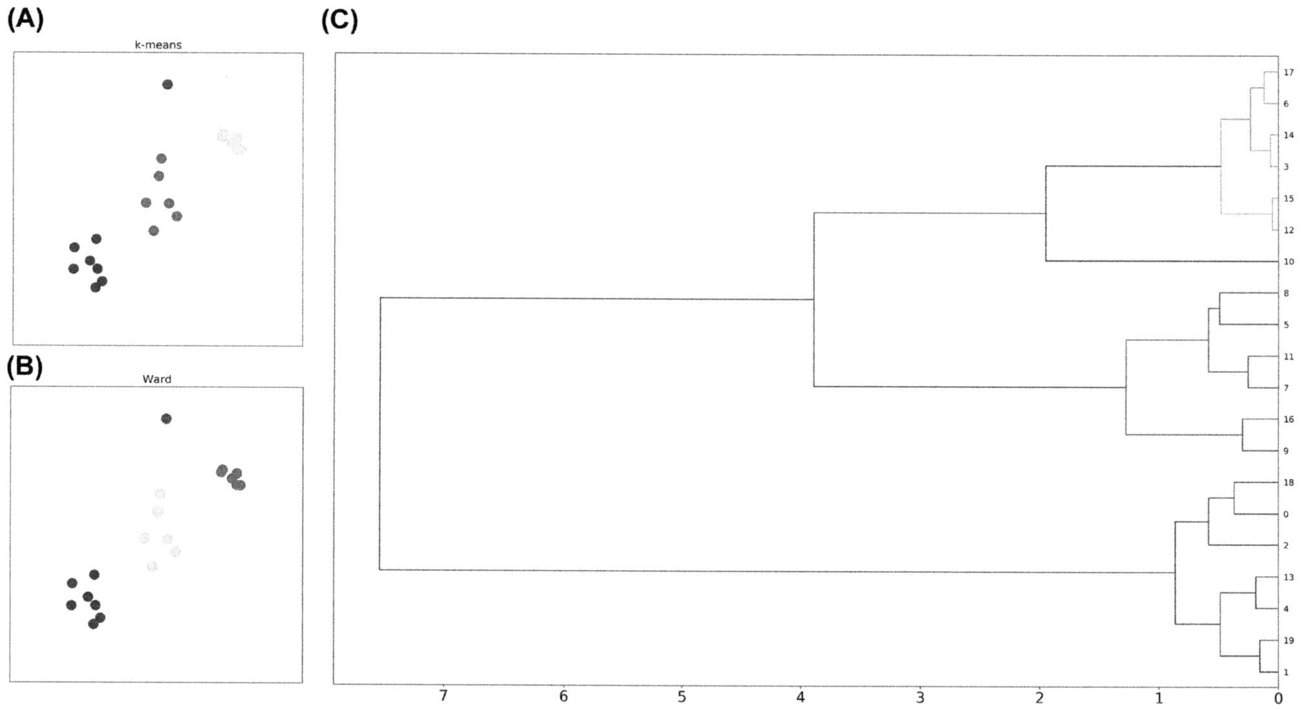

FIGURE 7.19

An illustration of a typical dendrogram that is used in hierarchical clustering.

7.3.5.5 Clustering performance evaluation

The main question here is how to estimate the optimal number of clusters, k, which will result in the best set of partitions for a clustering problem. Toward this direction, two strategies have been proposed. The first strategy uses prior knowledge regarding the true assignment of each observation to evaluate a set of clusters to quantify the level of "agreement" between them. The second strategy does not use any prior knowledge regarding the true labels and focuses on the identification of the optimal number of clusters in terms of homogeneous clusters.

Regarding the first type of strategy, several similarity metrics have been proposed to quantify the difference between the assigned clustering labels, say $\widetilde{C}$, and the true clustering labels, say C. An example of a widely used metric is the Rand index (RI) [124]:

$$RI = \frac{n+m}{C_2^N}, \tag{7.122}$$

where n is the number of pairs of observations that exist both in C and $\widetilde{C}$, m is the number of pairs of observations that have been assigned to different clusters in C and $\widetilde{C}$, and C_2^N is the total number of possible pairs of observations in the original dataset. Thus, the RI measures the level of "agreement" between the true set C and the assigned set $\widetilde{C}$. To overcome some bias effects that are introduced due to the randomness, the chance factor needs to be taken into consideration. Toward this direction, the adjusted RI (ARI) [125] was proposed to normalize the RI in the range [-1, 1], where 1 is denotes a precise match and -1 otherwise:

$$ARI = \frac{RI - E[RI]}{\max(RI) - E[RI]}, \tag{7.123}$$

where $E[RI]$ is the expected RI [125]. In addition, entropy-based measures have been also proposed to quantify the agreement between C and $\widetilde{C}$, such as the adjusted and the normalized mutual information (NMI) scores that quantify the information entropy that is shared between the sets C and $\widetilde{C}$. NMI is defined as follows [126]:

$$NMI = \frac{MI(C, \widetilde{C})}{mean(H(C), H(\widetilde{C}))}, \tag{7.124}$$

where $MI(C, \widetilde{C})$ is the mutual information between the sets C and $\widetilde{C}$, and $mean(H(C), H(\widetilde{C}))$ is the average entropy in the sets C and $\widetilde{C}$. The MI does not take into consideration the chance factor, such as the ARI.

Meanwhile, the adjusted MI can be also defined in this case as in

$$AMI = \frac{MI - E[MI]}{mean(H(C), H(\widetilde{C})) - E[MI]}, \tag{7.125}$$

where $E[MI]$ is the expected value as in Ref. [126]. Values close to 0 denote a large independence between the estimated and the true labels, whereas values close to 1 denote an increased level of "agreement." For a quick review on the MI and conditional entropies, see Section 7.2. Alternative methods that quantify the characteristics of the clusters include other entropy-based measures, such as the homogeneity score and the completeness scores. The homogeneity score [127], assume h, is defined as

$$h = 1 - \frac{H(C|\tilde{C})}{H(C)}, \tag{7.126}$$

and measures the level of "agreement" between C and $\tilde{C}$ in terms of the conditional entropy of C given the $\tilde{C}$, $H(C|\tilde{C})$, in the case where each cluster contains single class observations. On the other hand, the completeness score [128], assume c, is a similar measure which is defined as

$$c = 1 - \frac{H(\tilde{C}|C)}{H(\tilde{C})}, \tag{7.127}$$

and measures the level of "agreement" between $\tilde{C}$ and C in terms of the conditional entropy of $\tilde{C}$ given the C, $H(\tilde{C}|C)$, in the case where all the observations of a given class are assigned to the same cluster. A homogeneity value 1 denotes perfectly homogeneous clusters, whereas a homogeneity value 0 denotes the exact opposite. The same stands for the completeness score. No information, however, is given regarding the topological characteristics of the clusters and the chance factor (labeling randomness) especially in the case where the number of samples is small.

All the previous types of measures, however, are trivial as they require the true set of labels and do not provide a complete insight regarding the optimal number of clusters in the absence of the ground truth. So, what happens in the case where the true set of labels is unknown? Toward this direction, three measures are widely used in practice, including (i) the silhouette coefficient, (ii) the Davies–Bouldin (DB) index, and (iii) the Calinski-Harabasz (CH) index, where each similarity metric has its own value. For a given clustering set C, the silhouette index (SI) is defined as [129]

$$SI(C) = \frac{m - n}{max(m, n)}, \tag{7.128}$$

where (i) m is the average distance between an observation that belongs to cluster C_j, assume $x_i \in C_j, i = 1, 2, \ldots, n$, with its neighboring observations in that same cluster, i.e., $x_p \in C_j, p = 1, 2, \ldots, n$, where $p \neq j$, and is called the between-the-cluster distance, and (ii) n is the average distance between the same observation, x_i, with its neighboring observations from the next nearest cluster, i.e., $x_p \in C_q, p = 1, 2, \ldots, n$, $q = 1, 2, \ldots k$, and is called the within-the-cluster distance. The silhouette value is evaluated on a predefined number of clusters, say $1, 2, \ldots, k$, and the clustering number that achieves the highest silhouette score is the one that yields well-separated clusters in terms of density. Here, the term "well-separated clusters" refers to clusters with small between-the-cluster distance and large within-the-cluster distance. A silhouette value 1 denotes a highly dense and well-separated cluster, whereas a silhouette value -1 denotes the exact opposite. Values close to 0 denote the existence of overlaps between the resulting clusters.

DB index [130] measures the clustering similarity based on the average distance between the clustering centroids of a cluster with its most similar one from the clustering set. More specifically, for a given clustering set C, the DB index is defined as

$$DB(C) = \frac{1}{k}\sum_{i=1}^{k} max(R_{ij}), \tag{7.129}$$

where k is the number of clusters, and R_{ij} is the similarity score between cluster C_i and cluster C_j, where $j \neq i$ and $i, j = 1, 2, \ldots k$, which is given by

$$R_{ij} = \frac{d_i + d_j}{d_{ij}}, \tag{7.130}$$

where d_i is the average distance between the observations in cluster C_i from the cluster's centroid, d_j is the average distance between the observations in cluster C_j from the cluster's centroid, and d_{ij} is the distance between the centroids in clusters C_i and C_j. In practice, DB values close to 0 indicate a good clustering performance as, in that case, the average distance between the clustering centroids is large, which implies that the clusters are well-separated. The DB index is evaluated on a predefined number of clusters, say $1, 2, \ldots, k$, and the clustering number that achieves the lowest DB score is the one that yields well-separated clusters in terms of large distance between the clustering centroids.

The CH index [131] is an alternative clustering performance evaluation score that measures the BCV and WCV to quantify the cluster separation outcome. For a predefined number of clusters, assume k, BCV is given by:

$$BCV(k) = \sum_{i=1}^{k} n_i \|c_i - \mu\|^2, \tag{7.131}$$

where n_i is the number of samples in cluster C_i, c_i is the clustering centroid of C_i, μ is the mean of the samples, and $c_i - \mu^2$ is the Euclidean distance between c_i and μ. WCV is given by

$$WCV(k) = \sum_{i=1}^{k} \sum_{x \in C_i} x - c_i^2, \tag{7.132}$$

where c_i is the clustering centroid of cluster C_i, and $x - c_i^2$ is the Euclidean distance between the input samples x and the centroid c_i of cluster C_i.

Finally, the CH index is defined as

$$CH = \left(\frac{n-k}{k-1}\right) \frac{BCV(k)}{WCV(k)}, \tag{7.133}$$

where high CH scores denote well-separated clusters and 0 otherwise. Note that the CH index is also referred to as the variance ratio criterion.

7.4 Distributed learning

So, how can we develop supervised and unsupervised learning models for predicting disease outcomes and detecting similar patterns in the case where the data are stored in distributed environments (e.g., in different locations in the cloud)? Here, the term "distributed data" refers to clinical data that are stored in distributed clinical databases or private spaces in the cloud or decentralized clinical databases. In this section, two candidate strategies for developing supervised and unsupervised learning models across distributed clinical databases will be presented and compared, including the (i) incremental learning and (ii) stacked generalization (or stacking) using metalearners. To be able to describe these two methods, under the distributed learning schema, we will assume that the data that are stored in different locations are harmonized (i.e., have the exact same set of features) otherwise none of these methods are meaningful. In addition, the value of distributed learning is high in the case where a centralized database model is not feasible either due to the underlying legal and security threats [2,6] or due to the massive amounts of data.

The incremental learning strategy is a straightforward approach that focuses on updating a machine learning model on constantly upcoming data streams or batches. According to its original definition, incremental learning was designed as an approach to mine knowledge from big data that are not able to fit into the physical memory units of an average computer through a process known as out-of-core learning [132]. To do so, the big data are first divided into multiple subsets or batches, and the training process is then applied in a sequential manner. Assuming a big dataset $\boldsymbol{X}$ with size mxn, where m is the number of instances (rows) and n is the number of features (columns), the batches are formed by horizontally partitioning $\boldsymbol{X}$ into a predefined number of p-subsets, say $\boldsymbol{X}_1, \boldsymbol{X}_2, \ldots, \boldsymbol{X}_p$, where each subset (batch) has a size equal to m/p. For example, in the case where the dataset has 1,000,000 instances and 100,000 features, value p equal to 10 would yield 10 batches with a size equal to 100,000 per batch. What incremental learning does is that it first constructs a machine learning model that is trained on batch $\boldsymbol{X}_1$, say M_1, and then proceeds by sequentially updating M_1 on batch $\boldsymbol{X}_2$, reupdating M_2 on batch $\boldsymbol{X}_3$, and so on, until all the batches are used for training. It is then proven that the incremental update process yields a machine learning model M_p, which is equivalent to the "global" machine learning model $\boldsymbol{M}$, that is produced by training the machine learning algorithm on the whole dataset, $\boldsymbol{X}$ [28,36].

Based on the definition of incremental learning, we can extend its basis to support the development of machine learning models across data that are stored in different locations. Then, the problem can be formulated as follows: given a set of N-sites, say $\boldsymbol{T}_1, \boldsymbol{T}_2, \ldots, \boldsymbol{T}_N$, with a dataset existing on each site, say $\boldsymbol{X}_1, \boldsymbol{X}_2, \ldots, \boldsymbol{X}_N$, we first train a machine learning model $\boldsymbol{M}_1$ on the dataset $\boldsymbol{X}_1$ in site $\boldsymbol{T}_1$ and then update the model through the update function:

$$F(x) = F(x-1) + \beta h(x), \tag{7.134}$$

where $F(x)$ corresponds to the estimated mapper that is trained on the dataset in site $\boldsymbol{T_x}$, $F(x)$ corresponds to the estimated mapper that was trained on the dataset in the previous site $\boldsymbol{T_{x-1}}$, where $x \leq N$, and $\beta h(x)$ is the learner function on the dataset in site $\boldsymbol{T_x}$. As we have already discussed, the estimated mapper is usually a weight update function. The majority of the supervised and unsupervised machine learning schemas that were presented in Section 7.3 can be extended to support incremental learning. Indeed, the majority of the supervised learning algorithms use the SGD approach to minimize a loss function by solving the following generalized weight update process:

$$w(x) = w(x-1) - \beta(\nabla_w L(f(x_i), y_i) + a\nabla_w r(w)), \tag{7.135}$$

where $L(f(x_i), y_i)$ is a loss function given a score function $f(x_i)$ and a target y_i, $\nabla_w L(f(x_i), y_i)$ is the gradient of the loss function with respect to w, $r(w)$ is a regularization function, $\nabla_w r(w)$ is the gradient of the regularization function with respect to w, a is a hyperparameter, and β is a learning rate parameter. It is obvious that (7.15) is similar to the sequential update process in (7.14). By setting the loss function in (7.15) to (i) the logistic loss, we can obtain the logistic regression classifier, (ii) the hinge loss, we can obtain the SVM classifier, (iii) the linear loss, we can obtain the linear regression classifier, and (iv) the perceptron loss, we can obtain the perceptron (and MLP) classifier. The same stands for the tree ensembles that were described in Section 7.3.2.4. Regarding the Naïve Bayes models, the multinomial and Bernoulli models use an incremental version of the maximum likelihood estimation [80]. Regarding the unsupervised learning algorithms, the minibatch version of the k-means algorithm can be applied [112].

In general, incremental learning shares many similarities with the online learning approach [29,31,32], which updates an existing machine learning model on an upcoming set of batches by minimizing a global cost function. Examples of online learning algorithms include the SVMs [12], hybrid online learning methods based on nonlinear kernels [13], and SGD-based algorithms for convex optimization [11,14], among others. However, as its name implies, online learning is only executed in an "online" manner, where the training samples are constantly arriving. Incremental learning, on the other hand, can be seen as an "offline" approach, where the batches are already stored in different locations.

An alternative but rather ambiguous schema is stacked generalization [40,41]. Stacked generalization is based on the fact that classification outcomes of different machine learning algorithms can be combined by a metalearner, which learns from the outcomes through the "learning to learn" concept [40,41]. More specifically, according to the stacked generalization schema, a machine learning algorithm is applied on each one of the p-batches, yielding p-different machine learning models, $\boldsymbol{M_1}, \boldsymbol{M_2}, \ldots, \boldsymbol{M_p}$, each one with its own classification outcomes. The individual outcomes of the machine learning models are then combined by a metalearner to produce the final outcome. The machine learning algorithm that is applied on each batch can be different (e.g., logistic regression of the first batch, RF on the second batch, etc.). A metalearner [133] is a machine learning algorithm that is trained on the individual classification outcomes of two or more machine learning algorithms that were trained for a specific task to yield the final classification outcome. The overall scope of metalearning is vague as the combination of classification outcomes from different classifiers introduce biases during the assembly state. The difference between incremental learning and stacked generalization is depicted in Fig. 7.20. In the incremental learning approach, the machine learning algorithm is applied on the dataset $\boldsymbol{X}_1$ yielding model $\boldsymbol{M}_1$, which is reupdated on the dataset $\boldsymbol{X}_2$ yielding model $\boldsymbol{M}_2$, then reupdated on the dataset $\boldsymbol{X}_3$ yielding model $\boldsymbol{M}_3$, and finally reupdated on the dataset $\boldsymbol{X}_N$ yielding model $\boldsymbol{M}_N$, which outputs the outcome $\boldsymbol{Y}$. Each dataset $\boldsymbol{X}_1, \boldsymbol{X}_2, \boldsymbol{X}_3, \ldots, \boldsymbol{X}_N$ is stored in the site $\boldsymbol{T}_1, \boldsymbol{T}_2, \boldsymbol{T}_3, \ldots, \boldsymbol{T}_N$, respectively. In the stacked generalization case, a different (or common) machine learning algorithm (level 0 classifiers) is applied on $\boldsymbol{X}_1, \boldsymbol{X}_2, \boldsymbol{X}_3, \ldots, \boldsymbol{X}_N$, yielding the individual models $\boldsymbol{M}_{01}, \boldsymbol{M}_{02}, \boldsymbol{M}_{03}, \ldots, \boldsymbol{M}_{0N}$, where each individual model provides an outcome $\boldsymbol{Y}_1, \boldsymbol{Y}_2, \boldsymbol{Y}_3, \ldots, \boldsymbol{Y}_N$. The outcomes are then combined into $\boldsymbol{Y}$ through a metalearner (level 1 classifier).

A pseudocode implementation of the incremental learning process across N-sites is presented in Algorithm 7.6. The algorithm uses as input the number of sites, say N, and the supervised or unsupervised learning model that will be used for classification or clustering purposes, respectively, say $\boldsymbol{M}_o$. Then, the model $\boldsymbol{M}_o$ is trained on site 1 yielding the model $\boldsymbol{M}_1$. The model's weights are incrementally updated on site $\boldsymbol{X}_{i+1}$ yielding the model $\boldsymbol{M}_{i+1}$ and the update process is repeated $N-1$ times. The final model $\boldsymbol{M}_{N-1}$ is then returned as the incrementally trained model. The incremental learning strategy is by far more robust against stacked generalization where the aggregation of individual classification outcomes yield biases in the predicted outcomes. Moreover, in stacked generalization, each individual model provides local information about the training process, yielding outcomes with limited "horizon" instead of incremental learning where the model is sequentially updated on the "whole" data. However, because of its underlying "nature," incremental learning is a sequential process that evolves over time and thus requires more amount of time than parallel learning, especially in the case where the number of sites is large (e.g., $N > 100$) and/or when the individual datasets on each site are large scale. A solution to this can be offered by a properly constructed computing engine (e.g., a clustering computer or a virtual machine [VM] or server) that will be responsible for the coordination of the sequential model update process, where the time for the establishment of the communication links between the computing engine and each remote site can be significantly reduced through enhanced networking throughput services for transferring information at high network speed.

7.5 Visual analytics

Data visualization involves the mapping of various data types, such as integers, floating numbers, strings, etc., into visual components, such as dots, lines, bars, etc. The overall concept of visual analytics is based on the human perception.

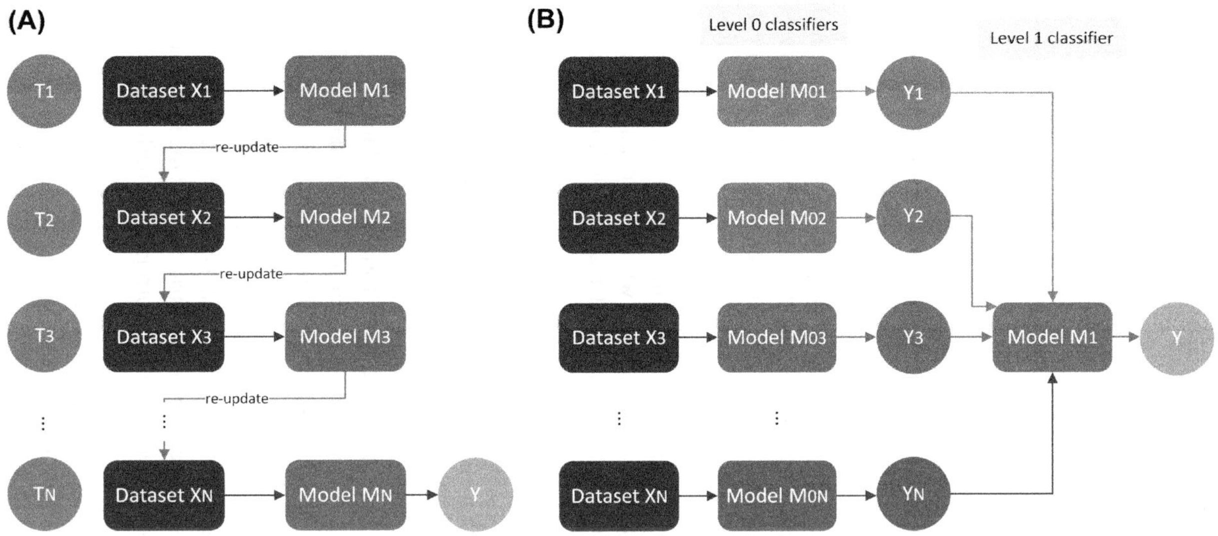

FIGURE 7.20

An illustration of the difference between incremental learning and stacked generalization for solving classification problems across multiple, dispersed datasets in *N*-remote locations (sites).

Algorithm 7.6 A PSEUDOCODE IMPLEMENTATION OF INCREMENTAL LEARNING ACROSS *N*-SITES.

```
define incremental_learning(N, M_0):
train M_0 on dataset X_1 in location 1 yielding the model M_1
for i = 1 : N − 1 do:
    compute the updated weights according to (7.135)
    update M_i on dataset X_{i+1} in location i + 1 yielding model M_{i+1} using (7.134)
    store the model M_{i+1} for updating on the next location
return M_{N−1};
```

Different visual analytics exist including line plots, bar plots, scatter plots, heatmaps, and network plots. The bar plots are useful for displaying the input values into the form of rectangular type of bars, where for a given feature $\boldsymbol{x}$, the total count of values (vertical axis) across different data ranges or intervals (horizontal axis) can be observed. Different variations of the bar plots currently exist in the literature, including the pie chart, which represents the aforementioned intervals in the form of pies, the histogram, which represents the frequency in the vertical axis instead of the count value, and the stacked bar chart, which stacks the values of multiple features across different data intervals on the same bar plot. The dot plots are the simplest form of data visualization, where the values of $\boldsymbol{x}$ are presented as points in the one-dimensional space. In the case where a second dimension exists, e.g., when the values of two features, say $\boldsymbol{x}$ and $\boldsymbol{y}$ are plotted, then the plot is called a scatterplot.

The line plot is a special case of the dot plot, where the points in the space are connected to formulate a line. This is useful especially when someone wishes to examine the distribution of the feature's values across a specific time interval (e.g., the electrical activity of the heart for a specific amount of time). The heatmaps are widely used to examine associations between two or more features. In a heatmap plot [43], the information is presented in the form of a rectangular matrix, where the cell (i,j) represents the association between features i and j using an appropriate color code, where the hue depends on the level of association between them, e.g., low, medium, or high. A heatmap, however, can be also visualized in the case of a single feature, where the cell (i,j) represents the association between the scale i (e.g., low, medium, high) and the data interval j of the input feature. This association is usually computed through a similarity metric, such as the inverse of the Euclidean distance or Pearson's product moment correlation coefficient with high values denoting features with strong associations (i.e., high similarity).

Because a heatmap depicts the pairwise relationships between two features, it can be projected to the two-dimensional space to yield a graph plot. According to graph theory, a graph [134] is defined as a set $\boldsymbol{G} = (\boldsymbol{V}, \boldsymbol{E})$, where $\boldsymbol{V}$ is the set of nodes and $\boldsymbol{E}$ is the set of edges. In a graph plot, the nodes correspond to the features, and the edge between two nodes represents the association or similarity score between them just like the cell (i,j) in the heatmap. A graph can either be directed on undirected. In a directed graph, the information flows only in one direction, whereas in an undirected graph, the information flows in both directions. This can be seen from the edges. In a directed graph, the edges have a direction, whereas in an undirected graph, the edges do not have a direction indicating a two-way relationship instead of a one-way relationship like in the former type of graph. Furthermore, a graph can be acyclic (loop-less) or cyclic. In an acyclic graph, no loops exist between the nodes (or self-loops), whereas in a cyclic graph loops are formed between the nodes. The importance of graphs in data visualization is crucial. Graphs can be used to identify nodes (features) with a large number of connected edges, which probably denote a hub, i.e., a node that is highly involved during the information flow within the network, as well as identify groups of nodes that are dense denoting a cluster and even detect outliers, i.e., nodes that are isolated from the rest of the nodes in the graph. Examples of graph applications include the detection of hubs in EEG and MEG studies [135,136], as well as the extraction of graph-related features that serve as input features for predicting disease outcomes [137]. A summary of all the previous data visualization methods is depicted in Fig. 7.21 using Python's seaborn package [138].

For the histogram, density, line, and dot plots, a random feature $\boldsymbol{x}$ was generated consisting of 200 samples that follow a normal distribution with a 0.8 variance. For the bar plot, the values of $\boldsymbol{x}$ were discretized. For the scatter and box plots, a new random feature $\boldsymbol{w}$ was generated with 0.2 variance. The index label in the line and dot plots indicate the ascending order of the samples in $\boldsymbol{x}$. For the heatmap plot, 10 random features were generated (200 samples per feature) using the Pearson correlation coefficient as a similarity measure. Regarding the graph plot, Python's Networkx package [139] was used to generate a random graph with 200 nodes and radius 0.4. The color of the node depends on the distance from the central node.

7.6 Popular machine learning frameworks and libraries

7.6.1 Apache Spark

Apache Spark [140] is a data analytics engine that provides a complete framework for large-scale data analytics. It provides highly flexible API services for (i) applying queries on structured data schemas based on the SQL (Structured Query Language) and the DataFrame technologies (Spark SQL) [141], (ii) building computationally efficient graphs for data management (GraphX) [142], (iii) streaming workflows for batch processing applications (Spark Streaming), and (iv) building machine learning models (MLib) [143]. The latter is a scalable machine learning algorithm that is interoperable with the Python and R programming languages. The machine learning algorithms that are currently supported by the Spark MLlib include both supervised and unsupervised machine learning algorithms. Examples of supervised machine learning algorithms include linear and logistic regression, Naïve Bayes, RFs, and gradient boosted trees, among others. Examples of unsupervised learning

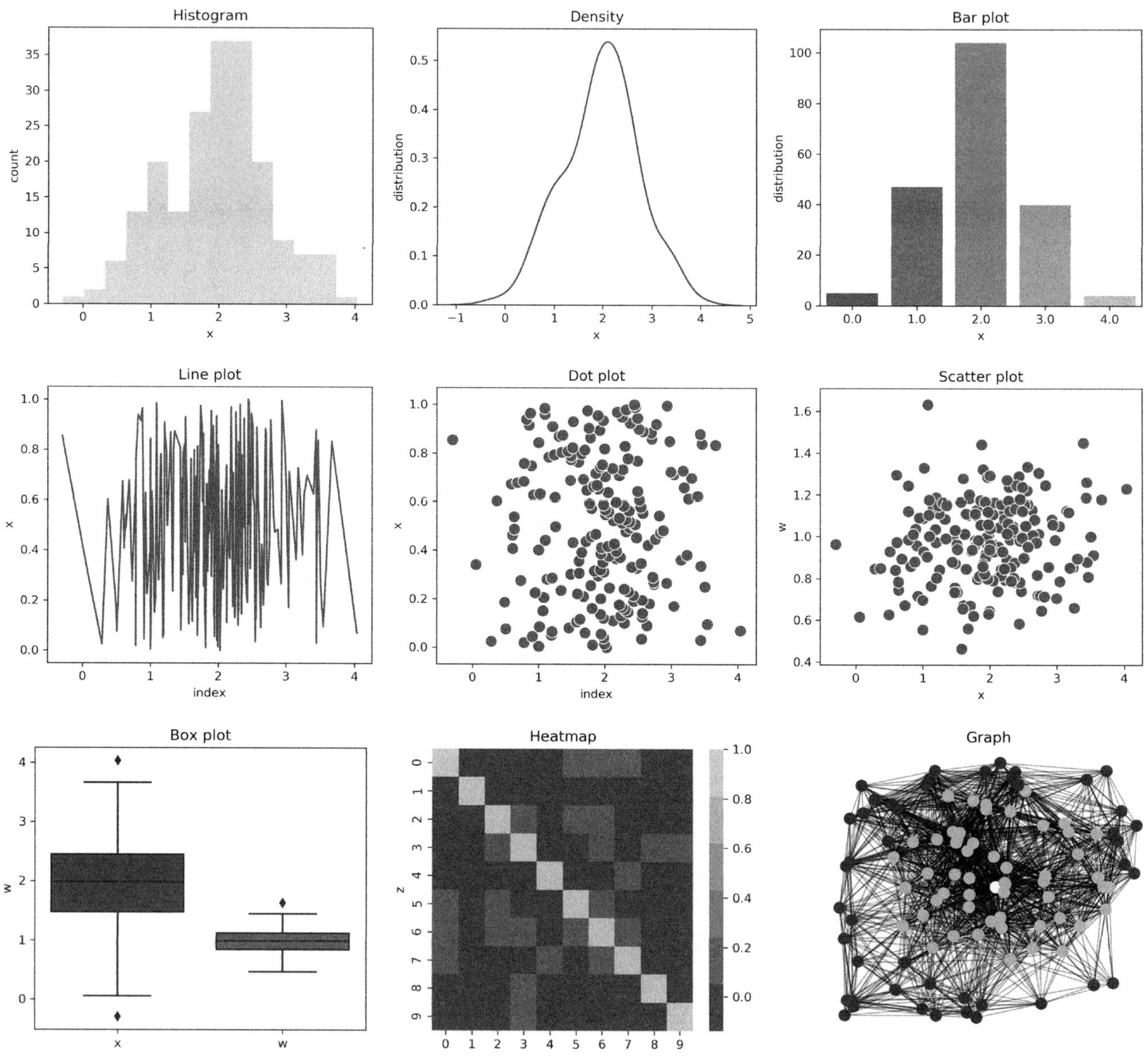

FIGURE 7.21

A summary of different types of data visualization methods.

algorithms include the k-means and related variants. The interesting thing about Spark MLib is the fact that is can be executed on Apache's Hadoop [144], a popular framework that offers the basis for the application of distributed analytics applications, where the large-scale input data are divided and distributed across a large number of available computer clusters (or servers).

The main idea behind Hadoop is based on separating a single workflow task (job) on a large-scale dataset into smaller workflows (jobs), which are executed in parallel on distributed subsets of the large-scale datasets that are stored into different computer clusters (engines). Toward this direction, the Apache Hadoop framework consists of four fundamental modules, namely the (i) Hadoop Common [145], which initializes the Hadoop services, (ii) Hadoop Distributed File System [146], which distributes the large-scale input dataset that is stored in a Hadoop database into subsets that are placed across the computer clusters using the Apache Hive [147] for management, (iii) Hadoop YARN (Yet Another resource Negotiator) [148], which schedules the jobs across the clusters, and (iv) Hadoop MapReduce [149], which enables the parallel processing across the clusters.

7.6.2 Scikit-learn

Scikit-learn [72] is an open-source machine learning library that is developed in Python and is distributed under the Berkley Software Distribution license. It provides a large variety of supervised learning and unsupervised learning algorithms for regression, prediction, clustering, and dimensionality reduction problems, along with data preprocessing and performance evaluation methods, as well as efficient resource management methods and strategies to computationally scale big data. Examples of supervised learning algorithms that are offered by the scikit-learn include the SVMs (e.g., linear SVM), decision trees and RFs, Naïve Bayes, neural networks (and multilayer neural networks), and ensemble methods, such as the AdaBoost and the gradient boosting classifier. Examples of unsupervised learning algorithms for clustering include the k-means (and minibatch k-means), spectral clustering, and hierarchical clustering, among others. Moreover, the scikit-learn offers methods for feature selection and dimensionality reduction, such as the information gain, the gain ratio, the principal component analysis, and for data preprocessing, such as data imputation, data discretization, data normalization, and data encoding. The performance evaluation strategies that are offered by the library include all the methods and metrics that were presented in Section 7.4.5. Finally, the scikit-learn supports incremental learning solutions for processing big data in the form of batches, as well as parallel and distributed computing using configuration switches and efficient resource management. Most of the scikit-learn functionalities have been also combined into an open-source framework known as Orange [150], which offers graphical user interface (GUI) that allows for the high-level construction of machine learning and data mining workflows, as well as for the execution of data visualization tasks. Orange is also provided in the form of a low-level Python library for further manipulation by the programmers.

7.6.3 Keras

Keras [46] is an open-source, easily scalable, deep learning software library, which has been developed in the Python programming language. It supports data mining using deep learning algorithms that are offered in the form of APIs based on neural networks with multiple hidden layers, such as CNNs and RNNs, including the popular LSTM implementation. To so, keras is executed on top of TensorFlow [151], a machine learning library that enables the mathematical computations of large workloads using graphs, where the nodes represent the mathematical formulations and the edges represent the information flow between them in the form of multidimensional data structures, which are known as tensors. TensorFlow provides numerous APIs that support multiple programming languages, such as Python, C, and Java. Its recent version Tensorflow Extended [152] goes one step further by providing the basis for the implementation of scalable machine learning pipelines that are able to deal with multiple components to solve high-performance machine learning workloads. The keras large-scale workloads can be also executed into the computer's graphical processing unit (GPU) instead of the CPU to enable faster execution. Furthermore, keras is also interoperable with two other frameworks namely the (i) Theano [153], which is a Python library that allows the definition, evaluation, and optimization of the mathematical expressions within the tensors of a graph, and (ii) Microsoft Cognitive Toolkit (CNTK) [154], which is an open-source deep learning toolkit that offers distributed deep learning algorithms, including CNNs, RNNs, and feedforward directed neural networks. CNTK uses graphs for solving optimization problems that support the parallelization across multiple GPUs and CPUs.

7.6.4 CRAN—machine learning

The Comprehensive R Archive Network (CRAN) [155] provides a large collection of machine learning packages that are implemented in the R programming language. The CRAN machine learning packages include a large number of algorithmic implementations that vary from SVMs and Bayesian models to RFs and boosted tree ensembles. The software packages also include implementations of single and multiple hidden neural networks for deep learning applications. The CRAN packages also include feature selection and performance evaluation methods, as well.

7.6.5 Weka

Another popular toolkit that offers a wide range of machine learning and data mining algorithms is Weka (Waikato Environment for Knowledge Analysis) [156]. Originally developed in Java, Weka includes a variety of tools for feature selection (e.g., gain ratio), data preprocessing (e.g., data normalization), classification (e.g., SVMs, RFs, decision trees, Naïve Bayes), regression (e.g., linear and logistic), and clustering (e.g., *k*-means), among many others. Although Weka is developed in Java, it is also interoperable with Python. In Weka, all the functionalities are offered by a user-friendly GUI that makes the application of data mining tasks easier for an average user.

7.6.6 Google artificial intelligence

Google offers a cloud-based artificial intelligence (AI) deep learning framework [157] that is built on preconfigured VMs that can be used to utilize machine learning and deep learning tasks by invoking the previous frameworks, such as scikit-learn and TensorFlow. The computational power from the VMs is then combined to yield high-performance data analytics with fast execution time through parallel execution. The amount of VMs to be invoked is usually defined by the user in terms of the processing cores that are needed to run a specific machine learning workload. The user is then charged according to the number of cores, as well as the amount of time that is needed to complete a machine learning task.

7.6.7 IBM watson machine learning

IBM Watson [157] offers machine learning solutions in multicloud environments, including public, private, and hybrid cloud computing environments. It enables the easier deployment and optimization of machine learning models that can be invoked by the Apache Spark libraries. The framework that is offered by IBM is scalable and computationally efficient as it supports fast model training hardware resources with easy management of pipelines through its modular architecture. IBM Watson also supports distributed model training using Apache Hadoop and model retraining in the case of constantly arriving data streams or batches (e.g., in the case of incremental learning) and finally enables the portability of machine learning models either inside or outside the cloud environment.

7.6.8 Amazon Web Services machine learning

Amazon Web Services offers a machine learning framework for building, training, and deploying machine learning models at small-, medium-, and large-scale datasets through the Amazon SageMaker [47]. The latter is a fully managed solution with built-in functionalities for data annotation and embedded machine learning technology toward the development of highly robust prediction models for applications in personalized medicine and logistics, among others. Fully customizable solutions in Python are also offered for the expert users supporting the automated configuration and high interoperability with popular software tools and libraries, including keras, TensorFlow, and Apache Spark ML, among others. Enhanced high throughput technology is offered for the model training process, where scalable, high-performance VMs (units) are deployed for model optimization and processing. Amazon SageMaker is addressed to data scientists and developers and currently supports the ability to effectively process up to petabytes of training data with more than 64 CPUs and 16 GB memory capacity per computing unit.

7.7 Examples of applications in the medical domain

The supervised learning algorithms constitute the most common approach for the construction of machine learning models toward the development of risk stratification models, as well as for predicting disease outcomes. Linear regression and logistic regression models have been widely used to predict a disease outcome given a set of independent features (observations). For example, logistic regression has been employed in Ref. [158] to develop a risk stratification model for determining the probability of lymphoma development in patients who have been diagnosed with primary Sjögren's syndrome. SVMs have been used for predicting diabetes mellitus [159] and readmission for patients with diabetes [160], as well as for the automated health status assessment from sensor data [161]. The Naïve Bayes approach has been employed for predicting chronic kidney failure [162], diabetes [163], and heart disease [164], among others, usually serving as an underrated method for prediction. Furthermore, as it has already been mentioned in Section 7.3.3, CNNs have been used to solve classification problems across large-scale image datasets including the detection of skin cancer [98] and diabetic retinopathy [99], as well as in omics studies for the prediction of chromatin marks in DNA sequences [100]. In addition, the LSTM has numerous applications in biomedical signal analysis for the prediction of ictal states in epileptic seizures [102], the early detection of heart failure onset [103], and the freezing of gait in Parkinson's disease [104]. As far as the clustering methods are concerned, the N-cut and spectral clustering methods have been widely used for image segmentation purposes, such as skull stripping [165], as well as for the identification of voxels with similar activation patterns in fMRI studies [115,166]. Hierarchical clustering has been applied in Ref. [167] to segment a population into five categories, namely "young, healthy," "middle age, healthy," "stable, chronic disease," "complicated chronic disease," and "frequent admitters" toward the validation of its discriminative ability on 4-year longitudinal healthcare utilization and mortality data.

The most common package for the application of deep learning algorithms, on large-scale datasets, is keras. The TensorFlow framework is most commonly deployed by the keras environment toward the efficient solution of complex mathematical calculations through graph theory. The Weka software, along with the CRAN machine learning and Python packages, is most commonly used for the application of machine learning algorithms for prediction and regression purposes, as well as for clustering purposes, with the CRAN package being widely used in bioinformatics and Python's scikit-learn being used as a general purpose framework for machine learning applications. Weka is an example of a software tool that is used for constructing regression, prediction, and clustering models. The well-established cloud vendors, such as IBM Watson, Google AI, and Amazon EWS, are often employed for high-performance computing purposes, where several computing clusters are combined to apply deep learning algorithms on big data.

7.8 Conclusions

The technical advancements toward the application of data analytics in healthcare have made a significant progress over the past years. Data analytics is more useful and more powerful than even before yielding high-performance machine learning and deep learning models for mining knowledge across massive amount of medical data. In general, a data analytics pipeline consists of three fundamental steps, namely (i) data curation for enhancing the quality of the medical data by removing outliers, incompatibilities, and inconsistencies, (ii) data mining using machine learning methods for mining useful knowledge across the medical data through the development of patient stratification models and the detection of biomarkers, among others, and (iii) evaluation of the performance of such models using various performance indicators.

Before distilling knowledge from medical data, it is necessary first to apply data preprocessing pipelines, including data curation, data discretization, and feature extraction, among others. As it is discussed in Chapter 3, data curation includes simple methodological steps for dealing with outliers, missing values, and incompatibilities within the medical data. Then, data discretization methods can be applied to transform the continuous features into categorical ones using various data binning methods, such as the equal frequency and equal width methods, as well as the entropy MDL method for automatically identifying the correct number of bins based on the information entropy. Furthermore, feature ranking methods, such as the information gain, the gain ratio, and the Gini impurity index, can be used to quantify the importance of each feature according to a target one that is usually a disease outcome or medical condition. On the other hand, feature selection methods, such as the FCBF selection method can be used to directly derive a specific subset of features that minimize the correlation between the rest of the features and maximizes the correlation with the target feature. This distinct set of important features can be considered as a set of prominent features for "describing" a disease outcome that serves as the backbone of understanding the disease's onset and progress, similar to the definition of the biomarkers.

The variety of machine learning and data mining algorithms for distilling knowledge across big medical data are vast varying from supervised learning algorithms for the development of robust classification models for predicting disease outcomes to unsupervised learning algorithms for clustering features in complex data structures with similar patterns (e.g., highly correlated features). The supervised and unsupervised learning algorithms that were presented in this chapter can be applied to different types of medical data, including omics, clinical, histological, and medical images, among others. In each case, the input features correspond to the values of the medical terms that are present in each type of medical data. For example, the input features in a clinical dataset can be a patient's medical examinations, history, demographics, etc., where the data types might be continuous or discrete (e.g., categorical, ordinal). In the domain of medical imaging, the input features quantify information regarding the pixels in the image slices (or voxels in the 3D space), such as the intensity, the gradient, etc. In general, the target feature is usually a disease outcome or a medical condition and can have either two values that correspond to the disease's (or condition's) presence (e.g., yes/no) or with multiple values regarding the levels of the presence (e.g., low, medium, high). In the first case, the classification problem is a binary one, whereas in the latter case, the problem is a multiclass. In both cases, the performance evaluation of the supervised learning algorithms can be performed by either separating the data into train and test subsets based on a ratio or by using the *k*-fold cross-validation process and its related variants (e.g., the repeated and the stratified *k*-fold cross-validation), as well as the LOO cross-validation, among others. In any case, various performance evaluation metrics, such as the accuracy, sensitivity, and specificity, are used to quantify the performance of the supervised learning model regarding the number of correctly classified and misclassified outcomes.

Toward this direction, a variety of machine learning algorithms were presented in Section 7.3, including supervised learning algorithms for dealing with binary and multiclass classification problems including regression algorithms, SVMs, Naïve Bayes, decision tree induction and ensembles, ANNs, and deep learning neural networks (DNNs). In regression analysis, the target feature is modeled as a linear combination of the input features using appropriate weights that are estimated in such a way so that a specific loss (cost) function is minimized. This loss function is usually the difference between the predicted value and the original value (e.g., the *l*1- or the *l*2-norm) and can be solved using gradient descent methods. Logistic regression uses the sigmoid function as a loss function to suppress the predicted values to be adapted to solve binary classification problems against linear regression where the outcomes are continuous values. The main drawback of regression analysis, however, is the fact that is assumes the input features to be linearly independent, which introduces large variance during the prediction process making them prone to overfitting. SVMs, on the other hand, follow a more generalized approach that focuses on the construction of hyperplanes in the multidimensional space (e.g., a line in the 1D space), which are able to solve multiclass classification problems by searching for projections that separate the input data in the high-dimensional spaces. An alternative and more probabilistic approach is the Naïve Bayes that uses conditional probabilities combined with the likelihood through the Bayes rule to quantify the probability of a data point belonging to a specific class. Optimization methods, such as MAP and the maximum likelihood, are used to derive the class with the highest a posterior probability or the likelihood of a data point being in that class.

Apart from the probabilistic approaches, more complex methods that are inspired from the biological processes in the neurons have been proposed in the past and optimized in the recent years. ANNs formulate a powerful family of machine learning algorithms that solve classification problems by simulating the information flow within a biological process that occurs in the neurons. The ANNs use an input layer, a hidden layer, and an output layer, where the information flow begins from the input layer and can flow either in a forward manner (FNNs) or in a recursive manner (recursive neural networks) within the hidden layer and the output layer. Regarding the FNNs, popular implementations, such as MLP, have been proposed to model the hidden nodes as combinations of the input layer using a specific loss function (e.g., a sigmoid) and yield classification outcomes based on an activation function (e.g., the ReLu or the *tanh*). The RNNs, on the other hand, are more complex and allow the backpropagation of the gradient weights in the previous layers making them more computationally demanding and complex. The RNNs belong to the family of the deep learning neural networks that use multiple hidden layers with backpropagation to distill knowledge from massive amounts of medical data. Popular implementations of the RNNs include the CNNs that are widely used in medical imaging for effectively solving image segmentation problems by hierarchically splitting the images into smaller parts and the LSTM neural network that solves the gradient "vanishing" problem through a forget gate that maintains the backpropagated weight gradients during the information flow within the RNN.

Data clustering approaches, on the other hand, are not able to solve prediction or classification problems due to the absence of information regarding the target feature. Thus, the data clustering algorithms are only limited to the detection of input features with similar patterns and the subsequent categorization (grouping or clustering) of such features into a predefined number of clusters. An example of a popular clustering algorithm is the k-means that starts by initializing random data points as centroids to define the clustering regions by computing a distance measure (usually the Euclidean distance) between each data point and the centroid. The centroid assignment process is then repeated in an iterative manner until the clustering regions remain the same (yield no change in the assignment process). A more sophisticated approach is spectral clustering that projects the input space using the Laplacian formula to a new space where the features can easily be clustered using the k- means. To do so, spectral clustering takes advantage of the fact that the eigenvectors of the Laplacian matrix are smooth in areas where the data points are dense and vary in areas where the data points are sparse. Hierarchical clustering is a hierarchical approach, as its name implies, which builds dendrograms either in a bottom-up or top-down manner. Each data point is first assigned to a unique cluster, and similar clusters are grouped according to a linkage function, such as the Euclidean distance, the Manhattan distance, or the Ward method. The process is repeated until the data points are assigned to a large cluster (i.e., the root of the dendrogram). The optimal number of clusters can be found by testing the performance of the clustering algorithm for a different number of clusters and picking up the number of clusters that minimize the between-the-cluster similarity and maximize the within-the-cluster similarity. Examples of such indices include SI, the DB index, and the CH index. In the case where the labels of the clusters is known before the clustering process, the clustering performance evaluation can be achieved by quantifying the level of "agreement" between the outcome labels and the original labels using popular measures, such as ARI, the completeness score, and the homogeneity score, among others.

Then, we have gone one step further to examine how we can apply supervised and unsupervised learning algorithms across clinical data that are distributed across multiple sites. A solution to this can be given by the incremental learning and stacked generalization strategies. Incremental learning focuses on updating an existing machine learning model on upcoming data streams or batches, similar to online learning, although the former one can also support the offline processing of medical data. This means that incremental learning can be used to train machine learning models on massive amounts of medical data by dividing them into subsets (batches), training the model on the first batch, and sequentially updating the initial model on the remaining batches. If we replace the batches with harmonized datasets that are stored in multiple sets and use a central engine that will coordinate the communication between the sites, we can use incremental learning to sequentially update a machine learning model on these sites. Apart from incremental learning, someone would adopt the stacked generalization strategy and instead apply an individual machine learning model on each harmonized dataset and simply combine the classification outcomes using a metalearner or a majority voting rule (e.g., weighted average) to yield the final ones. This approach, however, is prone to biases that are introduced during the assembly stage and limits the "horizon" of the training process as the individual models are trained on individual subsets.

As far as data visualization is concerned, a variety of data visualization methods exist, where each one offers a unique point of view toward the interpretation of complex and multidimensional data structures. Conventional data visualization methods, such as line plots, dot plots, bar plot,s and histogram plots, are able to quantify the descriptive profile of an input feature regarding the value range, the evolution over time, and the frequency of the values across different value ranges. But when it comes to multiple features and complex data structures where the hidden associations need to be examined, more sophisticated data visualization methods need to be provided. Toward this direction, the heatmaps and the graph plots offer a sophisticated way of representing the associations between the input features in the form of similarity matrices that are able to quantify the similarity (e.g., correlation coefficient) between two or more features and multidimensional graphs, where the edges reveal important information regarding the association of two nodes (each node corresponds to a feature) and the topology of the graphs reveals information regarding nodes (features) with increased number of incoming edges (e.g., hubs), as well as features concentrated in dense areas in the form clusters that reveal highly similar pairs of features.

Unlike simple statistics that work well with numerical data, do not involve any predictions, and support only descriptive and inferential approaches that test for hypotheses instead of taking decisions (nonheuristic), machine learning can handle both numerical and discrete data, generate predictions, solve classification, clustering, and regression problems, and identify hidden patterns (associations) in large-scale datasets using heuristic thinking with increased scalability and performance. So, we have come to a final question: can machine learning (and AI in general) be used to predict the future? Undoubtedly, machine learning and especially deep learning can shed light into hard and complex scientific problems varying from the prediction of rare disease outcomes and the detection of biomarkers and therapy treatment to the prediction of environmental disasters and economic breakthroughs. The existing technology offers the basis for distilling knowledge across huge amounts of generated data including built-in hyperparameter optimization methods, parallelized computing units, and high-throughput technologies. Although the benefits of AI are vast, so are the dangers of misusing it.

References

[1] Pezoulas VC, Kourou KD, Kalatzis F, Exarchos TP, Venetsanopoulou A, Zampeli E, et al. Medical data quality assessment: on the development of an automated framework for medical data curation. Comput Biol Med 2019;107:270–83.

[2] L'heureux A, Grolinger K, Elyamany HF, Capretz MA. Machine learning with big data: challenges and approaches. IEEE Access 2017;5:7776–97.

[3] De Sá CR, Soares C, Knobbe A. Entropy-based discretization methods for ranking data. Inf Sci 2016;329:921–36.

[4] Ghassemi M, Naumann T, Schulam P, Beam AL, Ranganath R. Opportunities in machine learning for healthcare. 2018. arXiv preprint 2018;arXiv:1806.00388.

[5] Chen M, Hao Y, Hwang K, Wang L, Wang L. Disease prediction by machine learning over big data from healthcare communities. IEEE Access 2017;5:8869–79.

[6] Mandl KD, Manrai AK. Potential excessive testing at scale: biomarkers, genomics, and machine learning. Jama 2019;321(8):739–40.

[7] Jung AA Gentle introduction to supervised machine learning. arXiv preprint 2018; arXiv:1805.05052.

[8] Gunst RF. Regression analysis and its application: a data-oriented approach. Routledge; 2018.

[9] Huang S, Cai N, Pacheco PP, Narrandes S, Wang Y, Xu W. Applications of support vector machine (SVM) learning in cancer genomics. Cancer Genomics Proteomics 2018;15(1):41–51.

[10] Breiman L. Classification and regression trees. Routledge; 2017.

[11] Kubat M. In: An introduction to machine learning, vol. 2. Switzerland: Springer International Publishing; 2017.

[12] Hertz JA. Introduction to the theory of neural computation. CRC Press; 2018.

[13] Chen JH, Asch SM. Machine learning and prediction in medicine—beyond the peak of inflated expectations. N Engl J Med 2017;376(26):2507–9.

[14] Simeone O. A brief introduction to machine learning for engineers. Found Trends® Signal Proces 2018;12(3–4):200–431.

[15] Jain AK. Data clustering: 50 years beyond K-means. Pattern Recognit Lett 2010;31(8): 651–66.

[16] Von Luxburg U. A tutorial on spectral clustering. Stat Comput 2007;17(4):395–416.

[17] Bouguettaya A, Yu Q, Liu X, Zhou X, Song A. Efficient agglomerative hierarchical clustering. Expert Syst Appl 2015;42(5):2785–97.

[18] Miotto R, Wang F, Wang S, Jiang X, Dudley JT. Deep learning for healthcare: review, opportunities and challenges. Briefings Bioinf 2017;19(6):1236–46.

[19] Che Z, Purushotham S, Khemani R, Liu Y. Distilling knowledge from deep networks with applications to healthcare domain. 2015. arXiv preprint, 2015;arXiv:1512.03542.

[20] Ronneberger O, Fischer P, Brox T. U-net: convolutional networks for biomedical image segmentation. In: Proceedings of the International Conference on medical image computing and computer-assisted intervention. Cham: Springer; 2015. p. 234–41.

[21] Sayed AH. Adaptation, learning, and optimization over networks. Found Trends® Mach Learn 2014;7(4–5):311–801.

[22] Obermeyer Z, Emanuel EJ. Predicting the future—big data, machine learning, and clinical medicine. N Engl J Med 2016;375(13):1216–9.

[23] Brown JS, Holmes JH, Shah K, Hall K, Lazarus R, Platt R. Distributed health data networks: a practical and preferred approach to multi-institutional evaluations of comparative effectiveness, safety, and quality of care. Med Care 2010;48:S45−51.

[24] Raghupathi W, Raghupathi V. Big data analytics in healthcare: promise and potential. Health Inf Sci Syst 2014;2:3.

[25] Krumholz HM. Big data and new knowledge in medicine: the thinking, training, and tools needed for a learning health system. Health Aff 2014;33(7):1163−70.

[26] Jochems A, Deist TM, van Soest J, Eble M, Bulens P, Coucke P, et al. Distributed learning: developing a predictive model based on data from multiple hospitals without data leaving the hospital - a real life proof of concept. Radiother Oncol 2016;121(3): 459−67.

[27] Deist TM, Jochems A, van Soest J, Nalbantov G, Oberije C, Walsh S, et al. Infrastructure and distributed learning methodology for privacy-preserving multi-centric rapid learning health care: euroCAT. Clin Transl Radiat Oncol. 2017;19(4):24−31.

[28] Bertsekas DP. Incremental gradient, subgradient, and proximal methods for convex optimization: a survey. Optimizat Mach Learn 2011;2010(1−38):3.

[29] Gepperth A, Hammer B. Incremental learning algorithms and applications. In: Proceedings of the european symposium on artificial neural networks (ESANN), Bruges, Belgium; 2016.

[30] Bekkerman R, Bilenko M, Langford J. Scaling up machine learning: parallel and distributed approaches. Cambridge University Press; 2011.

[31] Shalev-Shwartz S. Online learning and online convex optimization. Found Trends® Mach Learn 2012;4(2):107−94.

[32] Duchi J, Hazan E, Singer Y. Adaptive subgradient methods for online learning and stochastic optimization. J Mach Learn Res 2011;12:2121−59.

[33] Zheng J, Shen F, Fan H, Zhao J. An online incremental learning support vector machine for large-scale data. Neural Comput Appl 2013;22(5):1023−35.

[34] Si S, Kumar S, Yang L. Nonlinear online learning with adaptive nyström approximation. arXiv preprint 2018;arXiv:1802.07887.

[35] Bottou L. Large-scale machine learning with stochastic gradient descent. In: Proceedings of the COMPSTAT'2010. Physica-Verlag HD; 2010. p. 177−86.

[36] Bertsekas DP. Incremental proximal methods for large scale convex optimization. Math Program 2011;129(2):163−95.

[37] Zhang T. Solving large scale linear prediction problems using stochastic gradient descent algorithms. In: Proceedings of the 21st ACM International Conference on Machine learning; 2004. p. 116.

[38] Friedman JH. Greedy function approximation: a gradient boosting machine. Ann Stat 2001;29:1189−232.

[39] Christopher DM, Prabhakar R, Hinrich S. Introduction to information retrieval. Cambridge University Press; 2008.

[40] Zeng L, Li L, Duan L, Lu K, Shi Z, Wang M, Luo P. Distributed data mining: a survey. Int J Inf Technol Manag 2012;13(4):403−9.

[41] Tsoumakas G, Vlahavas I. Effective stacking of distributed classifiers. In: Proceedings of the ECAI'02 ACM international conference on artificial intelligence; 2002. p. 340−4.

[42] Prodromidis A, Chan P, Stolfo S. Meta-learning in distributed data mining systems: issues and approaches. Advances in distributed and parallel knowledge discovery, 3; 2000. p. 81−114.

[43] Senthilkumar SA, Rai BK, Meshram AA, Gunasekaran A, Chandrakumarmangalam S. Big Data in healthcare management: a review of literature. Am J Theor Appl Bus 2018; 4(2):57–69.

[44] Bruder V, Hlawatsch M, Frey S, Burch M, Weiskopf D, Ertl T. Volume-based large dynamic graph analytics. In: Proceedings of the 22nd international conference on information visualisation-biomedical visualization, visualisation on built and rural environments and geometric modelling and imaging; 2018. p. 210–9.

[45] Pedregosa F, Varoquaux G, Gramfort A, Michel V, Thirion B, Grisel O, et al. Scikit-learn: machine learning in Python. J Mach Learn Res 2011;12:2825–30.

[46] Gulli A, Pal S. Deep learning with keras. Packt Publishing Ltd; 2017.

[47] Amazon SageMaker, Link: https://aws.amazon.com/sagemaker/.

[48] Ramírez-Gallego S, García S, Mouriño-Talín H, Martínez-Rego D, Bolón-Canedo V, Alonso-Betanzos A, et al. Data discretization: taxonomy and big data challenge. Wiley Interdiscip Rev: Data Min Knowl Discov 2016;6(1):5–21.

[49] Grzymala-Busse J, Mroczek T. A comparison of four approaches to discretization based on entropy. Entropy 2016;18(3):69.

[50] Jović A, Brkić K, Bogunović N. A review of feature selection methods with applications. In: Proceedings of the 38th international convention on information and communication technology, electronics and microelectronics (MIPRO); 2015. p. 1200–5.

[51] Peng H, Long F, Ding C. Feature selection based on mutual information: criteria of max-dependency, max-relevance, and min-redundancy. IEEE Trans Pattern Anal Mach Intell 2005;(8):1226–38.

[52] Raschka S. Python machine learning. Packt Publishing Ltd; 2015.

[53] Yu L, Liu H. Feature selection for high-dimensional data: a fast correlation-based filter solution. In: Proceedings of the 20th international conference on machine learning (ICML-03); 2003. p. 856–63.

[54] Karegowda AG, Manjunath AS, Jayaram MA. Comparative study of attribute selection using gain ratio and correlation based feature selection. Int J Inf Technol Knowl Manag 2010;2(2):271–7.

[55] Liu H, Zhou M, Lu XS, Yao C. Weighted Gini index feature selection method for imbalanced data. In: Proceedings of the 15th international conference on networking, sensing and control (ICNSC); 2018. p. 1–6.

[56] Hira ZM, Gillies DF. A review of feature selection and feature extraction methods applied on microarray data. Adv Bioinformatics 2015;2015:198363.

[57] Leclercq M, Vittrant B, Martin-Magniette ML, Scott-Boyer MP, Perin O, Bergeron A, et al. Large-scale automatic feature selection for biomarker discovery in high-dimensional OMICs data. Front Genet 2019;10:452.

[58] Huang X, Zhang L, Wang B, Li F, Zhang Z. Feature clustering based support vector machine recursive feature elimination for gene selection. Appl Intell 2018;48(3): 594–607.

[59] Cai J, Luo J, Wang S, Yang S. Feature selection in machine learning: a new perspective. Neurocomputing 2018;300:70–9.

[60] Jain D, Singh V. Feature selection and classification systems for chronic disease prediction: a review. Egyp Inform J 2018;19(3):179–89.

[61] Olive DJ. Multiple linear regression. In: Linear regression. Cham: Springer; 2017. p. 17–83.

[62] Bun MJ, Harrison TD. OLS and IV estimation of regression models including endogenous interaction terms. Econom Rev 2018:1–14.

[63] McCullagh P. Generalized linear models. Routledge; 2019.
[64] Carroll RJ. Transformation and weighting in regression. Routledge; 2017.
[65] Liu Z, Li C. Recursive least squares for censored regression. IEEE Trans Signal Process 2016;65(6):1565–79.
[66] Dobriban E, Wager S. High-dimensional asymptotics of prediction: Ridge regression and classification. Ann Stat 2018;46(1):247–79.
[67] Ranstam J, Cook J. LASSO regression. Br J Surg 2018;105(10).
[68] Zhang Z, Lai Z, Xu Y, Shao L, Wu J, Xie GS. Discriminative elastic-net regularized linear regression. IEEE Trans Image Process 2017;26(3):1466–81.
[69] Dhar SK. Modeling with a bivariate geometric distribution. In: Advances on methodological and applied aspects of probability and statistics. CRC Press; 2019. p. 133–41.
[70] Hemmert GA, Schons LM, Wieseke J, Schimmelpfennig H. Log-likelihood-based pseudo-R 2 in logistic regression: deriving sample-sensitive benchmarks. Sociol Methods Res 2018;47(3):507–31.
[71] Bonte C, Vercauteren F. Privacy-preserving logistic regression training. BMC Medical Genomics 2018;11(4):86.
[72] Sci-kit learn: Machine learning in Python. Link: https://scikit-learn.org/stable/.
[73] Chen Y, Georgiou TT, Tannenbaum AR. Probabilistic kernel support vector machines. arXiv preprint. 2019. arXiv:1904.06762.
[74] Gu B, Quan X, Gu Y, Sheng VS, Zheng G. Chunk incremental learning for cost-sensitive hinge loss support vector machine. Pattern Recognit 2018;83:196–208.
[75] Guenther N, Schonlau M. Support vector machines. STATA J 2016;16(4):917–37.
[76] Meyer D, Wien FT. Support vector machines. The Interface to libsvm in package. 2015. e1071, 28.
[77] Gaudioso M, Gorgone E, Labbé M, Rodríguez-Chía AM. Lagrangian relaxation for SVM feature selection. Comput Oper Res 2017;87:137–45.
[78] Rish I. An empirical study of the naive Bayes classifier. In: IJCAI workshop on empirical methods in artificial intelligence, vol. 3; 2001. p. 41–6. 22.
[79] Bassett R, Deride J. Maximum a posteriori estimators as a limit of Bayes estimators. Math Program 2019;174(1–2):129–44.
[80] Maritz JS, Lwin T. Empirical bayes methods. Routledge; 2018.
[81] Jiang L, Wang S, Li C, Zhang L. Structure extended multinomial naive Bayes. Inf Sci 2016;329:346–56.
[82] Liu H, Cocea M. Induction of classification rules by gini-index based rule generation. Inf Sci 2018;436:227–46.
[83] Song YY, Ying LU. Decision tree methods: applications for classification and prediction. Shanghai Arch Psychiat 2015;27(2):130.
[84] Lee SJ, Xu Z, Li T, Yang Y. A novel bagging C4. 5 algorithm based on wrapper feature selection for supporting wise clinical decision making. J Biomed Inform 2018;78: 144–55.
[85] Pandya R, Pandya J. C5. 0 algorithm to improved decision tree with feature selection and reduced error pruning. Int J Comput Appl 2015;117(16):18–21.
[86] Mohamed WNHW, Salleh MNM, Omar AH. A comparative study of reduced error pruning method in decision tree algorithms. In: Proceedings of the 2012 IEEE International conference on control system, computing and engineering; 2012. p. 392–7.
[87] Baskin II, Marcou G, Horvath D, Varnek A. Bagging and boosting of classification models. Tutorials Chemoinformatics 2017:241–7.
[88] Pavlov YL. Random forests. Walter de Gruyter GmbH & Co KG; 2019.

[89] Schapire RE. Explaining adaboost. In: Empirical inference. Berlin, Heidelberg: Springer; 2013. p. 37−52.
[90] Chen T, Guestrin C. Xgboost: a scalable tree boosting system. In: Proceedings of the 22nd ACM SIGKDD international conference on knowledge discovery and data mining; 2016. p. 785−94.
[91] Schmidhuber J. Deep learning in neural networks: an overview. Neural Networks 2015;61:85−117.
[92] Stengel R. Introduction to neural networks!. Princeton University; 2017.
[93] Walczak S. Artificial neural networks. In: Advanced methodologies and technologies in artificial intelligence, computer simulation, and human-computer interaction. IGI Global; 2019. p. 40−53.
[94] Petersen J, Iversen HK, Puthusserypady S. Motor imagery based Brain Computer Interface paradigm for upper limb stroke rehabilitation. In: Proceedings of the 40th annual international conference of the IEEE engineering in medicine and biology society (EMBC); 2018. p. 1960−3.
[95] Alickovic E, Subasi A. Normalized neural networks for breast cancer classification. In: Proceedings of the international conference on medical and biological engineering. Cham: Springer; 2019. p. 519−24.
[96] Wei C, Lee JD, Liu Q, Ma T. On the margin theory of feedforward neural networks. arXiv preprint, 2018;arXiv:1810.05369.
[97] Lipton ZC, Berkowitz J, Elkan C. A critical review of recurrent neural networks for sequence learning. arXiv preprint, 2015;arXiv:1506.00019.
[98] Esteva A, Kuprel B, Novoa RA, Ko J, Swetter SM, Blau HM, et al. Dermatologist-level classification of skin cancer with deep neural networks. Nature 2017;542(7639): 115−8.
[99] Gulshan V, Peng L, Coram M, Stumpe MC, Wu D, Narayanaswamy A, et al. Development and validation of a deep learning algorithm for detection of diabetic retinopathy in retinal fundus photographs. J Am Med Assoc 2016;316(22):2402−10.
[100] Kelley DR, Reshef YA, Bileschi M, Belanger D, McLean CY, Snoek J. Sequential regulatory activity prediction across chromosomes with convolutional neural networks. Genome Res 2018;28(5):739−50.
[101] Dhillon IS. Stabilizing gradients for deep neural networks. 2018. arXiv preprint, 2018; arXiv:1803.09327.
[102] Tsiouris KM, Pezoulas VC, Zervakis M, Konitsiotis S, Koutsouris DD, Fotiadis DI. A Long Short-Term Memory deep learning network for the prediction of epileptic seizures using EEG signals. Comput Biol Med 2018;99:24−37.
[103] Maragatham G, Devi S. LSTM model for prediction of heart failure in big data. J Med Syst 2019;43(5):111.
[104] Zhao A, Qi L, Li J, Dong J, Yu H. LSTM for diagnosis of neurodegenerative diseases using gait data. In: Proceedings of the ninth international conference on graphic and image processing (ICGIP 2017). International Society for Optics and Photonics; 2018. 10615;106155B.
[105] Bengio Y, Grandvalet Y. No unbiased estimator of the variance of k-fold cross-validation. J Mach Learn Res 2004;5:1089−105.
[106] Kim JH. Estimating classification error rate: repeated cross-validation, repeated hold-out and bootstrap. Comput Stat Data Anal 2009;53(11):3735−45.

[107] Cawley GC. Leave-one-out cross-validation based model selection criteria for weighted LS-SVMs. In: Proceedings of the 2006 IEEE international joint conference on neural network proceedings; 2006. p. 1661−8.
[108] Zeng X, Martinez TR. Distribution-balanced stratified cross-validation for accuracy estimation. J Exp Theor Artif Intell 2000;12(1):1−2.
[109] Arlot S, Celisse A. A survey of cross-validation procedures for model selection. Stat Surv 2010;4:40−79.
[110] Powers DM. Evaluation: from precision, recall and F-measure to ROC, informedness, markedness and correlation. J Mach Learn Technol 2011;2(1):37−63.
[111] Obuchowski NA, Bullen JA. Receiver operating characteristic (ROC) curves: review of methods with applications in diagnostic medicine. Phys Med Biol 2018;63(7): 07TR01.
[112] Sculley D. Web-scale k-means clustering. In: Proceedings of the 19th international conference on World wide web. ACM; 2010. p. 1177−8.
[113] Müller AC, Guido S. Introduction to machine learning with Python: a guide for data scientists. O'Reilly Media, Inc.; 2016.
[114] Ding CH, He X, Zha H, Gu M, Simon HD. A min-max cut algorithm for graph partitioning and data clustering. In: Proceedings of the 2001 IEEE international conference on data mining; 2001. p. 107−14.
[115] Pezoulas VC, Michalopoulos K, Klados M, Micheloyannis S, Bourbakis N, Zervakis M. Functional connectivity analysis of cerebellum using spatially constrained spectral clustering. IEEE J Biomed Health Informat 2018;23(4):1710−9.
[116] Pezoulas VC, Zervakis M, Pologiorgi I, Seferlis S, Tsalikis GM, Zarifis G, et al. A tissue classification approach for brain tumor segmentation using MRI. In: Proceedings of the 2017 IEEE international conference on imaging systems and techniques (IST); 2017. p. 1−6.
[117] De Maesschalck R, Jouan-Rimbaud D, Massart DL. The mahalanobis distance. Chemometr Intell Lab Syst 2000;50(1):1−8.
[118] Xiang S, Nie F, Zhang C. Learning a Mahalanobis distance metric for data clustering and classification. Pattern Recog 2008;41(12):3600−12.
[119] Yim O, Ramdeen KT. Hierarchical cluster analysis: comparison of three linkage measures and application to psychological data. Quantitat. Methods Psychol 2015;11(1): 8−21.
[120] Murtagh F, Contreras P. Algorithms for hierarchical clustering: an overview, II. Wiley Interdiscip Rev: Data Min Knowl Discov 2017;7(6):e1219.
[121] Tran TN, Drab K, Daszykowski M. Revised DBSCAN algorithm to cluster data with dense adjacent clusters. Chemometr Intell Lab Syst 2013;120:92−6.
[122] Hang W, Chung FL, Wang S. Transfer affinity propagation-based clustering. Inf Sci June 20, 2016;348:337−56.
[123] Serdah AM, Ashour WM. Clustering large-scale data based on modified affinity propagation algorithm. J Artif Intell Soft Comput Res 2016;6(1):23−33.
[124] Yeh CC, Yang MS. Evaluation measures for cluster ensembles based on a fuzzy generalized Rand index. Appl Soft Comput 2017;57:225−34.
[125] Steinley D, Brusco MJ, Hubert L. The variance of the adjusted Rand index. Psychol Methods 2016;21(2):261−72.
[126] Amelio A, Pizzuti C. Correction for closeness: adjusting normalized mutual information measure for clustering comparison. Comput Intell 2017;33(3):579−601.

[127] Ozdemir A, Bolanos M, Bernat E, Aviyente S. Hierarchical spectral consensus clustering for group analysis of functional brain networks. IEEE (Inst Electr Electron Eng) Trans Biomed Eng 2015;62(9):2158–69.
[128] Brodinová Š, Zaharieva M, Filzmoser P, Ortner T, Breiteneder C. Clustering of imbalanced high-dimensional media data. Adv Data Analysis Class 2018;12(2):261–84.
[129] Starczewski A. A new validity index for crisp clusters. Pattern Anal Appl 2017;20(3): 687–700.
[130] Xiao J, Lu J, Li X. Davies Bouldin Index based hierarchical initialization K-means. Intell Data Anal 2017;21(6):1327–38.
[131] Lamirel JC, Dugué N, Cuxac P. New efficient clustering quality indexes. In: Proceedings of the 2016 international joint conference on neural networks (IJCNN); 2016. p. 3649–57.
[132] Kumar A, Boehm M, Yang J. Data management in machine learning: challenges, techniques, and systems. In: Proceedings of the 2017 ACM international conference on management of data; 2017. p. 1717–22.
[133] Finn C, Abbeel P, Levine S. Model-agnostic meta-learning for fast adaptation of deep networks. In: Proceedings of the 34th international conference on machine learning, vol. 70; 2017. p. 1126–35.
[134] Deo N. Graph theory with applications to engineering and computer science. Courier Dover Publications; 2017.
[135] Pezoulas VC, Zervakis M, Michelogiannis S, Klados MA. Resting-state functional connectivity and network analysis of cerebellum with respect to IQ and gender. Front Hum Neurosci 2017;11:189.
[136] Niso G, Carrasco S, Gudín M, Maestú F, del-Pozo F, Pereda E. What graph theory actually tells us about resting state interictal MEG epileptic activity. Neuroimage: Clin 2015;8:503–15.
[137] Tsiouris KM, Pezoulas VC, Koutsouris DD, Zervakis M, Fotiadis DI. Discrimination of preictal and interictal brain states from long-term EEG data. In: Proceedings of the IEEE 30th international symposium on computer-based medical systems (CBMS); 2017. p. 318–23.
[138] Seaborn: statistical data visualization. Link: https://seaborn.pydata.org/.
[139] NetworkX: software for complex networks. Link: https://networkx.github.io/.
[140] Zaharia M, Xin RS, Wendell P, Das T, Armbrust M, Dave A, Meng X, Rosen J, Venkataraman S, Franklin MJ, Ghodsi A. Apache spark: a unified engine for big data processing. Commun ACM 2016;59(11):56–65.
[141] Armbrust M, Xin RS, Lian C, Huai Y, Liu D, Bradley JK, et al. Spark sql: relational data processing in spark. In: Proceedings of the 2015 ACM SIGMOD international conference on management of data; 2015. p. 1383–94.
[142] Gonzalez JE, Xin RS, Dave A, Crankshaw D, Franklin MJ, Stoica I. Graphx: graph processing in a distributed dataflow framework. In: Proceedings of the 11th symposium on operating systems design and implementation; 2014. p. 599–613.
[143] Meng X, Bradley J, Yavuz B, Sparks E, Venkataraman S, Liu D, et al. Mllib: machine learning in Apache spark. J Mach Learn Res 2016;17(1):1235–41.
[144] Apache Hadoop. Link: https://hadoop.apache.org/.
[145] Holmes A. Hadoop in practice. Manning Publications Co.; 2012.
[146] White T. Hadoop: the definitive guide. O'Reilly Media; 2012.

[147] Ghazi MR, Gangodkar D. Hadoop, MapReduce and HDFS: a developers perspective. Proc Comp Sci 2015;48:45–50.
[148] Vavilapalli VK, Murthy AC, Douglas C, Agarwal S, Konar M, Evans R, et al. Apache hadoop yarn: yet another resource negotiator. In: Proceedings of the 4th annual symposium on cloud computing; 2013. p. 5.
[149] Taylor RC. An overview of the Hadoop/MapReduce/HBase framework and its current applications in bioinformatics. BMC Bioinforma BioMed Central 2010;11(12):S1.
[150] Demšar J, Curk T, Erjavec A, Gorup Č, Hočevar T, Milutinovič M, et al. Orange: data mining toolbox in Python. J Mach Learn Res 2013;14(1):2349–53.
[151] Abadi M, Barham P, Chen J, Chen Z, Davis A, Dean J, et al. Tensorflow: a system for large-scale machine learning. In: Proceedings of the 12th symposium on operating systems design and implementation; 2016. p. 265–83.
[152] Baylor D, Breck E, Cheng HT, Fiedel N, Foo CY, Haque Z, et al. Tfx: a tensorflow-based production-scale machine learning platform. In: Proceedings of the 23rd ACM SIGKDD international conference on knowledge discovery and data mining; 2017. p. 1387–95.
[153] Al-Rfou R, Alain G, Almahairi A, Angermueller C, Bahdanau D, Ballas N, et al. Theano: a Python framework for fast computation of mathematical expressions. 2016. arXiv preprint, 2016;arXiv:1605.02688.
[154] Seide F, Agarwal A. CNTK: microsoft's open-source deep-learning toolkit. In: Proceedings of the 22nd ACM SIGKDD international conference on knowledge discovery and data mining; 2016. p. 2135.
[155] Hothorn T. CRAN task view: machine learning & statistical learning. 2019.
[156] Hall M, Frank E, Holmes G, Pfahringer B, Reutemann P, Witten IH. The WEKA data mining software: an update. ACM SIGKDD explorations newsletter 2009;11(1):10–8.
[157] Chen Y, Argentinis JE, Weber G. IBM Watson: how cognitive computing can be applied to big data challenges in life sciences research. Clin Ther 2016;38(4):688–701.
[158] Fragkioudaki S, Mavragani CP, Moutsopoulos HM. Predicting the risk for lymphoma development in Sjogren syndrome: an easy tool for clinical use. Medicine 2016; 95(25).
[159] Zou Q, Qu K, Luo Y, Yin D, Ju Y, Tang H. Predicting diabetes mellitus with machine learning techniques. Front Genet 2018;9.
[160] Cui S, Wang D, Wang Y, Yu PW, Jin Y. An improved support vector machine-based diabetic readmission prediction. Comput Methods Progr Biomed 2018;166:123–35.
[161] Diez-Olivan A, Pagan JA, Khoa NLD, Sanz R, Sierra B. Kernel-based support vector machines for automated health status assessment in monitoring sensor data. Int J Adv Manuf Technol 2018;95(1–4):327–40.
[162] Devika R, Avilala SV, Subramaniyaswamy V. Comparative study of classifier for chronic kidney disease prediction using naive Bayes, KNN and random forest. In: 2019 3rd international conference on computing methodologies and communication (ICCMC); 2019. p. 679–84.
[163] Sisodia D, Sisodia DS. Prediction of diabetes using classification algorithms. Proc Comp Sci 2018;1132:1578–85.
[164] Bashir S, Khan ZS, Khan FH, Anjum A, Bashir K. Improving heart disease prediction using feature selection approaches. In: 2019 16th international bhurban conference on applied sciences and technology (IBCAST); 2019. p. 619–23.

[165] Pezoulas VC, Zervakis M, Pologiorgi I, Seferlis S, Tsalikis GM, Zarifis G, et al. A tissue classification approach for brain tumor segmentation using MRI. In: 2017 IEEE international conference on imaging systems and techniques (IST); 2017. p. 1–6.
[166] Craddock RC, James GA, Holtzheimer PE, Hu XP, Mayberg HS. A whole brain fMRI atlas generated via spatially constrained spectral clustering. Hum Brain Mapp 2012; 33(8):1914–28.
[167] Najjar A, Reinharz D, Girouard C, Gagné C. A two-step approach for mining patient treatment pathways in administrative healthcare databases. Artif Intell Med 2018;87: 34–48.

CHAPTER

Case studies

8

Chapter outline

8.1 Overview

Medical data sharing and harmonization have been prerequisites in a variety of clinical cohort studies that envisage to interlink multinational cohorts with heterogeneous data structures and schemas to address the unmet needs of various diseases and medical domains. As we have already mentioned in Chapter 3 and Chapter 5, the unmet needs include the detection of biomarkers and/or the validation of new ones to predict disease outcomes, the development of robust patient stratification models for the identification of high-risk individuals who are prone to developing a specific disease outcome, and the development of new, effective therapeutic treatments based on the disease's onset and progress. In this chapter, a variety of clinical cohort studies are presented across different medical domains including aging studies, studies on phenotypes, personality scores, cancer, obesity, autoimmune diseases to demonstrate the overall scientific impact and value of medical data sharing, and harmonization toward the integrative analysis of heterogeneous clinical data from various medical sources. Toward this direction, the tools and methodologies that have been already described in Chapter 5 are employed.

Numerous cases have been reported to demonstrate the applicability and heritability of medical data sharing and harmonization in healthcare. There is no doubt

Medical Data Sharing, Harmonization and Analytics. https://doi.org/10.1016/B978-0-12-816507-2.00008-6

that the value of medical data sharing and harmonization in addressing the clinical unmet needs across various medical domains has been validated both in large- and small-scale clinical cohort studies that have recently employed data harmonization as a "weapon" to enable the integrative analysis of heterogeneous medical data. More specifically, according to the existing literature, there are numerous reports [1–8] and emerging case studies [9–27] that strongly suggest that data harmonization can be used to interlink heterogeneous data across cohorts and identify common terms across vast patient records based on a reference (gold standard) model of common terms. The latter serves as a standard set of terms for mapping the parameters that share the same concept along with their related values having different measurement units and value ranges. The success story of medical data harmonization has been for long stated through the homogenization of personality scores (e.g., neuroticism and extraversion) [17], the identification of genetic variants in personality scores [16], the identification of risk factors and clinical phenotypes in aging and obesity [1–8], the investigation of the effects of age and gender during the execution of physical activities in the elder people [3], the identification of prominent determinants in aging, obesity and chronic diseases [1,4,7,8,21], and the extraction of harmonized variables (parameters) across well-established epidemiological databases [9–13], among many others.

The type of the case studies that are presented in the following sections are mainly cohort longitudinal studies involving both retrospective and prospective medical data, including genetic data (phenotypes), laboratory measures, patient history records, personality and depression scores, the presence of medical conditions and other diseases, lifestyle factors, demographics, and interventions, among others. All the different types of clinical case studies are further classified according to their scale or impact (e.g., as small- or large-scale depending on the number of cohorts that are included in each case study) and the medical domain (e.g., aging, obesity, epidemiology). The clinical unmet needs that were addressed by each individual study are also included.

8.2 Case studies

8.2.1 Aging studies

The Statistical Modeling of Aging and Risk of Transition (SMART) study [1] is a large-scale study that involves the harmonization of clinical data from 11 longitudinal cohorts on aging consisting of 11,541 individuals. The cohort data included information regarding phenotypes, medical history, and risk factors for different neuropathologies, such as the Alzheimer's disease, the mild cognitive impairment (MCI), the hippocampal sclerosis, and the dementia, among the elder individuals. The data sharing process fulfilled all the necessary legal and ethical requirements that were presented in Chapter 4. The data harmonization strategy that was adopted was similar to the stringent one (see Section 5.5.1) where the interlinked data followed a common, standardized format. The standardized format, which is referred to as a data template, was defined as the minimum set of parameters across at least two cohorts that are similar to the reference model that was already described in Section 5.4 and Section 3.4. The data template consists of different categories that describe the domain of interest, including information regarding demographics, medical history, interventions, genetic profiles, neuropathologies, autopsies, laboratory, and physical examinations. Considering the presence of high levels of variance during the heterogeneity of the data collection protocols that were used for recording the neurophysiological test scores across the cohort data, multivariate linear modeling methods were employed to overcome any biases to predict normalized baseline scores by taking into account multiple factors, such as the age, gender, ethnicity, and level of education, among others. This process was done to overcome the absence of MCI diagnostic reports in several cohorts. The SMART database can be used to investigate clinical phenotype and risk factors for various types of neuropathologies shedding light into the detection of genetic associations, comorbidities, and biomarkers in the field of neurological disorders and epidemiology.

The Comparison of Longitudinal European Studies on Aging (CLESA) Project [2] is a small-scale study that involves the harmonization of clinical data from 6 longitudinal cohorts (5 European and 1 Israeli) on aging consisting of 11,557 individuals who were separated into two aging categories to investigate statistical differences between the population groups. The data sharing process fulfilled all the necessary legal and ethical requirements that were presented in Chapter 4. The first aging category included the individuals from 65 to 74 years old and the second one included the individuals within the age interval from 75 to 84 years old. A common database harmonization schema was defined to enable the data harmonization process. The common schema consisted of a set of parameters that shared a common meaning across at least three of the participating cohorts to enable an unbiased statistical comparison between the measures. The domains that were recorded in the common schema included information regarding demographics, cognitive scores, depression scores, questionnaires, medical history, daily and instrumental activities, etc. Harmonization guidelines were developed regarding the demographics, health habits (e.g., smoking habits, alcohol consumption, physical activity, drug consumption), health status (e.g., hypertension, diabetes, stroke), physical functioning (e.g., disability), and social networking (e.g., loneliness, lifetime experiences) domains. Statistical methods, such as the generalized linear model, were applied to detect highly correlated terms across the data in each heterogeneous domain in the case where harmonization was not trivial. The final harmonized database schema consisted of 111 parameters and 11,557 individuals. The parameters were standardized based on the age and adjusted according to the sex during the statistical analysis process. Through the harmonized database, the CLESA Project has been able to identify multiple factors and conditions predicting the morbidity and mortality in the elderly people providing a great insight into the determinants of aging and chronic diseases, such as stroke, diabetes, and cardiovascular diseases.

The Healthy Ageing across the Life Course (HALCyon) research program [3] is a medium-scale study that involves the harmonization of clinical data from nine UK longitudinal cohorts (the Lothian birth cohort 1921 (LBC1921); the Hertfordshire Ageing Study; the Hertfordshire Cohort Study; the Caerphilly Prospective Study; the Boyd Orr cohort; the Aberdeen birth cohort 1936 (ABC1936); the English Longitudinal Study of Ageing; the MRC National Survey of Health and Development; and the National Child Development Study) on aging consisting of approximately 40,000 individuals. The data sharing process fulfilled all the necessary legal and ethical requirements that were presented in Chapter 4. The aim of HALCyon was to investigate age and gender differences on physical capability measures across various five-interval aging groups (from 50 to 90 years old) including the walking speed, grips strength, time to get up and go, and standing balance. The presence of each physical capability measure was first evaluated across the participating cohorts. The measures where then harmonized according to international standards (e.g., the walking speed was measured in meters/second across the cohorts) for each cohort in a manual manner. Descriptive statistics were then calculated for each harmonized measure and subsequently adjusted according to the gender, weight, and height (and the body mass index in some cases) factors. The ensuing statistical analysis revealed significant differences between males and females in the grip strength and the walking speed in small aging groups (e.g., around 50 years and above), as well as the absence of these differences in larger aging groups, yielding useful scientific evidence regarding the effect of gender and age in the physical capability levels across elder people.

The Consortium on Health and Ageing: Network of Cohorts in Europe and the United States (CHANCES) Project [4] is a large-scale, multinational study that involves the harmonization of clinical data from 14 longitudinal cohorts (from 23 European and 3 non-European countries) on aging (more than 60 years in the majority of the cases and more than 50 years in a few cohorts) consisting of 683,228 individuals. The data sharing process fulfilled all the necessary legal and ethical requirements that were presented in Chapter 4. The aim of CHANCES is to examine aging-related factors in different health conditions, including cardiovascular disease, diabetes, cancer, and cognitive impairment, among others, to identify the incidence of each health condition, the prevalence, and the aging and other genetic determinants (biomarkers), such as the risk factors, in the elderly individuals and/or validate the existing ones. The harmonization process involved the execution of a five-step procedure: (i) first, a list of exposures (e.g., lifestyle, medical history, dietary factors, demographics, blood biomarkers) and health outcomes (e.g., prevalence and incidence of cancer, osteoporosis, depression, cognitive impairment, disability) was defined by the clinical experts of the CHANCES, (ii) the parameters from the individual cohorts that were relevant to those from the list were then recruited and compared across the cohorts, where new parameters were also added where applicable, (iii) according to the level of agreement between each variable from the cohorts and the variables from the list, an algorithm was developed to assess the availability, comparability, and quality of the variables in each cohort in terms of the ability of each variable to generate it in the list. The list is similar to the reference model that was already described in Chapter 5. Up to now, the CHANCES Project has been able to harmonize 287 out of 409 variables across the 14 cohorts with the harmonization of the remaining ones being currently under development.

In the small-scale study of Bath [5], data harmonization was applied to interlink and homogenize the clinical data across two European longitudinal cohorts (two cohorts in the United Kingdom and one cohort in the Netherlands) on aging including 1768 patient records. The cohort data included information regarding the demographics (e.g., age, gender), physical (e.g., presence of rheumatism, diabetes, heart disease), behavioral, and mental health status (e.g., depression, anxiety), social activity (e.g., loneliness), and physical activity (e.g., walking, cycling), among others. The aim of this study was to develop guidelines for the harmonization of cohorts on aging and describe the challenges that were raised during the harmonization process. The cohorts were independently analyzed to detect terms with common meaning and/or coding and assess the comparability of the terms. The mental scales were standardized using logistic regression analysis based on the item response theory (IRT) approach (see Section 5.6.3) to detect significant associations between the measurement scales. The resulting harmonized schema consisted of 47 variables and 1768 records across different domains, including demographics, physical health, mental health, and physical activity. The harmonized data can be used to enhance the scientific impact of harmonization for conducting additional comparative studies to examine significant differences and associations between different scores in the elder people.

In the study of Gatz [6], an online questionnaire with multiple depression scales and measures of subjective health was created to generate a crosswalk sample to simulate the lack of common measures across heterogeneous datasets. The questionnaire with the depression scales was distributed to 1061 participants between 30 and 98 years old, and the questionnaire with the multiple measures of subjective health was distributed to 1065 participants. IRT analysis (see Section 5.6.3) approach was applied to compare the multiple depression scale scores across the population and harmonize them. As both of these scales share a common latent factor, it is possible to generate calibrated versions of these depression scales and thus harmonize the measurement scales. The IRT model was able to generate a common metric for translating between two measures of the same construct. Multiple factor analysis was also applied to combine multiple scales to generate new subjective health scales by taking into consideration the correlation between the combined factors.

8.2.2 Obesity

The BioSHaRE-EU Healthy Obese Project (HOP) [7,8] is a large-scale, multinational study that involves the harmonization of clinical data from 10 longitudinal cohorts on obesity (age group: 18–80 years; participating countries: Estonia, Germany, Finland, Italy, the Netherlands, Norway, the United Kingdom) consisting of 163,517 individuals (28,077 with obesity). The aim of the HOP is to address the unmet needs regarding the prevalence in metabolic syndrome and metabolically health obesity. A set of clinical variables was first identified by the clinical experts who are denoted as the target variables in the final harmonized schema along with the recruitment criteria for the discrimination of the metabolically obese patients (e.g., cutoff thresholds for the blood pressure, the triglyceride levels, the high-density lipoproteins, etc.). The final set of variables included mostly laboratory variables (e.g., systolic and diastolic blood pressure, fasting blood glucose, cholesterol levels, hypertriglyceridemia), demographics, behavioral and physical variables, medication history (e.g., myocardial infarction, stroke, diabetes), and prescribed medications, among others. This set of variables was selected according to the prevalence of the obese individuals, and the related lifestyle and risk factors, where for each variable the concept and its measurement unit are defined. The variables along with the descriptions were organized in the form of a DataSchema [28], which was subsequently used to quantify the level of agreement between the variables of each individual cohorts and those in the DataSchema (see Section 5.7). The DataSHaPER tool [29], which was presented in Section 5.7, was then applied to evaluate the potential of generating the variables of the DataSchema from the variables in each individual cohort using lexical matching methods based on a repository of medical terminologies. In a preliminary application, the harmonization process was able to match 96 DataSchema variables across 8 out of 10 cohort studies yielding a 73% matching accuracy. The matching accuracy was larger (i.e., 93%) in the case where the variables covered the medical history and interventions domains. The harmonized database schema enables the integrative analysis of international cohorts on obesity to apply statistical analysis for the detection of associations between the harmonized variables within a federated database system to address the unmet needs regarding the prevalence of the healthy obese individuals (i.e., those that are obese but do not exhibit any metabolic or cardiovascular disease) along with the lifestyle and behavioral risk factors.

8.2.3 Phenotypes and epidemics

In Ref. [9], six steps regarding the scope and application of data harmonization in cancer epidemiology are described. The primary questions for data harmonization were first defined to identify the parameters that efficiently describe the domain of interest, i.e., those that are involved in the relationship between mortality and body mass index. The concept of each parameter is then described in terms of an entity. For example, for a parameter, e.g., gender, a high-level class is usually defined to express this variable in a semantic way. Such entities must be carefully defined for all parameters to enable the high-level conceptual matching with similar ones. Once the complete set of parameters is selected along with the semantic expression of this set of parameters, one must then assess which of these parameters are present across the heterogeneous data. The semantic model must then be updated according to the remaining entities. The next step is to develop common data elements (CDEs) for each concept. The CDEs are defined according to the range of the values of the parameters that are present within the common data schema. For example, a CDE for the variable "gender" can be a categorical CDE with values 0 and 1, where, for example, 1 denotes a female and 0 denotes a male. Another example of a CDE can be the measurement unit of a specific laboratory test (e.g., in mg/day). At the final step, the CDEs between the set of parameters and the heterogeneous dataset are mapped. Inconsistencies between the data elements are also taken into consideration in the case where the value of one data element depends on the value of another. An example of an inconsistency occurs in the case where the patient has not conducted a biopsy, but there is a recorded value for a biopsy score (e.g., the number of foci in a biopsy tissue). The quality of the data harmonization process is also evaluated in terms of incorrect data types, erroneously parsed mappings, and the presence of missing values.

In Ref. [10], the detailed workflow that was developed by the Gene Environment Association Studies (GENEVA) initiative toward phenotype harmonization in genome-wide association studies (GWAS) is described. The phenotype studies are first collected through a data sharing process that determines the type of sharing, i.e., whether it will be partial or sole. The aim of GENEVA is to conduct a cross-study analysis given a specific phenotype. Toward this direction, the phenotype harmonization process begins with a detailed documentation of the available GWAS, the number of available patients, the inclusion and exclusion criteria, the data definitions, and the data distributions, among others. A set of common definitions and values are then specified similar to the reference model (see Chapter 5) to combine data elements into categories or classes and create new variables that summarize the information from two or more measurements, where applicable. Various covariates, such as the age and gender factors, are also identified at this stage for harmonizing these covariates. Toward this direction, a standard set of covariates is also developed to take into consideration the same set of covariates across the harmonized data. The harmonized genotype and phenotype data from each individual GWAS is then stored into a centralized relational database. A variety of data analytics methods can be applied to analyze the harmonized data for the application of statistical analysis on the extracted metadata and the examination of genotype–environment interactions for the construction of epidemiological models and the detection of genetic outliers through a genotype cleaning workflow [10]. Pitfalls introduced during the data harmonization process were also described including the diversity of the common phenotypes, the alignment of the genotype time frames, and the imputation of missing fields, among others.

An example of perhaps the largest initiative toward medical data sharing and harmonization is described in Ref. [11], where the DataSHaPER approach (see Section 5.7) was applied to harmonize risk factors and outcomes across 53 of the world's largest epidemiological studies including more than 6.9 million participants. The goal was to identify and quantify the importance and coverage of gold standard variables across multiple epidemiological research studies. The studies were selected according to ethical and legal guidelines that were provided by the P3G2 (Policy Partnerships Project for Genomic Governance), which offers a catalog [12] with existing frameworks for biobanks, genomic, and other related databases. The 53 selected studies fulfilled specific criteria regarding the existence of common biological samples, health outcomes, cognitive measures, data collection dates, demographics, and types of questionnaires. These criteria were separated into domains according to the DataSchema (see Section 5.7), which is similar to the reference model [8]. A DataSchema consisting of 148 variables (based on the ICD10 standard) was then constructed, where each variable belongs to a specific domain, such as the demographics, patient history, medical conditions, health-related questionnaires, etc. The percentage of coverage for each variable in the DataSchema was then assessed. Out of 148 variables, 38 were characterized as of high importance for epidemiological research, 45 variables were characterized as of medium importance, and 65 as low importance. According to the data harmonization process (see Section 5.7), 53% of the variables were characterized as partially and/or complete and 50% as either not possible or not applicable. Examples of harmonized variables include the systolic and diastolic blood pressure, the physical activity levels, the use of alcohol, and the body mass index, among others.

In Ref. [13], a large-scale study was conducted to harmonize epidemiological respiratory (spirometry) data across 9 longitudinal cohorts in the United States consisting of 65,251 individuals. The cohorts were part of the NHLBI (National Heart, Lung, and Blood Institute) Pooled Cohorts Study [14], which was initially designed for the purposes of cardiovascular epidemiology and then to measure spirometry measures (e.g., lung function) across individuals including chronic lower respiratory disease, chronic obstructive pulmonary disease, asthma, and dyspnea, among others. A set of main measures was first identified, including inhalational exposures, respiratory measures, medications, symptoms, and demographics, among others. A set of potentially harmonized variables was then identified on each cohort where a variable was defined as harmonized if it was present in more than two cohorts. Standardized formats and range values were then defined for each variable (e.g., "0" denotes "no" and "1" denotes "yes"). Rules were developed to match variables describing the same concept. Covariates, such as the age, ethnicity, and gender, were also taken into consideration during the statistical analysis process to measure the within- and between-subject variability on the individual spirometry measures. Furthermore, a quality control pipeline was applied before the harmonization process to remove outliers and handle missing values. According to the overall workflow, the subjects were categorized according to the number of valid spirometry examinations. During the data harmonization process, spirometry measures with reduced variability and outliers were removed to further reduce biases yielding harmonized data with reduced within-study measurement error, as well as between-study heterogeneity to minimize the classification error. The harmonized data are suitable for metaanalysis to advance the research in respiratory epidemiology and develop guidelines to leverage decades of research toward the development of standardized protocols and gold standard measures.

8.2.4 Autoimmune diseases

The European Project on OSteoArthritis (EPOSA) [15] is a small-scale study that involves the harmonization of clinical data from five European longitudinal cohorts (in Germany, Italy, the Netherlands, Spain, and the United Kingdom) on aging consisting of 10,107 individuals with mild and/or severe osteoarthritis. The data sharing process fulfilled all the necessary legal and ethical requirements that were presented in Chapter 4. The aim of EPOSA is to investigate the determinants of personal and societal factors of osteoarthritis on the elder European population in terms of the prevalence and impact on the quality of life. A postharmonization process was adopted, where the presence of each variable across at least three cohorts is first assessed. Then the variables that are present in less than three cohorts are ignored from the harmonization process. For each one of the remaining variables, the concept and the values are described and evaluated. If the concept of a variable matches with the concept of the recruitment criteria, the harmonization process is applied by standardizing the units yielding high-quality harmonized variables across several domains, including demographics, socioeconomic, chronic diseases, and physical performance, among others.

8.2.5 Personality scores

The Genetics of Personality Consortium [16] is a large-scale, multinational study that involves the harmonization of clinical data from 23 longitudinal cohorts (Europe, the United States, and Australia) on personality traits (neuroticism and extraversion scores) consisting of more than 160,000 individuals. The goal of the study was to harmonize relevant personality traits across nine different personality repositories. Toward this direction, IRT analysis (see Section 5.6.3) was used to link the individual cohort data to common metrics for the neuroticism and extraversion scores, respectively. In fact, the personality scores were estimated according to the given number of items on each repository based on the probability of a specific response in an item yielding comparable (nonbiased) personality scores. The differential item functioning factor was computed (as in Section 5.6.3) to test for the variance of an item across the cohorts. The estimated scores were then compared against their independence with the original scores, as well as the noninvariance (i.e., the measurement noninvariance) across the cohorts yielding independent and robust estimated values. The success of the IRT analysis toward data harmonization was accomplished by identifying a genetic variant that is associated with personality.

In Ref. [17], a small-scale study was conducted toward the harmonization and subsequent integration of two cohort studies, namely the Terman Life Cycle Study and the Hawaii Personality and Health Longitudinal Study on personality traits. More specifically, the goal of this study was to harmonize child personality traits across the two cohorts to generate child personality factors and finally integrate the harmonized factors to apply a personality–health model. The latter uses the child personality traits or factors (e.g., extraversion, agreeableness, conscientious, emotional stability, and intellect) to predict illness reports, mental adjustment, and self-rated health traits by taking into consideration health behavior factors along with educational factors (which serve as mediators), as well. The results across the two cohorts validate the success of data harmonization toward the prediction of personality and health factors to detect physiological dysfunction and mortality risk factors during the life span development.

8.2.6 Other case studies

In Ref. [18], an alternative, medium-scale study was conducted across six cohorts on cognitive impairment and dementia to prestatistically harmonize cognitive measures toward the investigation of the correlation between the blood pressure and the cognitive impairment and dementia, as well. The main scope of the study was to examine the statistical diversity that exists among the harmonized data, which might introduce measurement biases during the metaanalysis procedure of the psychometric variables. Toward this direction, concepts from the IRT (see Section 5.6.3) were used to reduce the measurement biases between the cognitive score ranges and the scoring procedures across the cohorts by determining the degree of equivalence between these scores. A notable amount of statistical variability was discovered in the format, content, administration, and scoring procedures, during the cognitive measurement, among the cohorts. Such statistical variabilities were reported to be taken into consideration before the harmonization process to reduce the variability between the cohort data to be harmonized.

The Mechanisms of the Development of ALLergy (MeDALL) Project [19] is a large-scale study that was designed to provide a great insight on the underlying structure of asthma and allergies across 15 birth cohorts (in 9 European countries) with 47,998 children. The goal of the project was to address the unmet needs in epidemics and especially in allergies regarding the allergy onset and progress over time and specifically on how the environmental factors affect the different allergy phenotypes. Toward this direction, genetic data, immunological data, and biological data were recruited to address these clinical needs. The data harmonization process was based on the DataSHaPER strategy according to which a DataSchema consisting of 137 variables was initially constructed by the experts in the previous medical data domains. The inferential equivalence of each variable was quantified in terms of "partial," "complete," or "impossible" with those from the DataSchema. Pairing rules (see Chapter 5) were defined to match variables that share a common conceptual basis across 3551 variables in total. Approximately 70% of the variables were matched as "complete," whereas the remaining ones were classified as "partial" or "impossible." The harmonized database schema can be used for the development of mathematical models based on biomarkers toward the early diagnosis, prevention, and targeted therapeutic treatment of allergies.

In Ref. [20], the importance of standardizing studies of preeclampsia is highlighted to understand the association between pregnancy and future cardiovascular health. Toward this direction, the CoLAB Preeclampsia angiogenic factor study [30] is presented as a promising platform that was designed to interlink laboratory measurements from 23 cohorts with pregnancy data (approximately 16,516 pregnancies) toward the validation of a new, potential biomarker for preeclampsia. The success of the data standardization process was validated as the subsequent integration of the standardized data was able to identify the PIGF as a prominent biomarker for preeclampsia, with increased statistical power and generalizability.

8.2.7 Ongoing projects

The Canadian Partnership for Tomorrow Project (CPTP) [21] is an ongoing project that aims to harmonize five Canadian cohorts (British Columbia Generations Project, Alberta's Tomorrow Project, the Ontario Health Study, CARTaGENE [Quebec], and the Atlantic Partnership for Tomorrow's Health) with more than 300,000 individuals toward the detection of prominent risk factors in cancer and other chronic diseases. The cohort data (12 datasets in total) included health and risk factor questionnaires, physical measures, and biological samples, among others. The data harmonization process was based on the guidelines of the Maelstrom Research group [22] (see Chapter 5) where a DataSchema with 694 variables was developed by the experts as the set of harmonized variables. For each DataSchema variable, additional information was also included, such as the meaning, the format, the class, and the measurement units. Tools that were part of the BioSHaRE Project [7] were then used to quantify the potential of each variable in the heterogeneous cohort data to generate each DataSchema variable. The DataSchema variables that were able to be generated by at least two cohorts were then maintained, whereas the rest of the variables were ignored. The evaluation process was able to identify 81.6% (6799 variables) of the DataSchema variables across the cohorts with a final number of 307,017 individuals. More than 60% of the variables were mapped with those from the DataSchema yielding a harmonized database schema of 432 terms (apart from one case where the number was 188). The database schema includes harmonized health and risk factors data that can be used to address the unmet needs in cancer and other chronic diseases.

The EUROlinkCAT (Establishing a linked European Cohort of Children with Congenital Anomalies) [23] is an EU-funded project and a promising data sharing initiative that envisages to interlink child cohorts and biobanks to provide a centralized data management platform with more than 200,000 children who have been diagnosed with congenital anomalies (CA). The EUROlinkCAT initiative makes use of the existing EUROCAT infrastructure [24] to interlink 21 medical registries including CA data from children (up to age 10) across Europe. The initiative has recently published a common data model (i.e., a reference model) [25] including standard descriptions and formats that can enable the coanalysis of the diverse coding terms that are present across multiple registries. The coanalysis is expected to yield risk factors and genetic variants for children with CA.

The ESCAPE-NET (European Sudden Cardiac Arrest network: toward Prevention, Education and NEw Treatment) [26] is an example of an EU-funded project where European scientific teams have been gathered to design sudden cardiac arrest prevention and treatment strategies by combining existing European databases. The goal of ESCAPE-NET is to identify the occurrence of sudden cardiac arrest (SCA) and develop related risk factors and resuscitation guidelines to prevent SCA and thus shed light into the survival chances and prevalence in patients who have been diagnosed with SCA. Toward this direction, 11 cohorts will be recruited with approximately 85,790 individuals with SCA. The cohort data consist of clinical measures and DNA samples, among others. Five prospective cohorts will be also included to increase the total number of the overall population by 53,060 individuals. The harmonized data will be analyzed to investigate the role of environmental stress in SCA occurrence and survival after SCA.

The LifeBrain Project [27] is another example of an ongoing EU-funded project, which aims to harmonize brain, cognitive, and mental health measures along with genetic data of more than 6000 individuals across 11 European cohorts. The goal of LifeBrain is to identify prominent risk and protective factors throughout the individual life span concerning the onset and progress of brain disorders including mental and cognitive disorders. Toward this direction, brain imaging datasets from European centers along with genetic data will be harmonized and subsequently integrated to increase the statistical power of the study. In addition, the project aims to develop new health policies and guidelines for new and effective therapies along with preventive strategies for the early diagnosis of brain disorders and prominent determinants. Methods for the BioSHaRE-EU–funded project [7] will be adopted to define the harmonized variables with standardized scales. Standard procedures, such as normalization and gray matter versus white matter volume adjustment, will be applied in the neuroimaging data to reduce measurement biases. Methods from the IRT analysis (see Section 5.6.3) have also been adopted to reduce the measurement bias across the clinical data.

The HarmonicSS–HARMONIzation and integrative analysis of regional, national, and international Cohorts on primary Sjögren's syndrome (pSS) project [31] is an ongoing EU-funded project that aims to harmonize 23 regional, national, and international longitudinal cohorts of patients diagnosed with pSS by taking into consideration ethical, legal, and privacy issues to construct a federated cloud-based cohort. On the latter, data mining, genetic data analytics, data governance, and visual analytics methods have been developed, as well as tools for clinical trial patient selection. The goal of HarmonicSS is to address the unmet needs in pSS toward the validation of existing biomarkers and/or the identification of new ones to shed light into the disease's onset and progress, the development of highly robust patient stratification models for detecting high-risk individuals along with lymphomagenesis models and the investigation of new, cost-effective therapies for effective treatment.

A reference model was developed by the clinical experts of the HarmonicSS consortium including a set of parameters that efficiently describe the requirements of the pSS domain knowledge. The pSS reference model [32] includes a set of more than 100 pSS-related parameters that are related to various laboratory tests, demographics, lifestyle factors, medical conditions, and interventions, among others. For the purposes of the data harmonization process, the pSS reference model has been transformed into an ontology using classes and subclasses to describe the parameters and enable the semantic interlinking of heterogeneous ontologies from different cohorts. In fact, HarmonicSS has adopted a semiautomated data harmonization schema (see Section 5.7) to enable the harmonization of ontologies. A data curation pipeline is first applied on the raw cohort data to remove outliers and deal with missing values and incompatibilities, as well as fix any inconsistent fields [33]. Metadata regarding the value range and data type of each feature are then extracted to formulate the basis for the application of the data harmonization process. The metadata are transformed into an ontology that is aligned with the reference pSS ontology in a semiautomated manner. All these functionalities have been incorporated into a GDPR-compliant, federated cloud computing platform fulfilling all the necessary legal and ethical requirements for data sharing and protection, where the cohort data are first evaluated for their compliance by a committee and stored in secure, distributed databases (i.e., private cloud spaces) to enhance the security of the sensitive data.

Until now, the cohort data harmonization process has been able to match more than 90% of the terms (which are relevant to the pSS reference model) in 21 out of 23 cohorts, where the pSS reference model is used as a gold standard to identify and match the pSS-related terms in each cohort. Distributed big data mining and genetic data analytics tools, along with patient selection tools for multinational clinical trials, have been already incorporated on the cloud computing platform to develop lymphomagenesis and lymphoma prediction models for patient stratification purposes, as well as to detect new biomarkers and/or validate the existing ones and propose more targeted therapeutic treatments to address the unmet needs in pSS. Machine learning algorithms varying from simple linear and logistic regression algorithms and Naïve Bayes to artificial neural networks and gradient boosting tree ensembles have been adopted and extended to support distributed learning through a process that is known as incremental learning (Section 7.4).

8.2.8 Summary

A summary of the aforementioned case studies in medical data sharing and harmonization is depicted in Table 8.1. For each case study, the following information is summarized: (i) the domain of the study, (ii) the cohorts involved in the study, (iii) the aim of the study, and (iv) the outcome and clinical unmet needs that were/will be addressed by each study along with the scientific outcomes.

8.3 Conclusions

Data harmonization bridges the gap between data sharing and data analytics in such a way so that the coanalysis of the shared data can yield scientific outcomes with increased statistical power. A metaanalysis of the harmonized data can then be applied to deal with the unmet needs in different medical domains. A notable

Table 8.1 A detailed summary of the case studies/projects in medical data sharing and harmonization.

Case study	Domain	Cohorts	Aim	Outcomes and unmet needs that were/will be addressed
[1]	Aging[c]	11 (11,541 individuals)	Harmonize genetic profiles, phenotypes, medical history, laboratory and physical examinations on different neuropathologies among the elder people (SMART Project)	A database that will be used to investigate clinical phenotype and risk factors for various types of neuropathologies toward the detection of genetic associations, comorbidities, and biomarkers in the field of neurological disorders and epidemiology
[2]	Aging[b]	6 (11,557 individuals)	Harmonize cognitive scores, depression scores, physical functioning, demographics and social networking data among the elder people (CLESA Project)	A final harmonized database schema consisting of 111 parameters that were used to identify multiple factors and conditions predicting the morbidity and mortality across elder people
[3]	Aging[b]	9 (40,000 individuals)	Investigate age and gender differences on physical capability measures across five aging groups (HALCyon Project)	A manual data harmonization process was applied along with a statistical analysis that revealed significant differences in the physical capability levels across elder people, as well as the effect of age and gender
[4]	Aging[c]	14 (683,228 individuals)	Examine aging-related factors in different health conditions (e.g.,	A highly qualified, harmonized database schema including 287 out of

Table 8.1 A detailed summary of the case studies/projects in medical data sharing and harmonization.—*cont'd*

Case study	Domain	Cohorts	Aim	Outcomes and unmet needs that were/will be addressed
			cardiovascular disease) to identify risk factors and biomarkers among the elder people (CHANCES Project)	409 variables (e.g., health outcomes, risk factors) across the cohorts who were able to generate the variables from a reference model
[5]	Aging[a]	2 (1768 patient records)	Interlink and harmonize data across two cohorts to develop guidelines for the harmonization of cohorts on aging, as well as the challenges that are posed during the harmonization process	A harmonized database schema consisting of 47 variables and 1768 records across different domains including demographics, physical activity, mental health, etc., which can be used to identify significant differences and associations between physical and mental health scores
[6]	Aging[a]	1065 individuals	Simulate the lack of common measures across heterogeneous datasets	IRT analysis was used to generate a common metric between measures of the same construct, and multiple factor analysis was used to combine multiple scales to generate new unbiased health scales
[7,8]	Obesity[b]	10 (163,517 individuals)	Address the unmet needs regarding the prevalence in metabolic syndrome and	A harmonized database schema with 96 DataSchema variables identified

Continued

Table 8.1 A detailed summary of the case studies/projects in medical data sharing and harmonization.—*cont'd*

Case study	Domain	Cohorts	Aim	Outcomes and unmet needs that were/will be addressed
			metabolically health obesity (BioSHaRE-EU Healthy Obese Project)	across 8 out of 10 cohorts that can be used to reveal lifestyle and behavioral risk factors among healthy obese individuals
[9]	Phenotypes and epidemics	–	Present guidelines and methodological steps for data harmonization along with methods for evaluating the quality of the harmonized data	Six methodological steps and examples of common data elements and their precise definition toward the success of the data harmonization process
[10]	Phenotypes and epidemics	–	Develop a cross-study methodology for the harmonization of phenotype and genotype data across epidemiological studies	Present the Gene Environment Association Studies (GENEVA) initiative that focuses on the development of guidelines and reference models toward the generation of high-quality, harmonized phenotype and genotype data
[11]	Phenotypes and epidemics[c]	53 (6.9 million individuals)	Identify and quantify the importance of gold standard variables across a large number of epidemiological studies (DataSHaPER)	A general purpose DataSchema for epidemiological studies consisting of 148 variables was generated and the data harmonization process was able to identify 53% of the variables as partially or complete and 50% as not possible

Table 8.1 A detailed summary of the case studies/projects in medical data sharing and harmonization.—*cont'd*

Case study	Domain	Cohorts	Aim	Outcomes and unmet needs that were/will be addressed
				or not applicable to be harmonized
[13]	Phenotypes and epidemics[b]	9 (65,251 individuals)	Harmonize epidemiological respiratory data for the purposes of respiratory epidemiology, e.g., asthma, dyspnea, chronic lower respiratory disease (NHLBI Pooled Cohorts Study)	A harmonized database schema along with a standard protocol where pairing rules were used to identify terms that share a common concept that can be used to advance research in respiratory epidemiology and identify risk factors
[15]	Autoimmune diseases[b]	5 (10,107 individuals)	Investigate the determinants of personal and societal factors of mild and/or severe osteoarthritis in terms of the prevalence and impact on the quality of life (EPOSA Project)	High-quality harmonized variables across different domains including demographics, chronic diseases, and physical performance, among others, which can be analyzed to study the prevalence and impact on the quality of life
[16]	Personality scores[c]	23 (>160,000 individuals)	Harmonize personality traits across different personality repositories to detect biomarkers in personality (GPC)	IRT analysis was able to reduce measurement biases and the variance among the variables yielding robust estimated values that were used to identify a prominent genetic variant associated with personality

Continued

Table 8.1 A detailed summary of the case studies/projects in medical data sharing and harmonization.—*cont'd*

Case study	Domain	Cohorts	Aim	Outcomes and unmet needs that were/will be addressed
[17]	Personality scores[a]	2	Harmonize child personality traits to generate child personality factors to apply a personality-health model	The integration of the data from the two cohorts was able to yield a personality-health model that detects physiological dysfunction and risk factors during the life span development
[18]	Cognitive impairment and dementia[b]	6	Examine the statistical diversity that exists among the harmonized data that might introduce measurement biases during the metaanalysis	Statistical variability was discovered in the format, content, administration, and scoring procedures during the cognitive measurement, which have been highlighted for further consideration
[20]	Pregnancies and cardiovascular heath[c]	23 (16,516 pregnancies)	Understand the relationship between pregnancy and cardiovascular health (CoLAB)	A platform that enables the subsequent integration of the standardized data that were able to identify a new biomarker for preeclampsia
Ongoing initiatives/projects				
[21]	Cancer and other chronic diseases[a]	5 (300,000 individuals)	Detect prominent risk factors in cancer and other chronic diseases (CPTP Project)	The DataSHaPER strategy was adopted, where a DataSchema consisting of 694 variables was used as the reference model (so far). Pairing rules were preliminary defined to harmonize terms

Table 8.1 A detailed summary of the case studies/projects in medical data sharing and harmonization.—*cont'd*

Case study	Domain	Cohorts	Aim	Outcomes and unmet needs that were/will be addressed
				where 60% of the variables were identified as complete to be matched with. The harmonized database schema will be used to detect prominent risk factors in cancer and other chronic diseases
[19]	Asthma and allergies[c]	15 (47,998 children)	Provide insight into the underlying structure of asthma and allergies regarding the onset and progress over time, as well as on how the environmental factors affect the allergy phenotypes (MeDALL Project)	The DataSHaPER strategy was adopted, where a DataSchema consisting of 147 variables was used as the reference model (so far). Pairing rules were preliminary defined to harmonize terms where 70% of the variables were identified as complete to be matched with. The harmonized database schema will be used to detect environmental factors that affect the allergy phenotypes
[23]	Congenital anomalies in children[c]	21 (>200,000 children)	Interlink medical registries with children who have been diagnosed with congenital anomalies	A common data model was recently published including standard descriptions and formats that enable the coanalysis of

Continued

Table 8.1 A detailed summary of the case studies/projects in medical data sharing and harmonization.—*cont'd*

Case study	Domain	Cohorts	Aim	Outcomes and unmet needs that were/will be addressed
			(EUROlinkCAT Project)	diverse coding terms among the heterogeneous medical registries
[26]	Sudden cardiac arrest (SCA)[c]	11 (35,790 individuals)	Design SCA prevention and treatment strategies (ESCAPE-NET Project)	The coanalysis of the harmonized data is expected to yield risk factors and resuscitation guidelines to prevent SCA and provide insight on the survival chances and prevalence in SCA
[27]	Mental and cognitive disorders[c]	11 (6000 individuals)	Identify risk and protective factors throughout an individual's life span regarding mental and cognitive disorders (LifeBrain Project)	Methods from the BioSHaRE EU–funded project will be adopted to create a DataSchema, as well as apply standard procedures regarding the adjustment of medical imaging data and the application of IRT analysis to reduce measurement biases and thus identify prominent determinants for the early diagnosis of mental and cognitive disorders
[31]	Autoimmune diseases[c]	23 (10,000 individuals)	Harmonize clinical and laboratory measures and coanalyze the harmonized data to address the unmet	• An innovative cloud computing platform for medical data sharing,

Table 8.1 A detailed summary of the case studies/projects in medical data sharing and harmonization.—*cont'd*

Case study	Domain	Cohorts	Aim	Outcomes and unmet needs that were/will be addressed
			needs in primary Sjögren's Syndrome including robust patient stratification, the identification of new therapies, and validation/detection of existing/new biomarkers (HarmonicSS Project)	harmonization, and analytics • A reference model with more than 100 terms related to the primary Sjögren's Syndrome was constructed as a gold standard protocol for the analysis of autoimmune diseases along with a semiautomated data harmonization tool that is used to quantify the level of agreement of each variable across the cohorts • The harmonized data will be analyzed to develop lymphoma prediction models and detect new biomarkers and cost-effective therapies

[a] *denotes a small-scale study (less than or equal to 5 cohorts).*
[b] *denotes a medium-scale study (more than 5 cohorts but less than or equal to 10 cohorts).*
[c] *denotes a large-scale study (more than 10 cohorts).*

progress has been made within our decade regarding the clinical value and scientific impact of medical data sharing and harmonization in healthcare to deal with the unmet needs in a variety of medical domains. The technical advances in the field of medical data harmonization are mainly based in the use of a gold standard model, which serves as a common (reference) data schema for the harmonization of heterogeneous data schemas. The fact that all the case studies adopt a common data schema, such as an ontology, is important not only for the high-level description of a disease's domain knowledge using a set of disease-related parameters, such as laboratory examinations, demographics, lifestyle factors, etc., but also for the application of computational methods, such as lexical and semantic matching methods to enable the semiautomated alignment of the common terms or concepts between two or more ontologies.

The majority of the case studies that were presented in this chapter have clearly demonstrated the value of medical data sharing and harmonization in addressing the unmet needs toward the identification of new biomarkers in pregnancy and cardiovascular health, aging, autoimmune diseases, phenotypes and epidemics, and personality scores. The harmonized database schemas and methodological advances that have been already developed in different case studies on aging [1–6], obesity [7,8], autoimmune diseases [15], personality scores [16,17], phenotype and epidemics [9–13], and cognitive and mental disorders [18], among others have been used to identify multiple factors and conditions in mortality across elder people [2], health outcomes, and risk factors in aging [1–6], as well as to study the effect of gender and age in the physical capability levels across elder people [3], to reveal prominent genetic variants that are associated with personality scores [16,17], and risk factors during the life span development [17], to identify risk factors in epidemiology [9–13], new biomarkers for preeclampsia [20], and finally phenotypes, risk factors, and biomarkers for different types of neuropathologies [1], among others.

Furthermore, the large number of ongoing initiatives and projects in the European Union and the United States regarding medical data sharing, harmonization, and analytics are very promising. Indeed, initiatives, such as (i) CPTP [21] that aims to detect prominent determinants in cancer and other chronic diseases, (ii) MeDALL Project [19] that aims to provide insight into the underlying structure of asthma and allergies to understand how environmental factors can affect the allergy phenotypes, (iii) the EUROlinkCAT Project [23] that focuses on the interlinking of medical registries with children who have been diagnosed with CA, (iv) ESCAPE-NET Project [26] that aims to develop prevention and treatment guidelines for SCA, (v) the LifeBrain Project [27] that focuses on the identification of risk factors regarding mental and cognitive disorders, and (vi) the HARMONIzation and integrative analysis of regional, national, and international Cohorts on pSS—HarmonicSS [31] Project that aims to address the unmet needs in pSS, envisage to promote medical data sharing, harmonization, and analytics into the broader scientific community. The majority of these initiatives have adopted a data harmonization strategy that was initially described into the BioSHaRE Project [7,8], a project that serves as pioneer in the field of medical data harmonization. The HarmonicSS project has launched a cloud computing platform that adopts the medical data sharing, harmonization, and analytics workflow that was presented in this book to deal with the unmet needs in pSS.

References

[1] Abner EL, Schmitt FA, Nelson PT, Lou W, Wan L, Gauriglia R, et al. The Statistical Modeling of Aging and Risk of Transition Project: data collection and harmonization across 11 longitudinal cohort studies of aging, cognition, and dementia. Obs Stud 2015;1(2015):56.

[2] Minicuci N, Noale M, Bardage C, Blumstein T, Deeg DJ, Gindin J, et al. Cross-national determinants of quality of life from six longitudinal studies on aging: the CLESA project. Aging Clin Exp Res 2003;15(3):187−202.

[3] Cooper R, Hardy R, Sayer AA, Ben-Shlomo Y, Birnie K, Cooper C, et al. Age and gender differences in physical capability levels from mid-life onwards: the harmonisation and meta-analysis of data from eight UK cohort studies. PLoS One 2011;6(11): e27899.

[4] Boffetta P, Bobak M, Borsch-Supan A, Brenner H, Eriksson S, Grodstein F, et al. The Consortium on Health and Ageing: network of Cohorts in Europe and the United States (CHANCES) project—design, population and data harmonization of a large-scale, international study. Eur J Epidemiol 2014;29(12):929−36.

[5] Bath PA, Deeg D, Poppelaars JAN. The harmonisation of longitudinal data: a case study using data from cohort studies in The Netherlands and the United Kingdom. Ageing Soc 2010;30(8):1419−37.

[6] Gatz M, Reynolds CA, Finkel D, Hahn CJ, Zhou Y, Zavala C. Data harmonization in aging research: not so fast. Exp Aging Res 2015;41(5):475−95.

[7] Doiron D, Burton P, Marcon Y, Gaye A, Wolffenbuttel BH, Perola M, et al. Data harmonization and federated analysis of population-based studies: the BioSHaRE project. Emerg Themes Epidemiol 2013;10(1):12.

[8] van Vliet-Ostaptchouk JV, Nuotio ML, Slagter SN, Doiron D, Fischer K, Foco L, et al. The prevalence of metabolic syndrome and metabolically healthy obesity in Europe: a collaborative analysis of ten large cohort studies. BMC Endocrine Disorders 2014; 14(1):9.

[9] Rolland B, Reid S, Stelling D, Warnick G, Thornquist M, Feng Z. Toward rigorous data harmonization in cancer epidemiology research: one approach. Am J Epidemiol 2015; 182(12):1033−8.

[10] Bennett SN, Caporaso N, Fitzpatrick AL, Agrawal A, Barnes K, Boyd HA, et al. Phenotype harmonization and cross-study collaboration in GWAS consortia: the GENEVA experience. Genet Epidemiol 2011;35(3):159−73.

[11] Fortier I, Doiron D, Little J, Ferretti V, L'Heureux F, Stolk RP, et al. Is rigorous retrospective harmonization possible? Application of the DataSHaPER approach across 53 large studies. Int J Epidemiol 2011;40(5):1314−28.

[12] Policy Partnerships Project for Genomic Governance (P3G2). http://p3g2.org/.

[13] Oelsner EC, Balte PP, Cassano PA, Couper D, Enright PL, Folsom AR, et al. Harmonization of respiratory data from 9 US population-based cohorts: the NHLBI Pooled Cohorts Study. Am J Epidemiol 2018;187(11):2265−78.

[14] National Heart, Lung, and Blood Institute (NHLBI). https://www.nhlbi.nih.gov/.

[15] Schaap LA, Peeters GM, Dennison EM, Zambon S, Nikolaus T, Sanchez-Martinez M, et al. European Project on OSteoArthritis (EPOSA): methodological challenges in harmonization of existing data from five European population-based cohorts on aging. BMC Musculoskelet Disord 2011;12(1):272.

[16] Van Den Berg SM, De Moor MH, McGue M, Pettersson E, Terracciano A, Verweij KJ, et al. Harmonization of neuroticism and extraversion phenotypes across inventories and cohorts in the genetics of personality consortium: an application of item response theory. Behav Genet 2014;44(4):295–313.
[17] Kern ML, Hampson SE, Goldberg LR, Friedman HS. Integrating prospective longitudinal data: modeling personality and health in the terman life Cycle and Hawaii longitudinal studies. Dev Psychol 2014;50(5):1390.
[18] Briceño EM, Gross AL, Giordani B, Manly JJ, Gottesman RF, Elkind MS, et al. Pre-statistical harmonization of cognitive measures across six population-based cohorts: ARIC, CARDIA, CHS, FHS, MESA, and NOMAS. Alzheimer's & Dement 2018; 14(7):P1611–2.
[19] Benet M, Albang R, Pinart M, Hohmann C, Tischer CG, Annesi-Maesano I, et al. Integrating clinical and epidemiologic data on allergic diseases across birth cohorts: a harmonization study in the Mechanisms of the development of allergy project. Am J Epidemiol 2018;188(2):408–17.
[20] Staff AC, Redman CW, Williams D, Leeson P, Moe K, Thilaganathan B, et al. Pregnancy and long-term maternal cardiovascular health: progress through harmonization of research cohorts and biobanks. Hypertension 2016;67(2):251–60.
[21] Fortier I, Dragieva N, Saliba M, Craig C, Robson PJ. Harmonization of the health and risk factor questionnaire data of the Canadian partnership for Tomorrow project: a descriptive analysis. CMAJ open 2019;7(2):E272.
[22] Fortier I, Raina P, Van den Heuvel ER, Griffith LE, Craig C, Saliba M, et al. Maelstrom Research guidelines for rigorous retrospective data harmonization. Int J Epidemiol 2017;46(1):103–5.
[23] Rankin J, Glinianaia S, Morris J, Loane M, Garne E. EUROlinkCAT: establishing a linked European cohort of children with congenital anomalies. A collaborative project of European registries of congenital anomalies. BJOG-AN Intl J Obstet Gynaecol 2019; 126(S1):131.
[24] Morris JK, Wellesley DG, Barisic I, Addor MC, Bergman JE, Braz P, et al. Epidemiology of congenital cerebral anomalies in Europe: a multicentre, population-based EUROCAT study. Arch Dis Child 2019. pii: archdischild-2018-316733. doi:10.1136/archdischild-2018-316733.
[25] Loane M, Densem J, Morris J, Tan J. EUROlinkCAT: common data model. Intl J Popul Data Sci 2018;3(2).
[26] Tan HL, Dagres N, Böttiger BW, Schwartz PJ. European sudden cardiac arrest network: towards prevention, education and new effective treatments (ESCAPE-NET): a major European Horizon 2020 project focused on cardiac arrest. Eur Heart J 2018;39(2): 86–8.
[27] Walhovd KB, Fjell AM, Westerhausen R, Nyberg L, Ebmeier KP, Lindenberger U, et al. Healthy minds 0–100 years: optimising the use of European brain imaging cohorts ("Lifebrain"). Eur Psychiatry 2018;50:47–56.
[28] Kourou KD, Pezoulas VC, Georga EI, Exarchos TP, Tsanakas P, Tsiknakis M. Cohort harmonization and integrative analysis from a biomedical Engineering perspective. IEEE Rev Biomed Eng 2018;12:303–18.
[29] Fortier I, Burton PR, Robson PJ, Ferretti V, Little J, L'Heureux F. Quality, quantity and harmony: the DataSHaPER approach to integrating data across bioclinical studies. Int J Epidemiol 2010;39(5):1383–93.
[30] Global Pregnancy Collaboration (CoLab). https://pregnancycolab.tghn.org/.

[31] HARMONIzation and integrative analysis of regional, national and international Cohorts on primary Sjögren's Syndrome (pSS) towards improved stratification, treatment and health policy making. Link: https://www.harmonicss.eu/.

[32] Pezoulas VC, Exarchos TP, Andronikou V, Varvarigou T, Tzioufas A, De Vita S, Fotiadis DI. Towards the establishment of a biomedical ontology for the primary Sjögren's Syndrome. In: Proceedings of the IEEE engineering in medicine and biology conference; 2018. p. 4089−92.

[33] Pezoulas VC, Kourou KD, Kalatzis F, Exarchos TP, Venetsanopoulou A, Zampeli E, et al. Medical data quality assessment: on the development of an automated framework for medical data curation. Comput Biol Med 2019;107:270−83.

CHAPTER

Conclusions and future trends

9

Chapter outline

9.1 Conclusions

Why is medical data sharing necessary? What is data harmonization in the clinical domain? How can we ensure the security and quality of the medical data? What is the scientific value of a cloud infrastructure in health care? What techniques can be recruited to coanalyze the harmonized data? These are some of the questions which were adequately answered within the context of this book. As we have already seen, everything begins from data sharing. The interlinking of medical data from different medical data sources is the basis for the development of a data sharing framework which is able to address the unmet needs in various diseases, including the development of robust patient stratification models, the development of algorithms for biomarker detection, and novel methods for proposing cost-effective therapies and treatment monitoring. The types of medical data are many (Chapter 2), varying from biosignals and medical images to omics and laboratory tests, which can be subsequently found in a variety of medical data sources varying from patient registries, health sensors, and electronic health records, to genome registries, clinical trials, cohorts, and clinical claims. These massively accumulated types of data, on a daily basis, have given rise to the ambiguous term "big data," which overcomes the traditional definition of the ordinary data by extending the "volume" property into more dimensions (i.e., velocity, veracity, and variety) yielding large data elements which are capable of addressing the clinical unmet needs across different medical domains.

Every data sharing framework, however, must first take into account any legal and ethical issues which are posed during the sharing of sensitive personal data. Toward this direction, crucial data protection legislations (Chapter 4), such as the General Data Protection Regulation (GDPR) in the European Union and the Health Insurance Portability and Accountability Act (HIPAA) in the United States, along with their intersection, i.e., the EU–US Privacy Shield and popular data sharing initiatives, such as the United Nation's resolution 68/167, the Privacy Guidelines

Medical Data Sharing, Harmonization and Analytics. https://doi.org/10.1016/B978-0-12-816507-2.00009-8

issued by the Organisation for Economic Co-operation and Development (OECD) on the Protection of Privacy and Transborder Flows of Personal Data and those from the Council of Europe Convention 108 and the International Data Protection Commissioner's (IDPC) Initiative, have been extensively described to provide great insight into the legal basis along with the strengths and weaknesses of such initiatives. Seven principles (i.e., qualified data, security safeguards, data minimization, respect to the individual rights, open actions and policies, lawful processing, purposes of processing), which lie in the intersection of these data sharing initiatives, were identified to design a strategy for the responsible sharing of medical data which can be adopted by any global data protection framework worldwide. More specifically, the data must be qualified in terms of accuracy, completeness, adequacy, and relevance, as well as be up to date. Security safeguards must be also taken into consideration before data sharing. The individuals must have the right to object, acquire, and edit their personal data. Any data processing action must be conducted under the purpose of processing which must be described before the sharing of sensitive data. Finally, any unnecessary information must be reduced (data minimization) and the actions and policies must be open with respect to the personal data.

The data must be qualified before the data sharing process. Toward this direction, great emphasis must be given on the development of methods for enhancing the quality of the shared data in terms of accuracy (how accurate the data are in terms of incompatibilities), relevance (how relevant the data are based on the purpose of the analysis and the research questions that are made), and completeness (how complete the data are in terms of missing values and the purpose of the analysis). In Chapter 3, a computational workflow for medical data quality control was presented to deal with outliers, missing values, and any further incompatibilities and inconsistencies which are present within the clinical data. Both univariate methods, such as the z-score, the interquartile range and multivariate methods, such as the local outlier factor, the elliptic curves, and the isolation forests, were used to detect outliers with increased accuracy. In addition, the importance of the reference model was highlighted to enhance the relevance and completeness of the medical data which serves as a gold-standard template that describes the domain knowledge of a disease of interest. Lexical matching methods, such as the Jaro distance score, the Jaro–Winkler distance score, and the Levenshtein distance score, can be used to quantify the lexical similarity between a pair of medical terms to increase the relevance of the medical data, as well as to standardize terms with the same concept but different measurement units to enable data harmonization.

Medical data sharing, however, is meaningless without any attempt to interlink the shared data which implies the existence of an overwhelming scientific need toward the successful integration of the shared data with minimum information loss. So, the main question here is how to integrate the interlinked medical data with heterogeneous data structure due to the underlying differences in the data recording and data collection protocols across the clinical centers? As we have already seen in Chapter 5, the answer is medical data harmonization. Data harmonization can either be flexible or stringent, depending on whether we wish to harmonize data with common data collection protocols or retrospective data with different data collection protocols. As someone would expect, retrospective flexible harmonization is of great importance. In general, data harmonization requires the existence of a predefined, gold-standard set of parameters (a reference model or a standard template), which describes the domain knowledge of a disease of interest. Then, lexical and semantic matching methods are applied to seek for common terms between the terms of a heterogeneous dataset and those from the reference model. Lexical matching methods involve the application of string matching methods which quantify the similarity between two strings in terms of common characters. Examples of such methods include the Jaro distance, the Levenshtein distance, the Hamming distance, and the Dice distance scores, among others. Semantic matching methods are more robust because they combine lexical matching methods and semantic models to detect terms that might not be lexically identical but share a common concept. Apart from lexical and semantic matching methods, we have also gone through a second approach, namely the item response theory (IRT) analysis. The concept of IRT is based on the estimation of new values given a set of items and latent factors which are able to reduce the measurement bias which is introduced during the measurement of the same parameter (item) with different measurement units that represent the same thing. IRT analysis requires a predefined knowledge regarding the parameters to be matched and thus focuses more on data standardization instead of data harmonization. Besides, IRT analysis is a statistical method which is more or less trivial because of the hypothesis testing procedure that it adopts.

The number of case studies presented in Chapter 8 regarding the scientific impact and value of medical data sharing, harmonization, and analytics toward the development of effective strategies and mechanisms to deal with the crucial clinical unmet needs in different medical domains, including aging, obesity, phenotypes and epidemics, personality scores, and autoimmune diseases, among others, is very promising. Indeed, the harmonized database schemas, which have been already generated across national, international, regional, and longitudinal cohorts on the above medical domains, enable the integrative analysis of the heterogeneous cohort data toward the development of novel tools and methods for reducing (and even vanishing) measurement biases across heterogeneous datasets and dealing with the detection of prominent biomarkers, as well as the development of standard database schemas or reference models. Such models, which have been developed by the clinical experts across the aforementioned medical domains, provide a substantial basis for comprehending the domain knowledge of a disease toward the development of global data collection and data recording protocols, the adoption of which can overcome measurement biases and structural heterogeneities.

More specifically, data processing methods, such as data quality assessment, data discretization, and feature selection, can be applied to enhance the quality of the data, encode the data values according to their frequency, and identify prominent features given a target one, respectively. As far as the detection of biomarkers is concerned, feature ranking algorithms, such as the InfoGain, the Gain Ratio, and the Gini impurity index, and feature extraction algorithms, such as the fast correlation–based feature selection algorithm, can be applied to either rank the features or extract a specific subset of features according to a target feature. Regarding the development of patient stratification models, machine learning algorithms are usually applied to predict a specific disease outcome (i.e., a target feature) given a set of training features. Both conventional and complex supervised learning algorithms, such as the regression models, the decision trees, the random forest, the support vector machines, the Bayesian models along with tree ensemble methods, such as the AdaBoost algorithm and the extreme gradient boosting trees, can be applied on the harmonized data to solve multiclass classification problems. On the other hand, without the presence of a target feature, unsupervised learning methods, such as the *k*-means, the spectral clustering, and the hierarchical clustering, can be applied to group features with similar patterns (e.g., highly correlated features).

What happens in the case where the amount of data is extremely large and/or when the data are stored in distributed databases? In that case, the incremental learning strategy can be applied to develop machine learning models which are able to adapt in such environments through a process known as additive learning. As we have already seen, the majority of the supervised and unsupervised learning algorithms which were mentioned before can support the incremental learning property which is similar to the online learning approach. Furthermore, incremental learning can be used to apply machine learning models in a batch processing manner which occurs in the challenging case where the data are too big to fit into the memory (i.e., out-of-core learning). In that case, the big data are separated into smaller subsets, which are known as batches, and then the machine learning algorithm is incrementally applied on the batches to yield the final supervised or unsupervised machine learning model as if it was applied on the whole data. The number of batches determines the complexity of the training process. In the case where the data are stored in distributed databases, the data can be seen as batches and the machine learning model can be incrementally updated on each individual site yielding the final supervised or unsupervised machine learning model as if it was applied on the data that lie in different locations.

So, how can we combine all these data sharing, data harmonization, and data analytics methods together to offer them to the users, on demand? The answer is through a health cloud infrastructure. Also known as a cloud computing platform, a cloud infrastructure can offer tremendous opportunities for providing data management and data analytics services, on demand, to different users including data providers (e.g., clinicians), data processors (e.g., software engineers, cloud administrators), and other healthcare stakeholders (e.g., health policy-makers, pharmaceutical companies). Through a health cloud computing platform, a clinician can share patient data through GDPR and HIPAA compliant mechanisms; a healthcare stakeholder, such as a pharmaceutical company, can conduct clinical trials to develop cost-effective therapeutic treatments; and a data processor can execute data analytics services to harmonize the data, develop machine learning models for predicting disease outcomes and detecting new biomarkers. A typical cloud computing platform consists of three layers, namely the IaaS (infrastructure as a service) layer, the PaaS (provider as a service) layer, and the SaaS (software as a service) layer. Each cloud layer contributes the most toward the realization of a cloud platform. More specifically, the IaaS layer provides the hardware components which are required toward the development of the hypervisor and the virtual machines. The SaaS layer provides the software components which are necessary for the development of the data management and data analytics services which are established in the final layer, the PaaS layer.

Security protocols and guidelines like those described in Chapter 6, such as the guidelines from the Institute of Electrical and Electronic Engineers (IEEE), the European Network and Information Security Agency (ENISA), and the National Institute of Standards and Technology (NIST), regarding the adoption of standard communication protocols and services, must be carefully taken into consideration to ensure the legal and ethical compliance of the cloud computing platform. The different types of cloud platform, such as the public cloud platform, the private cloud platform, and the hybrid cloud platforms, must adopt these protocols and guidelines to ensure their legal and ethical transparency.

9.2 Future trends

Significant scientific interest will be noted in the field of semiautomated data harmonization toward the development of machine learning methods which will be able to reduce the "semi" term yielding much more automated (less semiautomated) data harmonization mechanisms [1,2]. The main idea behind this is to maintain a "smart" ontology repository, which will serve as a training set for the application of a machine learning model which will be able to learn from the available information regarding the semantic matching process for each ontology to semantically match upcoming ontologies in an automated manner. More specifically, assuming a set of available matches for a specific laboratory term, one could use the existing mapping information to automatically detect an existing match with the highest probability. For example, possible matches for the laboratory term "hemoglobin" could be the term "haemoglobin" with a high probability, as well as the terms "HGB," "HgB," and "Hgb," with less probability, instead. Thus, in the case where one of these terms appears in an upcoming ontology, the data harmonization algorithm shall be able to generate knowledge from the repository to provide the mapping suggestion with the highest probability (in the previous case the term "hemoglobin"). As a matter of fact, the semantic model can significantly reduce the loss of information by merging classes or subclasses (and thus variables) which share a common conceptual basis. Thus, emphasis must be given on the development of robust and accurate ontologies for a disease of interest through the cooperation of both technical and clinical experts.

Medical data curation remains a controversial challenge [3,4]. The development of methods and frameworks, which can effectively deal with the quality of medical data in terms of accuracy, relevance, completeness, conformity, validity, and timeliness, is a challenging task. Conventional methods, such as outlier detection and data imputation, need to be extended to cover different types of medical data. The most challenging part in data quality assessment is the development of data curation methods for genetic data, where the structure and the coding of such data differs a lot from the conventional clinical datasets and requires high-level clinical expertise and additional guidance regarding the format of the data. The genetic data are usually recorded as symbols (e.g., the single-nucleotide polymorphisms—SNPs—genotypes which can be recorded using the CC and CT symbols), a fact that hampers the application of the conventional numerical methods for outlier detection and data imputation, among others. Furthermore, the interdependency between the variables in the data is another challenge. For example, a male patient should not have any recorded information regarding the pregnancy status or a nonsmoker patient should not have any recorded information regarding the number of cigarettes per day. Medical data quality assessment comes in hand with medical data standardization and the existence of an accurate reference model which can be used to normalize the data as a preharmonization step. Clinical guidance is, however, necessary to validate the existence of outliers and the efficacy of the methods toward data imputation. In general, data imputation is an ambiguous method, as the imputation process fills the missing values using virtual patient data, which, however, is not widely acceptable by the overall scientific community, especially when the concept of data imputation is applied on clinical data, where the presence of medical conditions and interventions, as well as the existence of normal or abnormal levels of laboratory measurements, is absent and cannot be arbitrarily estimated. Methodological steps, however, for imputing missing data according to the original population (distribution) are often applied by the majority of the common machine learning workflows to reduce the loss of information during the data integration process.

Further interest is demonstrated on the development of out-of-core learning algorithms to train machine learning and deep learning models across big data [5–8]. The massive amounts of daily generated medical data with an outrageous volume size hamper the development of machine learning models because the data do not “fit” into any conventional RAM. As its name implies, out-of-core learning or external memory learning will undoubtedly be the future in machine learning. An external memory is similar to the conventional RAM with the only difference that an additional cache memory exists to enable faster data fetching and therefore the generated machine learning models are often referred to as cache-aware models as the cache memory directly affects the model training process. The big data are divided into smaller data streams (batches) and are stored in different locations into the external memory. These subsets can be seen as horizontal partitions of the original dataset. Each batch is then fetched through the external memory into the RAM to initiate the sequential (or incremental) learning process. The development of incremental learning algorithms is based on the sequential minimization of a cost function (e.g., the mean square error between the predicted value and the original value) so that the minimization process can be updated incrementally. The stochastic gradient descent method is an example of such an approach where the gradient of the cost function is minimized to simplify the mathematical solution. The interesting part with the incremental learning strategy is that it can be extended to train machine learning models across data which are stored in distributed databases. In that case, a central memory can be used to incrementally fetch the data to sequentially train a machine learning from each database. A serious limitation of the incremental learning process, however, is the fact that the data on each site must be harmonized to be treated as horizontal partitions, like in the previous case regarding the big data training process.

Furthermore, in the case of big data, where the size of the data varies from terabytes to even petabytes of daily generated data, the size and the number of batches in turn can be extremely large and thus batch processing can be a really time-consuming task. In addition, the reduced number of machine learning algorithms that support incremental learning has led the scientific community to the development of strategies which combine computational resources to achieve high-performance computing. The latter requires a large number of computational units which are combined into a parallel computational environment to scale down computationally hard problems, such as the training of deep learning algorithms on big data to develop highly robust models for predicting disease outcomes. Toward this direction, the initial computational task is split up into a predefined number of tasks which are distributed across computing clusters and then subsequently executed in a parallel manner. A computing cluster is a set of computational units which are combined to form an integrated unit which combines the computational power of each individual unit. When several computing clusters are combined together, the resulting multicluster architecture can solve high-performance computing tasks for the analysis of big data. The development of algorithms that support this option is challenging, although as we have already seen in Chapters 6 and 7, there are several popular cloud computing vendors which are able to support high-performance computing applications through the combination of several virtual machines on the cloud computing platform. The distribution of high-performance cloud computing services, on demand, will definitely be the point of view in the next years because it not only avoids the installation of any hardware and/or software resources locally (in premises) but also is much more cost-effective in terms of increased computational power over cost and less time-consuming.

A global data protection and data sharing framework still remains a major scientific challenge in health care [9,10]. The absence of preliminary knowledge regarding the underlying legal and ethical requirements that obscure the sharing of personal sensitive data is a major challenge that needs to be taken into consideration by the existing healthcare infrastructures through the development of GDPR and HIPAA compliant data sharing strategies and cloud computing infrastructures, in general. Furthermore, the absence of a common data collection protocol (under a medical domain of interest), which could be adopted by all medical centers and clinical facilities, worldwide, to reduce measurement biases and structural heterogeneities, hampers the integration of the shared medical data under a specific medical domain and thus emphasis must be given on the construction of gold-standard (i.e., reference) data models for describing the domain knowledge of various diseases, as well as on the development of clinical guidelines and common practices for data collection and data recording protocols in different medical domains. The idea of a sustainable and federated health cloud infrastructure, which will be able to interlink clinical databases from different clinical centers across the world, fulfilling all the national and international legal and ethical requirements posed during the sharing of sensitive data, remains another major scientific challenge. Security protocols, global guidelines, and related strategies along with straightforward user access management technologies, such as intrusion detection methods, need to be effectively adopted by all cloud vendors in health care to provide robust data management and data analytics services and solutions with respect to the rights of the individuals and the rights of processing sensitive patient data. Besides, medical data harmonization and data analytics are pointless without the existence of a well-established data protection and sharing framework, which envisages to enable the interconnection of international clinical databases and registries.

References

[1] Lee JSH, Kibbe WA, Grossman RL. Data harmonization for a molecularly driven health system. Cell 2018;174(5):1045–8.

[2] Abedjan Z, Boujemaa N, Campbell S, Casla P, Chatterjea S, Consoli S, et al. Data science in healthcare: benefits, challenges and opportunities. In: Data science for healthcare. Cham: Springer; 2019.

[3] Brenner SE, Bulyk M, Crawford DC, Mesirov JP, Morgan AA, Radivojac P. Precision Medicine: improving health through high-resolution analysis of personal data. In: Pacific symposium on biocomputing, vol. 24; 2019. p. 220–3.

[4] Chen PHC, Liu Y, Peng L. How to develop machine learning models for healthcare. Nat Mater 2019;18(5):410.

[5] Graham SA, Depp CA. Artificial intelligence and risk prediction in geriatric mental health: what happens next? Int Psychogeriatr 2019;31(7):921–3.

[6] Wang Y, Kung L, Byrd TA. Big data analytics: understanding its capabilities and potential benefits for healthcare organizations. Technol Forecast Soc Chang 2018;126:3–13.

[7] Beam AL, Kohane IS. Big data and machine learning in health care. Jama 2018; 319(13):1317–8.

[8] Farahani B, Barzegari M, Aliee FS. Towards collaborative machine learning driven healthcare internet of things. In: Proceedings of the international conference on omni-layer intelligent systems (ACM); 2019. p. 134–40.

[9] Stark Z, Dolman L, Manolio TA, Ozenberger B, Hill SL, Caulfied MJ, et al. Integrating genomics into healthcare: a global responsibility. Am J Hum Genet 2019;104(1): 13–20.

[10] Cook-Deegan R, Majumder MA, McGuire AL. Introduction: sharing data in a medical information commons. J Law Med Ethics 2019;47(1):7–11.

Index

Note: 'Page numbers followed by "*f*" indicate figures and "*t*" indicate tables and "*b*" indicate boxes'.

D

H

K

L

M

N

O

P

Q

R

S

T

X

9780128165072